Electric Motors and Transformers
Theory and Practice

AUTHOR

Dr. Sanjay B. Bodkhe
Professor in Electrical Engineering and Registrar,
RAMDEOBABA UNIVERSITY
(Formerly, Shri Ramdeobaba College of Engineering and Management)
Nagpur, Maharashtra 440013

REVIEWER

Dr. R. K. Srivastava
Professor
Department of Electrical Engineering,
Indian Institute of Technology (BHU),
Varanasi, Uttar Pradesh 221005

KHANNA BOOK
PUBLISHING CO. (P) Ltd.
4C/4344, Ansari Road, Darya Ganj,
New Delhi, India

All India Council for
Technical Education
Nelson Mandela Marg, Vasant Kunj,
New Delhi, India

Dear Readers,

To prevent the piracy, this book is secured with HIGH SECURITY HOLOGRAM on the front title cover. In case you don't find the hologram on the front cover title, please write us to at contact@khannabooks.com or whatsapp us at +91-99109 09320 and avail special gift voucher for yourself.

Specimen of Hologram on front Cover title:

Moreover, there is a SPECIAL DISCOUNT COUPON for you with EVERY HOLOGRAM.

How to avail this SPECIAL DISCOUNT:

Step 1: Scratch the hologram

Step 2: Under the scratch area, your "coupon code" is available

Step 3: Logon to www.khannabooks.com

Step 4: Use your "coupon code" in the shopping cart and get your copy at a special discount

Step 5: Enjoy your reading!

ISBN: *978-93-55388-45-2*

Electric Motors and Transformers Theory and Practicals

by Dr. Sanjay B. Bodkhe

First Edition: 2025

Originally Published by:
All India Council for Technical Education, Ministry of Education, GOI

Exclusively Published & Distributed by:
Khanna Book Publishing Co. (P) Ltd.
4C/4344, Ansari Road, Daryaganj,
New Delhi (INDIA)

Visit us at: **www.khannabooks.com**
Write us at: **contact@khannabooks.com**

To view complete list of books, Please scan the QR Code:

Printed in Bharat.

प्रो. टी. जी. सीताराम
अध्यक्ष
Prof. T. G. Sitharam
Chairman

अखिल भारतीय तकनीकी शिक्षा परिषद्
(भारत सरकार का एक सांविधिक निकाय)
(शिक्षा मंत्रालय, भारत सरकार)
नेल्सन मंडेला मार्ग, वसंत कुंज, नई दिल्ली-110070
दूरभाष : 011-26131498
ई-मेल : chairman@aicte-india.org

ALL INDIA COUNCIL FOR TECHNICAL EDUCATION
(A STATUTORY BODY OF THE GOVT. OF INDIA)
(Ministry of Education, Govt. of India)
Nelson Mandela Marg, Vasant Kunj, New Delhi-110070
Phone : 011-26131498
E-mail : chairman@aicte-india.org

FOREWORD

Engineers are the backbone of any modern society. They are the ones responsible for the marvels as well as the improved quality of life across the world. Engineers have driven humanity towards greater heights in a more evolved and unprecedented manner.

The All India Council for Technical Education (AICTE), have spared no efforts towards the strengthening of the technical education in the country. AICTE is always committed towards promoting quality Technical Education to make India a modern developed nation emphasizing on the overall welfare of mankind.

An array of initiatives has been taken by AICTE in last decade which have been accelerated now by the National Education Policy (NEP) 2020. The implementation of NEP under the visionary leadership of Hon'ble Prime Minister of India envisages the provision for education in regional languages to all, thereby ensuring that every graduate becomes competent enough and is in a position to contribute towards the national growth and development through innovation & entrepreneurship.

One of the spheres where AICTE had been relentlessly working since past couple of years is providing high quality original technical contents at Under Graduate & Diploma level prepared and translated by eminent educators in various Indian languages to its aspirants. For students pursuing 2nd year of their Engineering education, AICTE has identified 87 books, which shall be translated into 12 Indian languages - Hindi, Tamil, Gujarati, Odia, Bengali, Kannada, Urdu, Punjabi, Telugu, Marathi, Assamese & Malayalam. In addition to the English medium, books in different Indian Languages are going to support the students to understand the concepts in their respective mother tongue.

On behalf of AICTE, I express sincere gratitude to all distinguished authors, reviewers and translators from the renowned institutions of high repute for their admirable contribution in a record span of time.

AICTE is confident that these outcomes based original contents shall help aspirants to master the subject with comprehension and greater ease.

(Prof. T. G. Sitharam)

ACKNOWLEDGEMENT

This book is an outcome of various suggestions of AICTE members, experts and authors who shared their opinion and thoughts to further advance the engineering education in our country. The author is grateful to the authorities of AICTE, particularly Prof. T. G. Sitharam, Chairman; Dr. Abhay Jere, Vice-Chairman; Prof. Rajive Kumar, Member-Secretary; and Dr. Sunil Luthra, Director, Training and Learning Bureau for their planning to publish the book on '**Electric Motors and Transformers: Theory and Practicals**'.

For me, writing this book is a blessing of Shri Saibaba of Shirdi and Shri Vijaybaba of Nagpur. I have no words to express my gratitude but simply a prayer to them and the almighty.

I sincerely acknowledge the valuable contributions of the reviewer of the book

Dr. R. K. Srivastava, Professor, IIT (BHU) Varanasi and thank him for the painstaking efforts to make this book fruitful for the students.

I am grateful to the Hon'ble Chairman, General Secretary, Chancellor and Vice-Chancellor of Ramdeobaba University, Nagpur for providing a conducive environment and support.

I would also like to thank all my friends, colleagues and students for boosting my perseverance and encouraging me in this noble endeavour. I acknowledge with gratitude, the support and love of my family– my parents Mr. Bhaurao and Mrs. Meera; my wife Jayashree and my beloved daughters Ishika and Poorva. They all kept me going, and this book would not have been possible without them.

Finally, acknowledgements are due to the different contributors and industries in this field whose published books, review articles, papers, photographs, footnotes, references and other valuable information enriched me at the time of writing the book.

Dr. Sanjay B. Bodkhe

PREFACE

Outcome based education (OBE) as the name imply focuses on the outcomes rather than the output. It talks of the abilities developed in the students to: explore, learn and apply the fundamental concepts; analyse the given complex situations; evaluate and; develop probable solutions. It assigns less weightage to the marks, grade and rank secured by students. OBE is already in existence in many developed countries since more than 50 years. In India also, the institutes of national importance are practicing it since the same period. However, penetration of OBE in other universities and institutes started roughly a decade before when the National Board of Accreditation (NBA) revised its manual in the year 2013. The first step towards implementation of OBE is to build an outcome-based curriculum and outcome-based course content. Efforts in this direction are facilitated when textbooks and reference books having OBE approach are also available.

Numerous books on electric machines written by different authors exist in the market. This AICTE Technical book on electric motors and transformers is an attempt to rewrite the concepts in line with OBE approach. It will be translated by AICTE in different Indian languages and therefore will also aid to assist the students from socio-economically disadvantaged areas who have completed their schooling with non-English medium of instruction.

Except the basic concepts of electrical engineering studied at first year level; there is no other pre-requisite for understanding this book. Unit-1 and 2 deal with dc machines, Unit-3 deals with single –phase transformer while the last two units cover three-phase transformer and special purpose transformers. The content of this book is drafted in a systematic and orderly manner throughout the volume keeping in mind the target beneficiaries, wide coverage of essential information and the topics recommended by AICTE. Efforts are made to explain the concept of every topic in the simplest possible way. Innovative illustrations with detail labelling are presented to make the discussions self-explanatory. The units also include medium and higher level problems solved in a step-wise manner and followed by multiple choice questions, short and long questions and unsolved problems with final answers for self-assessment by students. At the end of every unit, additional information (which is not a part of the syllabus) is provided for curious minds in the 'Know More' sections.

I sincerely hope that this book will inspire the students to learn, discuss, apply and analyse the concepts related to electric motors and transformers and make it one of their favourite courses in the complete program. I shall be thankful for all the suggestions, which will contribute to the improvement of future editions of the book. It gives me immense pleasure to place this book in the hands of the teachers and students. It was indeed an immense pleasure to work on the different aspects covered in the book. Happy reading, fruitful learning!

Dr. Sanjay B. Bodkhe

OUTCOME BASED EDUCATION

For the implementation of Outcome Based Education (OBE), the first requirement is to develop an outcome-based curriculum and incorporate an outcome-based assessment in the education system. By going through the outcome-based assessments, evaluators will be able to evaluate whether the students have achieved the outlined standard, specific and measurable outcomes. With the proper incorporation of OBE there will be a definite commitment to achieve a minimum standard for all learners without giving up at any level. At the end of the program running with the aid of OBE, a student will be able to arrive at the following program outcomes.

Programme Outcomes (POs) are statements that describe what students are expected to know and be able to do upon graduating from the program. These relate to the skills, knowledge, analytical ability attitude and behaviour that students acquire through the program. The POs essentially indicate what the students can do from subject-wise knowledge acquired by them during the program. As such, POs define the professional profile of an engineering diploma graduate.

National Board of Accreditation (NBA) has defined the following seven POs for an Engineering diploma graduate:

PO-1. Basic and Discipline specific knowledge: Apply knowledge of basic mathematics, science and engineering fundamentals and engineering specialization to solve the engineering problems.

PO-2. Problem analysis: Identify and analyses well-defined engineering problems using codified standard methods.

PO-3. Design/ development of solutions: Design solutions for well-defined technical problems and assist with the design of systems components or processes to meet specified needs.

PO-4. Engineering Tools, Experimentation and Testing: Apply modern engineering tools and appropriate techniques to conduct standard tests and measurements.

PO-5 Engineering practices for society, sustainability and environment: Apply appropriate technology in context of society, sustainability, environment and ethical practices.

PO-6. Project Management: Use engineering management principles individually, as a team member or a leader to manage projects and effectively communicate about well-defined engineering activities.

PO-7. Life-long learning: Ability to analyse individual needs and engage in updating in the context of technological changes.

COURSE OUTCOMES (CO)

The theory, practical experiences and relevant soft skills associated with this course are to be taught and implemented, so that the students demonstrate the following industry oriented COs associated with the above-mentioned competency:

CO-1: Analyse and compare different types of dc generator; discuss their construction and operation; derive mathematical relations and apply them to calculate the unknown quantities from given data; and dismantle a dc machine to validate the theoretical understanding.

CO-2: Apply knowledge of basic mathematics, science and engineering to understand the working of dc motor; calculate the unknown quantities or parameters; discuss the methods of starting and speed control and perform experiments on dc motors in laboratory using engineering tools to test the performance and controllability.

CO-3: Discuss the types, construction and material used in single-phase transformer, analyse its working at No-load and at On-load conditions; understand the phasor-diagram, equivalent circuit and conditions for parallel operation; determine the performance in terms of efficiency and regulation; and conduct experiments in laboratory to test the performance and correlate with theoretical concepts.

CO-4: *Differentiate* power and distribution transformers; compare three-phase transformer with a bank of three, single-phase transformers; understand cooling methods, Scott connection and parallel operation; refer to Indian Standards to understand the connections, specifications and criteria for selection of distribution transformer and identify the phase-wise windings of a three-phase transformer when they are masked.

CO-5: Understand the need, construction and operation of different special purpose transformers; discuss the effect of non-linear load on transformer performance and test these special purpose transformers using modern engineering tools.

Course Outcomes	EXPECTED MAPPING WITH PROGRAMME OUTCOMES (1- Weak Correlation; 2- Medium correlation; 3- Strong Correlation)						
	PO-1	PO-2	PO-3	PO-4	PO-5	PO-6	PO-7
CO-1	3	3	1	3	1	–	1
CO-2	3	3	1	3	1	–	1
CO-3	3	3	1	3	1	–	1
CO-4	3	3	1	3	2	–	1
CO-5	3	3	1	3	2	–	1

GUIDELINES FOR TEACHERS

To implement Outcome Based Education (OBE) knowledge level and skill set of the students should be enhanced. Teachers should take a major responsibility for the proper implementation of OBE. Some of the responsibilities (not limited to) for the teachers in OBE system may be as follows:

- Within reasonable constraint, they should manoeuvre time to the best advantage of all students.
- They should assess the students only upon certain defined criterion without considering any other potential ineligibility to discriminate them.
- They should try to grow the learning abilities of the students to a certain level before they leave the institute.
- They should try to ensure that all the students are equipped with the quality knowledge as well as competence after they finish their education.
- They should always encourage the students to develop their ultimate performance capabilities.
- They should facilitate and encourage group work and team work to consolidate newer approach.
- They should follow Blooms taxonomy in every part of the assessment.

Bloom's Taxonomy

Level	Teacher should Check	Student should be able to	Possible Mode of Assessment
Creating	Students ability to create	Design or Create	Mini project
Evaluating	Students ability to Justify	Argue or Defend	Assignment
Analysing	Students ability to distinguish	Differentiate or Distinguish	Project/Lab Methodology
Applying	Students ability to use information	Operate or Demonstrate	Technical Presentation/ Demonstration
Understanding	Students ability to explain the ideas	Explain or Classify	Presentation/Seminar
Remembering	Students ability to recall (or remember)	Define or Recall	Quiz

GUIDELINES FOR STUDENTS

Students should take equal responsibility for implementing the Outcome Based Education (OBE). Some of the responsibilities (not limited to) for the students in OBE system are as follows:

- Students should be well aware of each Unit Outcome (UO) before the start of a unit in each and every course.
- Students should be well aware of each Course Outcome (CO) before the start of the course.
- Students should be well aware of each Program Outcome (PO) before the start of the program.
- Students should think critically and reasonably with proper reflection and action.
- Learning of the students should be connected and integrated with practical and real life consequences.
- Students should be well aware of their competency at every level of OBE.

ABBREVIATIONS AND SYMBOLS

List of Abbreviations

General Terms

Abbreviations	Full form	Abbreviations	Full form
PMBLDC	Permanent Magnet Brush Less Direct Current motor	LV	Low voltage side
BLDC	Brush Less Direct Current motor	BCC	Body centered cubic structure of grains
NdFeB	Neodymium iron boron magnets	FCC	Face centered cubic structure of grains
PV	Photo Voltaic	IS	Indian Standard
GNA	Geometrically Neutral Axis	AN	Air Natural cooling
PMSM	Permanent magnet synchronous motor	ONAN	Oil-immersed Air Natural cooling
IGBT	Insulated-gate bipolar transistor	OFAN	Oil-immersed, Forced oil circulation with Natural cooling
PLC	Programmable logic controllers	OFAF	Oil-immersed, Forced oil circulation with Forced air cooling
IEEE	Institute of Electrical and Electronics Engineers	OFWF	Oil-immersed, Forced oil circulation with Forced water cooling
EMI	Electromagnetic interference	OLTC	ON-load tap changers
SCR	Silicon controlled rectifier	UPS	uninterruptible power supply
EV	Electric vehicle	MNA	Magnetic Neutral Axis
BHP	Brake-horse-power	VFD	Variable Frequency Drive
HC	Holding Magnet or Holding Coil in 3-Point starter	OLR	Over Load Release in 3-Point starter
CT	Current transformer	SMPS	Switch Mode Power Supply
PT	Voltage transformer	PM	Permanent magnets
CRGO	Cold Rolled Grain Oriented steel	CRNGO	Cold Rolled Non-Grain Oriented steel
HRS	Hot Rolled Steel	HV	High voltage side

List of Units and Symbols

Symbols	Description	Symbols	Description
emf	Electromotive force	K	Transformation ratio of transformer
B	Flux density (Wb/m2)	R_1	Primary winding resistance per phase (Ω)
λ	Flux linkage (volt.second)	R_2	Secondary winding resistance per phase (Ω)
ϕ	Magnetic flux (Wb)	X_1	Primary winding leakage reactance per phase (Ω)
ϕ_m	Flux per pole in dc machines	X_2	Secondary winding leakage reactance per phase (Ω)
ϕ_m	Peak value of flux in transformer (Wb)	X_1	Primary winding impedance per phase (Ω)
ϕ_a	Armature flux (Wb)	X_2	Secondary winding impedance per phase (Ω)
ϕ_{sh}	Flux produced by shunt field winding (Wb)	R_o	No load resistance (Ω)
ϕ_{se}	Flux produced by series field winding (Wb)	X_o	Magnetizing reactance (Ω)
f	Frequency (Hz)	S	Apparent power (VA)
N	Rotor speed (rpm)	V_{ph}	Phase voltage (V)
ω	Rotor speed (rad/s)	V_L	Line voltage (V)
P	Number of poles	I_{ph}	Phase current (A)
Z	Total number of armature conductors	I_L	Line current (A)
θ_e	Electrical angle (degree)	Y	Star connection without neutral terminal (HV side)
θ_m	Mechanical angle (degree)	D	Delta connection (HV side)
Y_A	Front pitch	Z	Zigzag connection (LV side)
Y_B	Back pitch	YN	Star connection with neutral terminal brought out (HV side)
A	Number of parallel paths in dc machines	ZN	Zigzag connection with neutral terminal brought out (HV side)
E_g	Emf generated across armature winding in dc generator	y	Star connection without neutral terminal (LV side)
E_b	Back emf induced in armature winding in dc motor	d	Delta connection (LV side)
F_a	Armature mmf (AT)	z	Zigzag connection (LV side)
F_d	Demagnetizing component of Fa (AT)	yn	Star connection with neutral terminal brought out (LV side)
F_c	Cross-magnetizing component of Fa (AT)	zn	Zigzag connection with neutral terminal brought out (LV side)
V_f	Voltage across field winding (V)	V_1	Voltage across primary winding

I_f	Field current (A)	V_2	Voltage across secondary winding
R_f	Field winding resistance (Ω)	E_1	Emf induced in primary winding (V)
V_a	Voltage across armature winding (V)	E_2	Emf induced in secondary winding (V)
I_a	Armature current (A)	I_1	Primary winding current (A)
R_a	Armature winding resistance (Ω)	I_o	Primary winding current at No-load condition (A)
R_{sh}	Shunt field winding resistance (Ω)	I_μ	Magnetizing component of Io(A)
R_{se}	Series field winding resistance (Ω)	I_w	Working component of Io (A)
R_c	Critical resistance value of shunt field winding (Ω)	I_2	Secondary winding current (A)
N_c	Critical speed value in dc shunt generator (rpm)	I'_2	Load component of secondary winding current (A)
I_L	Load current (A)	N_1	Primary winding number of turns per phase
P_h	Hysteresis loss (W)	N_2	Secondary winding number of turns per phase
P_i	Eddy current loss (W)	Vt	Supply voltage connected across armature winding (V)
P_i	Iron loss or Core loss (W)	Ta	Electromagnetic torque developed on armature winding (Nm)
P_{cu}	Copper loss or Ohmic loss (W)	Tsh	Torque available at the shaft terminal (Nm)
η	Efficiency	TL	Load torque (Nm)
K_f	Field constant	Ke	Voltage constant

LIST OF FIGURES

LIST OF TABLES

CONTENTS

1 DC Generator

UNIT SPECIFICS

Through this unit we have discussed the following aspects:

- Construction of different parts of direct current (dc) generator, their functions and materials used;
- Principle of operation of dc generator;
- Fleming's Right Hand Rule and its application to determine the direction of generated emf;
- Concept of slip-ring, split-ring and commutator;
- Lap and wave type armature windings with steps for their development;
- Mathematical formulation of (speed induced) motional emf;
- Commutation process and method for its improvement;
- Armature reaction process, its effects and remedies;
- Types and applications of dc generator;
- Overview of different types of measuring instruments.

The above topics are discussed in detail with sufficient number of illustrations and solved examples to enable the reader get an insight of the processes and understand better.

Besides giving many multiple-choice questions as well as questions of short and long answer types in two categories following the lower and higher order of Bloom's taxonomy, assignments through numerical problems, list of references and suggested readings are also provided for practice. A dynamic QR code is provided at the end which can be scanned for relevant supportive knowledge.

After the related laboratory experiment based on the content, is a "Know More" section. It has been designed to provide opportunity to the readers to explore more on the topic.

RATIONALE

Electrical machine is an important and inevitable element in human life. Right from the generation of electricity to the rolling of wheels, avoiding a rotating electrical machine in mechanical systems of manufacturing industries, locomotion, Electric Vehicle (EV) and Hybrid Electric Vehicle (HEV) is not possible. DC generators are the first dynamic electromechanical systems that were used for the generation of electricity. Due to the popularity of alternating current (ac) right from the generation to utilization, bulk dc power generation using dc generators is now no more used. However dc generators are used in certain control machinery and lab electrics-for the determination of output power and torque. Homopolar dc generator working on the principle of Faraday's Disk Generator delivering low voltage high current dc is used for certain specific applications. The basic principles and the engineering concepts involved in it are very much useful to a student of electrical engineering and they serve as a logical and convenient beginning towards the study of conventional dc motors, ac generator, ac motors and special purpose machines. The control of dc machines is simpler than that of ac machines. Many phenomena occurring in ac / dc machines can be easily explained with dc machines.

PRE-REQUISITES

Basics of Electrical Engineering

UNIT OUTCOMES

List of outcomes of this unit is as follows:

U1-O1: Describe the construction, operation and functions of different parts of dc machine
U1-O2: Discuss commutation, armature reaction, their issues and solutions
U1-O3: Calculate winding data, emf, efficiency and other quantities of dc generator
U1-O4: Apply Fleming's right hand rule to determine the direction of induced current
U1-O5: Discuss electrical measuring instruments and their types
U1-O6: Dismantle a dc machine and identify the different parts

Unit-1 Outcome	EXPECTED MAPPING WITH COURSE OUTCOMES (1 - Weak Correlation; 2 - Medium Correlation; 3 - Strong Correlation)				
	CO-1	CO-2	CO-3	CO-4	CO-5
U1-O1	3	-	-	-	-
U1-O2	3	-	-	-	-
U1-O3	3	-	-	-	-
U1-O4	3	-	-	-	-
U1-O5	3	-	-	-	-
U1-O6	3	-	-	-	-

1.1 INTRODUCTION

Electrical energy is rarely available in nature to be used directly. The electrical energy dissipated during lightning strike is available naturally but due to its transient and uncertain nature, it cannot be utilized or stored for later use. All electrical appliances and systems that we use actually convert the input electrical energy into some other usable form like heat, light, sound, displacement (linear or angular) etc. The physical machines used in large manufacturing industries and electrical traction / locomotion are seldom static and are supposed to perform useful work. To enable them for doing so, they are coupled with electric motors which impart them the required mechanical energy in rotational form. Electrical traction locomotive, cement industry, textile industry, steel industry, paper industry, food industry and pharmaceutical industry are some of the examples. This indicates that there has to be some devices which can convert energy from non-electrical form to electrical form and vice-versa.

An electromechanical device which converts mechanical energy into electrical energy is known as *electric generator* whereas the one which converts electrical energy into mechanical energy is known as *electric motor*. The generator receives mechanical energy in rotational form from a device generally referred as *prime mover*. This prime mover can be either a steam turbine, gas turbine, hydraulic turbine or wind turbine etc. The electric power available for utilization are of two type – *Direct Current (dc)* and *Alternating Current (ac)*. Generator and motor can be similarly classified as *dc* type and *ac* type. Schematic block diagrams of generator and motor are shown in Fig. 1.1 (a) & (b) respectively.

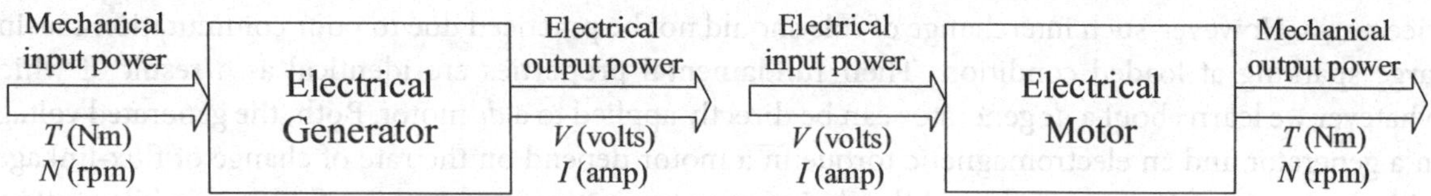

Fig. 1.1(a): Block diagram of electrical generator **Fig. 1.1(b):** Block diagram of electrical motor

The history of generation of electricity as a modern industrial revolution date back to 18th century. By the end of this century, storage batteries were invented. In 1820, it was discovered that, the electric currents are associated with magnetic field. A year later, it was found that a force is exerted on a current carrying conductor when it is placed in the magnetic field. In 1831, the Michael Faraday's laws of induction made it clear that an inter relationship exists between electric and magnetic circuits. During the same period, a parallel research led to the invention of an electromechanical dc source of constant amplitude to recharge the batteries. In the second half of the 19th century the invention of commutator facilitated the way for the birth of dc generator and motor.

Since 20th century, *ac* power is widely adopted for electricity generation, transmission, distribution as well as utilization. The reasons are obvious. The *dc* motor and generator work on the physical electrical contacts between rotating commutator and stationary carbon brushes, leading to electrical spark and related Electro Magnetic Interference (EMI) problems to modern equipment. Due to sparking at the trailing edge of brush and commutator segments, there is chance of explosion when hazardous gases are present in atmosphere, like in underground mines. As such, the application of *dc* machines in underground mines is totally forbidden. The commutator of a large *dc* machines are made with Copper Silver alloy. Due to these reasons the procurement, control and maintenance of *dc* machine are not cost effective as that of an equivalent ac machine. With to the advent of powerful permanent magnets like NdFeB and Hall effect (Integrated Circuit -IC) sensor / latch, currently Permanent Magnet Brush Less Direct Current (PMBLDC) Motors are becoming more popular. Therefore practically, the use of conventional *dc* machine as *dc* generator for bulk power generations is rare, except in certain specific applications.

In certain applications like traction, the *dc* motor operates as a generator for short time periods in regenerative braking mode. The conventional aircrafts and ships that are isolated from land based *ac* grid employ *dc* sources including *dc* generators and batteries. Currently these are being replaced using *ac* machines and modern power electronics. Homopolar *dc* generators based on Faraday's disk capable of delivering low voltage and extremely high *dc* current are in use for metal extraction and in defence sector. In modern times, due to the availability of power semiconductor technology, the available *ac* power is converted into *dc* using a solid-state rectifier wherever *dc* power is needed. Also, the solar Photo Voltaic (PV) system is preferred to generate the *dc* power directly. It has merits like static operation, environment friendly and minimum maintenance. Solar PV generation and Fuel cell technology do not require prime mover for *dc* power generation.

1.2 DC MACHINE

DC machine can function as a (a) generator (b) motor and (c) brake. The constructions of both, *dc* generator and *dc* motor are similar therefore, any *dc* generator can be ideally operated as a motor and vice-versa. However, such interchange of role should not be practiced due to poor commutation causing large sparking at loaded condition. Their fundamental properties are identical as a result of which whatever we learn about a *dc* generator can be directly applied to a *dc* motor. Both, the generated voltage in a generator and an electromagnetic torque in a motor depend on the rate of change of flux-linkages with the winding. DC generators can be designed to generate power ranging from a few kilo-watts to hundreds of kilo-watts. DC motors are constructed in many sizes from 1.5 – 3 V motor required in toys to a huge 100 MW (or more) motor used in rolling mill application.

1.2.1 DC Generator: Principle of operation

The working of a *dc* generator is based on Faraday's law of electromagnetic induction. A *motional (Speed Induced) electromotive force (emf)* is induced in the conductor when it moves in a stationary magnetic field such that it cuts the field at right angles to the direction of flux. If the flux density of stationary magnetic field is B Wb/m^2, conductor active length is l meters, and its uniform peripheral velocity is v m/s than, the motional *emf* induced in the conductor is given by,

$$e = Blv \text{ volts} \qquad \text{... (1.1)}$$

To understand this principle, imagine a pair of stationary magnets placed a distance apart with opposite North and South poles facing towards each other as shown in Fig. 1.2. Let there be a cylindrical drum type rotor having two diametrically opposite slots and placed between the two magnets. The rotor assembly is free to rotate about its shaft axis and is called an *armature.* Now consider a rectangular insulated copper coil ABCD wound on the rotor core such that the two coil sides AB and CD are housed in the two slots as shown in Fig 1.2. Note that, the lengths AB and CD of the coil shall be able to cut the magnetic flux of the stationary magnets when the coil is being rotated. This results in change in flux linkage and generates speed induced *emf* in the coil sides. The remaining lengths BC and DA will not have any such contribution, but they help in addition or summation of voltages produced in AB and CD. *Therefore, the lengths AB and CD only are regarded as active lengths and are commonly termed as Conductors AB and CD.* These are placed in armature winding such that AB and CD at any instant of time while in rotation face opposite flux density (due to North and South poles) and result in opposite speed induced *emfs* in them, which get added.

The electrical power generated in the coil can be utilized only when an electric load (say a resistor R) is connected across it. To connect the rotating armature coil to a stationary electric load, some interfacing device will be required which will transfer the generated electrical power from rotating coil to the stationary load. In *dc* machines, this task is performed by a device called *commutator* and *a pair of brushes.* Details about commutator and brushes are provided ahead in Sections 1.3.4 and 1.3.5.

To understand the wave shape of generated emf, assume that instead of commutator, two slip-rings X_1 and X_2 are used. Slip-rings are the copper rings that are tightly fixed on the rotor shaft but electrically insulated from the shaft as well as from each other. End terminals of the coil ABCD are connected (internally) to these slip rings. Let the terminal A of the coil be connected to X_1 and terminal D be connected to X_2. Now consider two solid blocks of carbon that are called as carbon brushes. Let them be mounted on some other stationary object but riding on slip-rings, thereby maintaining an electrical contact with rotating slip rings. The external load (resistor R) is connected across these brushes.

Now, rotate the rotor with the help of prime mover. The conductors AB and CD will cut the main flux created by North and South poles and result into flux-linkages. By Faraday's law, *emf* will be generated in both conductors whose magnitude will be directly proportional to the rate of change of flux linkages.

$$e = -\frac{d\lambda}{dt} = -N_t \frac{d\phi}{dt} \text{ volts}$$

Where, dλ is the change in flux linkage (volt.second), dϕ is the change in flux (weber), is the change in time (second) and N_t is the number of turns in the coil (In Fig. 1.2, N_t is assumed equal to 1). The minus sign takes into account the direction of induced *emf* as given by Lenz's law. The total *emf* generated in the coil will be the summation of speed induced *emfs* in each conductor and it will appear across the pair of brushes. Since the armature coil forms a close circuit through the load resistor R, the generated *emf* will force a current in the load. Thus, electrical power is generated and is delivered to the load. Eq. (1.1) and (1.2) represent the motional *emf* and transformer action *emf* respectively but they give identical results though expressed in different manner based on two different processes, one is dynamic and the latter one is static. To prove this equality, refer to Fig. 1.2 again. Assume that the rotor is being rotated by the prime mover and let the infinitesimal distance travelled by any conductor of length '*l*' (say conductor AB or CD) in time '*dt*' is '*dx*'. While doing so, it is cutting the magnetic field of density '*B*'. The change in flux linkage experienced by the conductor will be,

$$d\lambda = B(\text{area of plane travelled by the conductor in time } `dt' = B(l\,dx)$$

Divide both sides of above equation by '*dt*'.

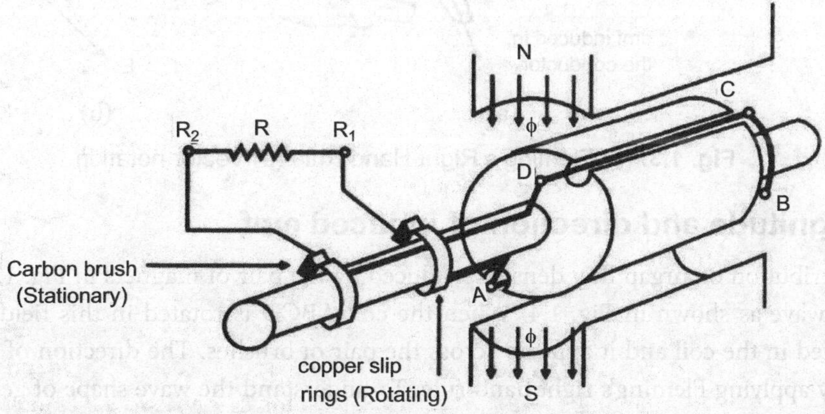

Fig. 1.2: Schematic diagram of a basic dc generator with one armature coil and slip-rings

Thus the above equation shows that, in magnitude, the voltages given by Eqn. (1.1) and (1.2) are identical.

1.2.2 Fleming's Right Hand Rule

This rule is employed to determine the direction of motional or speed induced *emf* (*an electrical quantity*) induced in a conductor having flux linkages with magnetic field. It involves three different domains i.e. *electric, magnetic, and physical* and can be applied whenever a relative motion exists between the conductor and the magnetic field. For example, we can apply it when,

- A moving conductor is cutting a steady magnetic field or
- A moving field is cutting a stationary conductor

To apply this rule, three fingers of right hand i.e., *first finger, thumb and middle finger* are stretched away from each other such that each of them gets aligned perpendicular to the remaining two.

- *First finger is oriented in the direction of magnetic field* (North to South) where the conductor is placed.
- *Thumb is oriented in the direction of motion of conductor* (relative to the direction of motion of magnetic field if the field is also moving)
- *Middle finger will give you the direction of emf or current induced in the conductor.*

Mathematically, the induced *emf* or electric field intensity is given as . These are shown in Fig. 1.3(a) & (b).

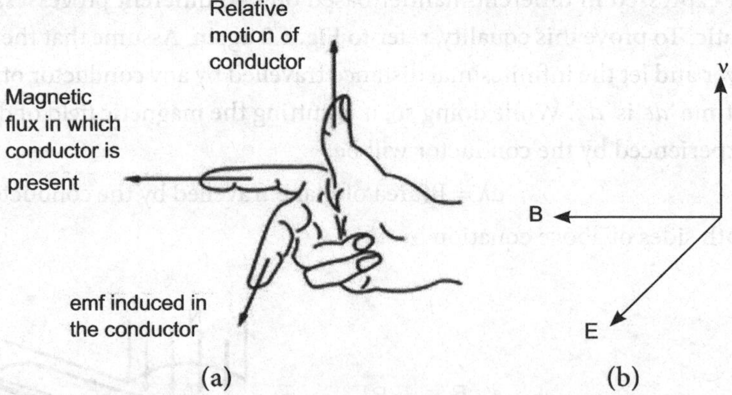

Fig. 1.3: (a) Fleming's Right Hand Rule (b) Vector notation

1.2.3 Magnitude and direction of induced emf

The space distribution of airgap flux density produced by the pair of magnets in Fig. 1.2 approximates a flat-topped wave as shown in Fig. 1.4. When the coil ABCD is rotated in this field, an alternating emf is generated in the coil and it appears across the pair of brushes. The direction of this emf can be determined by applying Fleming's right hand rule. To understand the wave shape of generated emf, let us consider various angular positions of the coil. Imagine that the rotor is rotated in counter-clockwise direction. Refer to Fig 1.5(a). The angular positions are marked as 0°, 90°, 180° and 270°.

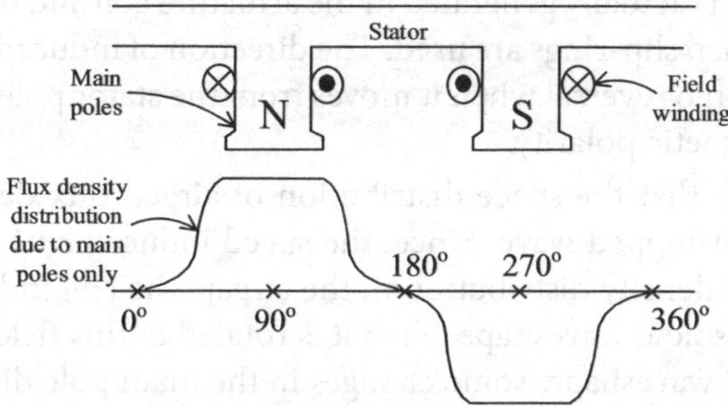

Fig. 1.4: Space distribution of air-gap flux density

(a) When conductor AB is at 0° and CD is at 180°:

Refer to Fig. 1.5(a). The conductors are moving along the lines of main flux established by the North and South poles and therefore, the speed *emf* induced in them will be zero.

(b) When conductor AB is at 90o and CD is at 270°:

Refer to Fig. 1.5(a) & (b). The coil has rotated by 90° angle. The two conductors are cutting the main flux, moving at right angle to the flux and therefore, speed *emf* will be induced. Applying Fleming's Right Hand Rule, the direction of generated current '*I*' will be found as B-A-X_1-load-X_2-D-C-B. In the load resistor 'R', the current is flowing from R_1 to R_2.

(c) When conductor AB is at 180° and CD is at 0°:

This position will be identical to position (a) above and the *emf* generated in both conductors will be zero.

(d) When conductor AB is at 270° and CD is at 90°:

Refer to Fig. 1.5 (a) & (c). The coil has further rotated by 90° physical angle. Applying Fleming's right hand rule again, the direction of generated current is found to be C-D-X_2-load-X_1-A-B-C. This time, the current in load resistor has reversed and is flowing from R_2 to R_1.

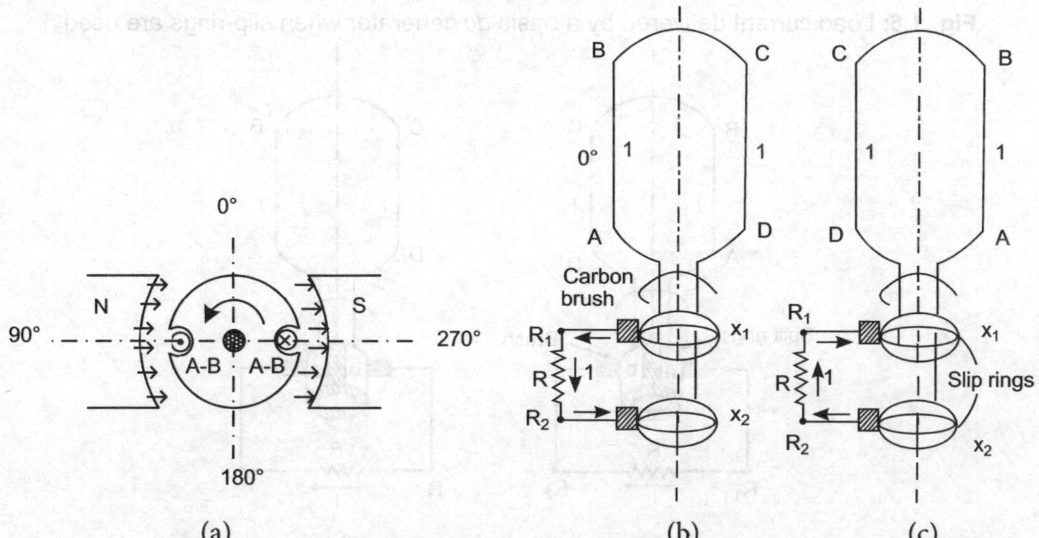

Fig. 1.5: Load current at different positions of armature coil when slip-rings are used

Thus the *emf* (and current) actually generated in the armature coil and delivered to the external load are of alternating type (*ac*) when slip-rings are used. The direction of induced current in every conductor of the armature coil will undergo reversal when it moves from the stator pole of one magnetic polarity to the next pole of opposite magnetic polarity.

In Fig. 1.4, it was shown that the space distribution of airgap flux density produced by the pair of magnets approximate a flat-topped wave. Since, the speed induced *emf* has similar waveshape with respect to time as that of flux density distribution in the airgap, the *emf* induced in the coil ABCD will also attain a similar non-sinusoidal waveshape when it is rotated in this field. Therefore, to generate an *emf* and current of sinusoidal waveshape, some changes in the main pole dimensions are required. The pole faces are shaped in such a way that the flux density distribution in the airgap becomes sinusoidal. Consequently, the current induced in coil ABCD that is delivered to the load becomes sinusoidal. This is shown in Fig. 1.6. Thus it is clear that, the *dc* generator actually generates alternating current (*ac*).

The *ac* nature of generated *emf* and current can be converted into *dc* if we replace the slip-*rings* by *split-ring*. Split-ring is a copper ring cut into two segments Y_1 and Y_2 such that both segments are semi-circular as shown in Fig. 1.7. They are tightly mounted on the shaft but electrically insulated from the shaft as well as from each other. End terminals of the coil ABCD are connected (internally) to these two segments of the split-ring. Let the terminal A of the coil be connected to Y_1 and terminal D be connected to Y_2. Two stationary carbon brushes for electrical connection to rotating armature are fixed on the stationary end-cover and are made to rock on the rotating split-ring as shown. The external load (resistor R) is connected across these brushes.

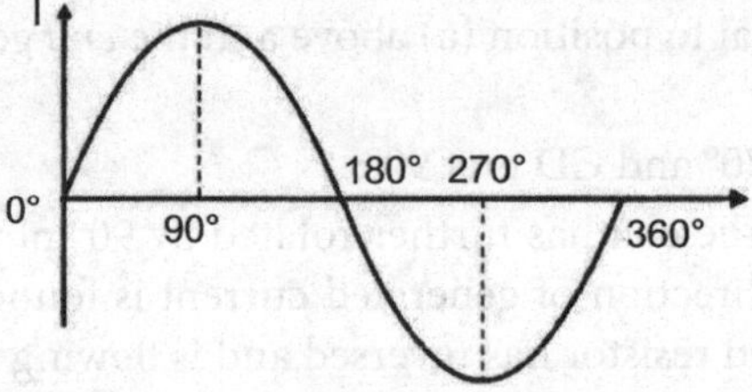

Fig. 1.6: Load current delivered by a basic *dc* generator when slip-rings are used

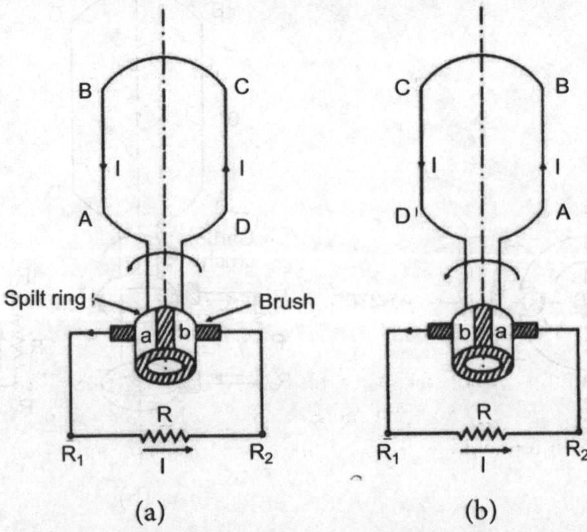

Fig. 1.7: Load current at different positions of armature coil when split-rings are used

Let a prime mover rotate the rotor in counter clockwise rotation and consider same four positions of the coil ABCD as discussed above. At positions (a) and (c), when the conductors AB and CD will be moving along the lines of main flux, the cutting action between them will be absent and the *emfs* induced will be zero. Whereas, in the other two positions when the conductors are cutting the flux and moving almost at right angles to the flux, *emf* will be induced. This is the same as was happened when slip-rings were used above. Let us now determine the direction of induced current flowing in the loop when the induced *emf* is not zero. You may apply Fleming's Right Hand Rule again.

(a) Conductor AB is at 90° and CD is at 270°:

Refer to Fig. 1.7(a). The direction of generated current in the loop will be B-A-Y_1-load-Y_2-D-C-B. Note that the current in load resistor is from R_1 to R_2.

(b) Conductor AB is at 270° and CD is at 90°:

Refer to Fig. 1.7(b). Due to 180o rotation, conductor AB has acquired the previous position of conductor CD and vive-versa. Also, the split-ring segments Y1 and Y2 have interchanged their positions. The direction of generated current is found to be C-D-Y2-load-Y1-A-B-C. Note carefully that, the direction of current in the load resistor has not changed. It is still from R1 to R2.

From Fig. 1.7 (a) & (b), it is clear that due to the use of split-ring, the direction of emf and current in the load does not change. It is unidirectional but of pulsating nature as shown in Fig. 1.8. Thus, though the actual generation is of ac type, the split-ring and brushes have rectified it (i.e. converted from ac to dc) for the load.

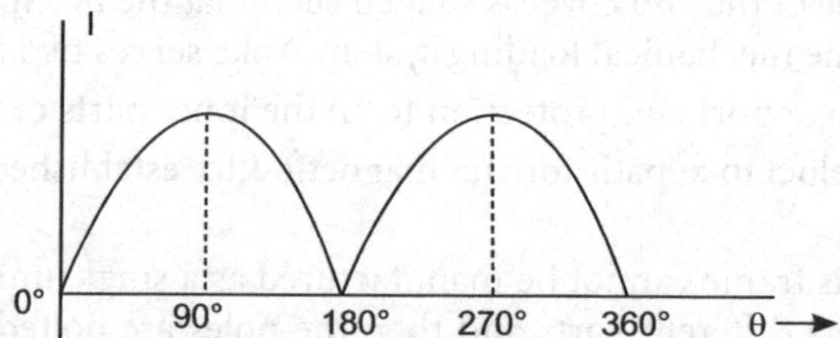

Fig. 1.8: Load current delivered by a basic dc generator when split-rings are used

A practical *dc* generator has many coils on the armature similar to the coil ABCD considered above. Therefore a split-ring arrangement with only two segments Y_1 & Y_2 will be insufficient. The number of copper segments should be equal to the number of armature coils. Such a device is then called as '*Commutator*'. If the armature is provided with many coils and a commutator than, the *dc* current in the load is almost steady and not of pulsating type. Thus, a commutator acts as a mechanical rectifier in a *dc* generator.

1.3 CONSTRUCTION OF DC MACHINE

The schematic view of a practical dc machine is shown in Fig. 1.9. An elementary dc machine consists of following major parts.

(i) Yoke	(ii) Main poles	(iii) Field winding
(iv) Armature core	(v) Commutator	(vi) Armature winding

The yoke, main poles and field winding together makes the stationary part called as *Stator.* Armature core, armature winding, shaft and commutator together represents the rotational part called as *Armature.* Additionally, pair(s) of stationary *carbon brushes*, a pair of *end-covers* and other mechanical parts are also present. Details about each of them are presented below.

Example 1.1: Consider a rectangular armature coil ABCD of single turn as shown in Fig. 1.2. Let the maximum flux density due to stator poles be 1.2 Wb/m² and the rotor speed be 600 rpm. If the cylindrical drum type rotor core has an axial length of 40 cm and a diameter of 120 cm, calculate the maximum value of emf induced in one conductor.

Data:

$B = 1.2\ Wb/m^2$, N = 600 rpm, l = 40 cm = 0.4m, D = 120 cm = 0.12m, E/conductor = ?

Solution: The coil ABCD has two conductors AB and CD.

Rotor speed in rps = speed in rpm/60 = 600/60 = 10 rps.

Rotor circumference = peripheral distance over the rotor surface

$= \pi D = 3.14 \times 0.12 = = 0.3768$ m

Peripheral velocity v = (πD) x speed in rps = 3.768 m/s

Maximum emf induced in one conductor = $1.2 \times 0.4 \times 3.768 = 1.8$ V

1.3.1 Yoke

Yoke, also termed as *Outer body* or *Frame* is the cylindrical outer shell within which all other parts are housed. It is closed at both ends by two *end-covers* that also support the bearings to assist smooth rotation. Since the flux in Yoke almost remains constant, *cast steel* is used for its fabrication. Most of the *dc* machines are *foot mounted*, which rest on a plane horizontal surface. Another type of mounting is the '*flange mounting*' in which one of the end cover is shaped such that the machine can be directly fixed on the vertical plane surface of the mechanical loading system. Yoke serves two functions:

- It provides physical support and protection to all the inner parts of the machine and
- It serves as a low reluctance path for the magnetic flux established by the poles fixed to the yoke.

In large *dc* machines, this frame cannot be manufactured as a single unit and therefore fabrication is done by suitably welding its different parts and then the poles are bolted to the inner periphery of the frame. In small size special machines, stack of thin laminations are fastened together to form a solid structure of the frame. Also, in smaller machines the field poles and yoke are shear cut in single stamping. Typical example is the motor used in small house-hold mixers.

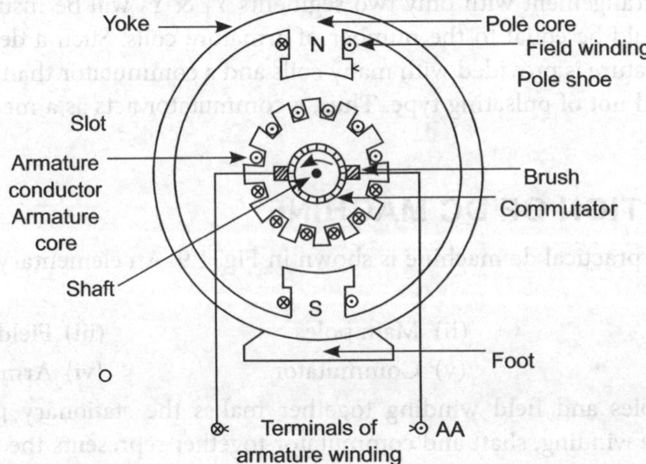

Fig. 1.9: Constructional details of a practical dc machine

1.3.2 Main Poles and Field winding

Solid projecting poles made from fabricated steel are fastened on the inner periphery of the Yoke with the help of bolts and are always even in number. The poles have pole core and pole shoe as shown in Fig. 1.9. The pole core is wound with a winding called as *Field winding.* When this winding is excited (means, when *dc* current is passed through it) it establishes the main magnetic field that directs its flux ϕ_m through the air-gap to the armature core and then to the next pole. The direction in which the field winding is wound on the poles should be such that when excited, opposite polarities (North and South) will be formed on the adjacent poles. Right hand thumb rule can be used to determine the direction of flux produced by the field winding (See appendix).

The pole shoes are generally laminated which means, they are a stack of thin sheets of magnetic material cut in the required shape. Every sheet is called as *lamination* or *stamping* and is electrically insulated from all others. Such structure using thin insulated laminations is preferred to minimize the *eddy current loss* in the pole shoes. Sometimes the pole core and pole shoe, both are made from same laminations. They either form an integral unit or are separately attached to each other. The enlarged shape of pole shoe helps to cover more area of the armature to come under the influence of main flux. The face of pole shoes is shaped in such a way that there is slightly increased airgap at their tips. This is done to create a sinusoidal nature of airgap flux density distribution in space (close to armature surface) which then ensures that the induced *emf* and current in the armature winding are also sinusoidal. Flux density in space can be measured using Hall Effect Gaussmeter. The stator poles serve three functions:

- Provide physical support to the field winding and establishes main flux .
- Reduces the reluctance of magnetic path and
- Spreads out the flux for better uniformity.

The choice of number of stator poles depends on many factors including the physical dimensions of the armature core. One of them is the frequency of *emf* reversal in the armature which is given by,

$$f = \frac{PN}{120} \text{ Hz} \tag{1.3}$$

where P is number of stator poles and N is the rotor speed in rpm. It is known that, the armature of *dc* machine faces alternating magnetic poles when being rotated and results in cyclic change in the direction of speed induced *emf* in an armature conductors. This was discussed in Section 1.2.3. If we choose a large number of poles, the frequency of reversals will be high and consequently it will result into excessive *iron loss* in the armature core. Normally, the value of f lie between 25 and 50 Hz. A typical laminated pole and the symbol of field winding are shown in Fig. 1.10(a) and (b). The stators of 2-pole and 4-pole *dc* machine are shown in Fig. 1.11(a) and (b).

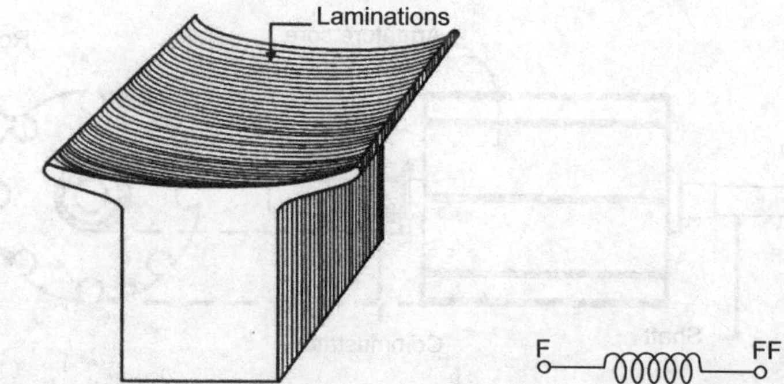

Fig. 1.10: (a) Actual Field pole without field winding (b) Symbol of field winding

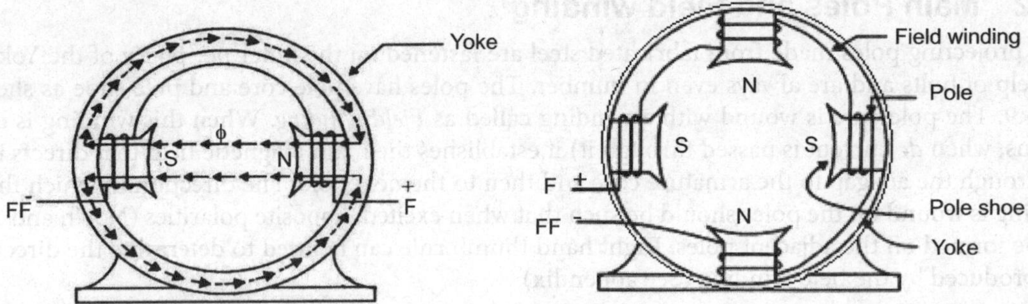

Fig. 1.11(a): Stator of a 2-pole dc machine **Fig. 1.11(b):** Stator of a 4-pole dc machine

1.3.3 Armature

Armature is the rotating part of a *dc* machine. It primarily consists of a shaft, drum type magnetic core and armature winding. The armature magnetic core is cylindrical, laminated and has semi-closed or open slots on its outer periphery. The central bore of the armature core carries a long shaft which extends either on both sides or on single side through end covers. To insert a shaft inside the core, a continuous hole along with a key-way is punched at the centre. The magnetic core is laminated to minimize the *eddy current loss*. Every lamination (also called as stamping) is around 0.5 mm thick of non-grain oriented electrical grade Silicon sheet steel or hot rolled electrical grade Silicon sheet steel. These laminations are also perforated for axial air ducts which permit the axial circulation of air and helps to improve cooling. In large machines radial ducts are also provided for cooling. For radial ducts a gap of around 10 mm is provided between two packets of several stampings of certain thickness, this allows the cooling air to flow in radial direction.

The slots on the periphery of armature core are either die-cut or punched in required shape. Their prime purpose is to house and hold the armature coils against the centrifugal forces that are developed on the coils when being rotated. One of the major functions of the magnetic core is to provide a low reluctance path through itself to the main flux ϕ_m established by stator poles. Therefore, the material used for it should have high permeability and high saturation flux density. Schematic view of an armature assembly is shown in Fig. 1.12.

The ends of shaft are supported by two bearings and helps in uniform rotation of the rotor. A cylindrical object called commutator is mounted on the shaft on one side of the core as shown. Armature windings terminals are brazed to the commutator segments. The extended portion of shaft is connected to prime mover or other rotating machines using mechanical coupling or flexible coupling, which enables both the machines to run at identical speeds if directly coupled.

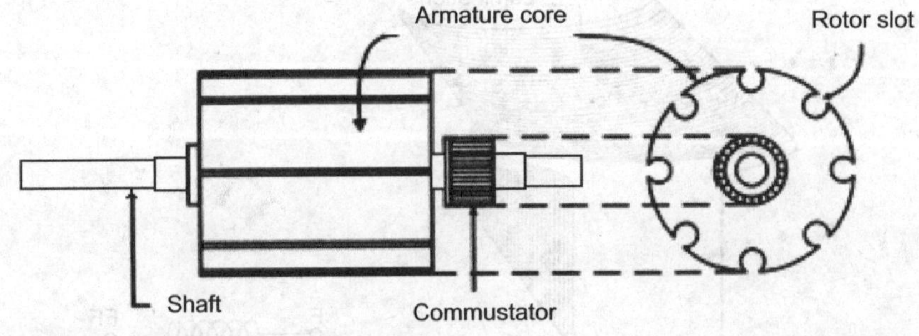

Fig. 1.12: Armature

1.3.4 Commutator

The armature winding of a *dc* machine is composed of *many number of coils* identical to the coil ABCD discussed in Section 1.2.1 and 1.2.3. In Section 1.2.3, only one coil was considered so obviously, the split-ring having two copper segments of semi-circular shape was appropriate to connect the two terminals of the coil (See Fig.1.7). However, with more number of coils, two segments would be insufficient and therefore, the split-ring is replaced by a device called '*Commutator*' that performs the same function as that of the split-ring but for more number of coils.

Constructionally, it is a cylindrically shaped device placed on the same shaft but at one end of the armature core. It consists of a number of segments or bars separated from each other by a suitable insulating material. Each segment has riser which is electrically jointed by brazing to the armature conductors. The commutator segments are made from hard drawn copper or copper-silver alloy. The thin insulation between adjacent segments is usually 0.8 mm thick and is made from mica or micanite. Different components of commutator assembly are shown in Fig. 1.13.

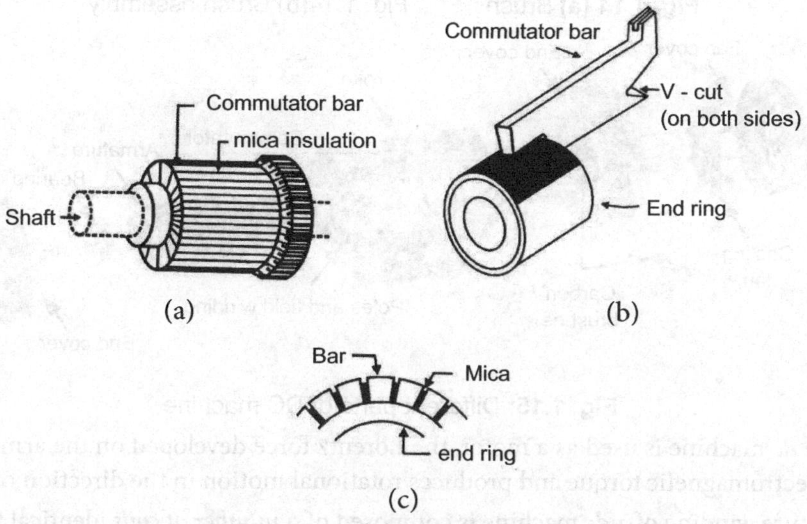

Fig. 1.13: Commutator assembly

1.3.5 Brush assembly and brushes

Brush assembly consists of brush holder and brush. It is fixed on the stationary end cover and arranged concentrically around the commutator. Brushes made from either hard carbon, natural graphite, electro-graphite or metal graphite are housed in the brush holders. They are made to rest on the commutator segments by applying suitable pressure with the help of stainless steel springs. Flexible conductors called *pig tail* are connected to the brushes which also serve as terminals of armature winding. Schematic view of a brush is shown in Fig. 1.14. Figure 1.15 shows the different parts of a *dc* machine.

1.4 ARMATURE WINDING

Armature winding is one of the most important constituent of any rotating machine. When generator is being rotated using a prime mover at certain speed, a speed induced *emf* gets generated and appears on its armature terminals. This generated *dc* voltage supplies *dc* current and *dc* power when a resistive load is connected across the armature terminals. The armature current results in the development of counter torque on the same rotor which opposes the prime mover torque and as result the speed of prime mover may drop with rise in load. The speed of prime mover is normally regulated using governors in any power plant and other speed control methods in standalone systems.

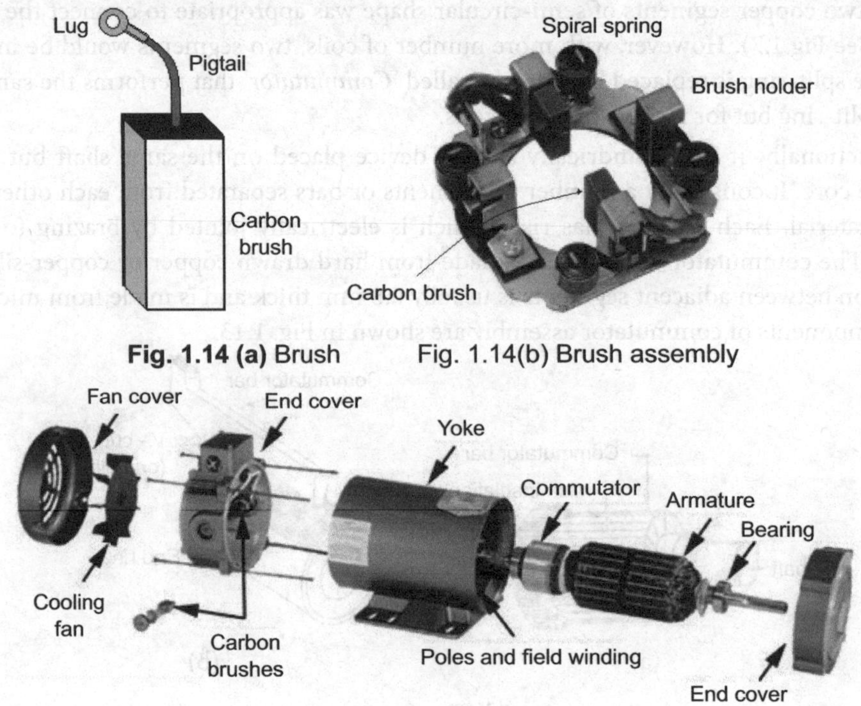

Fig. 1.14 (a) Brush   Fig. 1.14(b) Brush assembly

Fig. 1.15: Different parts of DC machine

When the *dc* machine is used as a motor, the Lorentz force developed on the armature conductors results in an electromagnetic torque and produces rotational motion in the direction of torque.

The armature winding of a *dc* machine is composed of *a number of coils* identical to the coil ABCD considered in Section 1.2.1 and 1.2.3. Before we discuss the armature winding and its types, it will be appropriate to begin with the elementary terms involved in it. They are (i) Armature coil, conductor and end connection, (ii) Coil span, (iii) Pole pitch, (iv) Full pitched and fractional pitched coil, (v) Front pitch, back pitch, resultant pitch, average pitch and, (vi) Commutator pitch. Fig. 1.16(a) shows a section of the developed (unfolded) armature winding to understand these terms. Fig 1.16(b) shows the symbol of armature winding.

1.4.1 Armature Coil, Conductor and End Connection

An armature coil ABCD was discussed in Section 1.2.1. The two coil-sides AB and CD that lie in the slots of armature core are termed as *conductors.* If the coil has only one turn, the number of conductors will be two. For '*n*' number of turns, the number of conductors will be 2*n* per coil. That portion of the coil which is not present in the slots but it completes the loop at front and rear sides is called as '*End connection*' or '*winding overhang*'. Thus, conductors and end connections together forms a coil. The terminals of all armature coils are suitably connected (brazed) to the respective commutator segments.

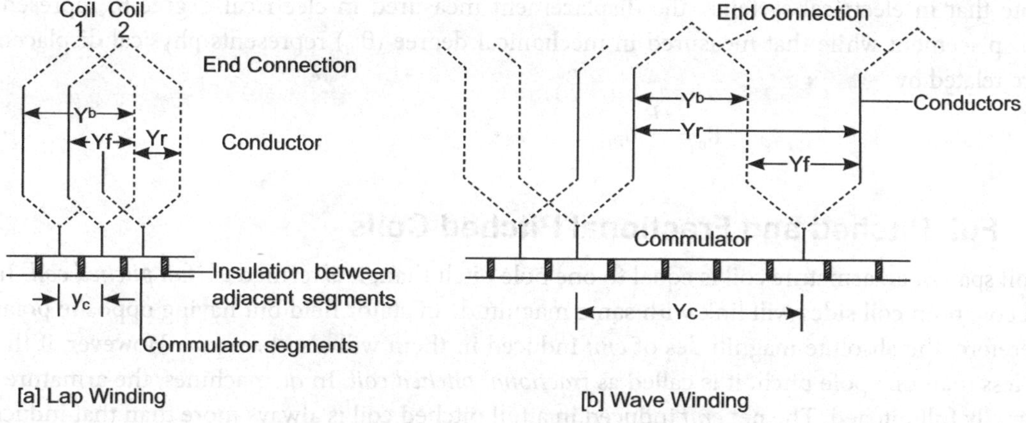

Fig. 1.16 (a): Unfolded view of armature windings

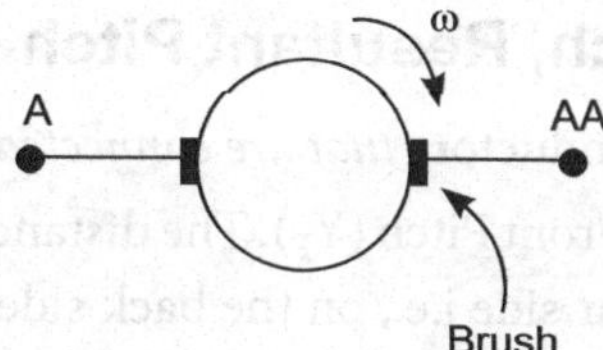

Fig. 1.16 (b): Symbol of rotating armature

1.4.2 Coil Span or Coil Pitch

The distance between two sides of an armature coil measured in terms of number of armature slots is called as coil span.

1.4.3 Pole Pitch

The number of armature slots or armature conductors coming under the influence of one stator pole is called as pole pitch. It can also be defined as the distance between adjacent poles but expressed in terms of number of armature slots or armature conductors. If 'Z' represents the total number of armature conductors (considering all armature coils) and 'P' represents the total number of stator poles than,

$$\text{Pole pitch} = \frac{Z}{P} \tag{1.4}$$

If we imagine an armature coil placed such that, it has one conductor under the centre of North Pole and other conductor under the centre of adjacent South Pole than, such a coil is called as full pitched coil. In a full pitched coil, the direction of emf induced in its two conductors will be of opposite polarity due to the influence of opposite polarity poles. In other words, the two emfs shall be 180° electrical apart when seen externally. This can be verified by applying Fleming's Right Hand Rule. Hence, one pole pitch is said to be of 180° electrical degrees or π radians. However, internally within the coil the two emfs will appear as aiding each other and the total emf across the coil will be the sum of these two emfs belonging to two different conductors.

Note that in electrical systems, the displacement measured in electrical degree θ_e represent the phase displacement while that measured in mechanical degree (θ_m) represents physical displacement. They are related by,

$$\theta_e = \frac{P}{2}\theta_m \tag{1.5}$$

1.4.4 Full Pitched and Fractional Pitched Coils

If the coil span of an armature coil is equal to one pole pitch than, it is termed as '*full pitched coil*.' In full pitched coil, both coil sides will link with same magnitude of stator field but having opposite polarities and therefore, the absolute magnitudes of *emf* induced in them will also be equal. However, if the coil span is less than one pole pitch, it is called as *fractional pitched coil*. In *dc* machines, the armature coils are normally full pitched. The net *emf* induced in a full pitched coil is always more than that induced in a fractional pitched coil.

1.4.5 Front Pitch, Back Pitch, Resultant Pitch and Average Pitch

The distance between two successive conductors *that are connected together on the commutator* side i.e., on the front side of rotor is called as Front Pitch (Y_F). The distance between two successive conductors *of the same coil* but measured on the rear side i.e., on the back side of rotor is called as Back Pitch (Y_B). The distance between the beginning of one coil and the beginning of next coil on the armature is called as Resultant Pitch (Y_F). The average pitch is given by,

$$Y_{av} = \frac{Y_F + Y_B}{2} \tag{1.6}$$

1.4.6 Commutator Pitch and Number of Commutator Segments

The distance between two copper segments of a commutator to which two terminals of the same armature coil are connected, is called as Commutator Pitch. It is measured in terms of number of commutator segments. The total number of commutator segments are equal to the number of coils.

1.4.7 Types of Armature Winding

Depending on the difference in arrangement of end connections at commutator side, the armature winding is classified as:

(i) *Lap winding*

(ii) *Wave winding.*

Selection amongst them depends on various factors like, required current rating, required voltage rating, size of the machine, available space, cost etc. Both lap and wave windings can be developed as

(i) Single layer winding

(ii) *Multilayer winding.*

In single layer winding, each armature slot consists of only one coil side whereas in double layer winding, two coil sides of different coils (one side per coil) are housed in each slot. For the satisfactory arrangement of end connections, double layer winding is commonly used. Lap and wave windings can also be designed and developed as

(i) *Simplex lap, Simplex wave,*

(ii) *Duplex lap, Duplex wave* and

(iii) *Triplex lap, Triplex wave* windings.

A duplex winding has two simplex windings in parallel whereas a triplex winding has three simplex windings in parallel. Usual practice to prefer the simplex windings.

While deciding the slots for coil sides, it is important to ensure that the two sides of every coil are placed under the stator poles of respective opposite polarities (North and South) so that, the *emf* induced in the two coil sides will always aid each other internally within their close circuit. Also, both lap and wave windings should form a close loop in themselves so that if we start from any point on one coil and traverse over all conductors of other coils, we should be able to reach back at the same point without any break or discontinuity in between.

1.5 SIMPLEX LAP WINDING

Lap winding is preferred for those applications where the *dc* machine is required to have low voltage and higher current rating. The lap wound *dc* generators are used as exciter or pilot generators supplying large *dc* field current at low voltage to the field winding of either three phase alternator or synchronous motor. The *dc* motors operating with battery supply may also have lap winding. In this winding, the two terminals of every armature coil are connected to the *adjacent* segments of commutator. Also, the finishing terminal of one coil is connected to the starting terminal of the next coil on the same commutator segment. Lap winding can be classified as,

(i) *Progressive lap* and

(ii) *Retrogressive lap winding*

The retrogressive type lap winding is usually preferred due to shorter span of the coil which also reduces the winding cost. Two coils wound as lap winding are shown in Fig. 1.16(a).

In Lap winding:

- Y_F and Y_B are always odd in number but never equal to each other ($Y_F \neq Y_B$). It means that and will differ by atleast 2.
  If $Y_F > Y_B$ than, $Y_F = (Y_B + 2)$ and it is called as *'Retrogressive Lap winding'*
  If $Y_F < Y_B$ than, $Y_B = (Y_F + 2)$ and it is called as *'Progressive Lap winding'*
- The average of and equals to one pole pitch.

$$Y_{av} = \frac{Y_F + Y_B}{2} = \frac{Z}{P}$$

- Resultant pitch $Y_R = (Y_B - Y_F)$.
- Commutator pitch = ±1.
- Number of *Parallel paths* (*A*) in the armature winding equals to the number of poles (*A* = *P*).
- Number of carbon brushes equals to the number of poles (P)
- Number of commutator segments = number of coils

Example:

For better understanding of Simplex lap winding, let us develop it for a *dc* machine having the following requirements: $P = 4$, armature coils = 12, double layer progressive winding, 1 turn per coil. The winding data is calculated as under. Since every coil has only one turn, conductor can also be termed as coil side.

- Total number of armature conductors Z = (12 × 2) = 24
- Pole pitch = $\frac{Z}{P} = \frac{24}{4}$ = 6 coil sides
- Average pitch $Y_{av} = \frac{Y_F + Y_B}{2} = \frac{Z}{P} = 6$, ∴ $Y_F + Y_B = 12$
- For Progressive lap winding, $Y_B = (Y_F + 2)$, ∴ $Y_F = 5$ coil sides, $Y_B = 7$ coil sides
- Being a double layer winding, number of slots = number of coils = 12
- Number of commutator segments = number of coils = 12
- Commutator pitch = ±1.

With above data, develop a winding diagram using the following steps as shown in Fig. 1.17.

Step 1: Calculate the winding data.

Step 2: Draw 12 commutator segments, C1, C2, C3,..........C12

Step 3: Draw 12 slots, S1, S2, S3,..........S12 by thin lines.

Step 4: Draw 2 North and 2 South poles (polarity of adjacent poles to be opposite) such that there will be 12/4 = 3 slots under each pole

Step 5: Draw two coil sides in each slot in the form of short parallel lines and number them as 1,2,3,.......24. If all odd numbers (1,3,5,7,9,11,13,15,17,19,21,23) represent coil sides in upper layer, the remaining with even number will represent those in lower layer.

Step 6: Assume and indicate the direction of current which will flow in each coil side by arrows. If current in coil sides under N-Pole is assumed downwards than in coil sides under S-Pole, it will be upwards.

Step 7: Make end connections on commutator side using the value of front pitch $Y_F = 5$. Coil side No.1 should be connected to Coil side No. (1 + 5) = 6. Similarly, Coil side No. 3 should be connected to Coil side No. (3 + 5) = 8 and so on.

Step 8: Make end connections at rear side using the value of back pitch $Y_B = 7$. Coil side No.1 should be connected to Coil side No. (1 + 7) = 8. Similarly, Coil side No. 3 should be connected to Coil side No. (3 + 7) = 10 and so on.

Step 9: Thus, the simplex, progressive, double layer, lap type armature winding with 12 coils is developed.

Step 10: Now let us determine the location of brushes on the commutator. For that, observe the direction of current in coil sides that were shown in Step 6. You will notice that at commutator segment C2, the currents from both coil sides 1 and 6 are incoming. Same is the case with C8. Similarly, at commutator segment C5, the currents of both coil sides 7 and 12 are outgoing. Same is the case with C11.

Therefore, for rotor position as shown in this Fig. 1.16, brushes should be located at C2, C5, C8 and C11. At C2 and C8, the brushes will receive current from the commutator whereas at C5 and C11, the current will leave the brushes and enter into the commutator. Thus 4 brushes shall be required which proves that in Lap winding, the number of brushes equals the number of stator poles.

Step 11: Brushes terminals at C2, C8 are joined together to form one terminal of armature winding and brush terminals at C5, C11 are joined together to form the other terminal.

1.5.1 Parallel Paths (A)

In Fig. 1.17, you will observe that the lap winding consisting of 12 coils got divided into 4 parallel paths with 3 coils per parallel path which proves that '*the number of parallel paths in a lap winding equals to the number of stator poles*'. This is simplified in Fig. 1.18. If '*E*' and '*I*' are the *emf* and current in each coil side than the net *emf* and current in every coil will be '2*E*' and '*I*' respectively. From Fig. 1.18, the *emf* across each parallel path and current in it will be '6*E*' and '*I*' respectively.

Emf generated across armature winding terminals will be same as the *emf* across parallel paths. By Kirchoff's current law (KCL), total output current of armature winding will be equal to the sum of currents in all parallel paths.

$$\therefore \quad E_g = 6E \text{ volts and } I_a = 4I \text{ amperes}$$

Fig. 1.17: Simplex lap winding for P = 4, 12 armature coils, double layer progressive type

1.6 SIMPLEX WAVE WINDING

In this type, the armature winding progresses in one direction from one coil to the next coil around the armature core just like waves. The two terminals of every armature coil are connected to the commutator segments a distance apart equal to the average pitch. Like lap winding, the wave winding can be classified as

(i) Progressive type

(ii) *Retrogressive type.*

A section of simplex wave winding is shown in Fig. 1.16 (b).

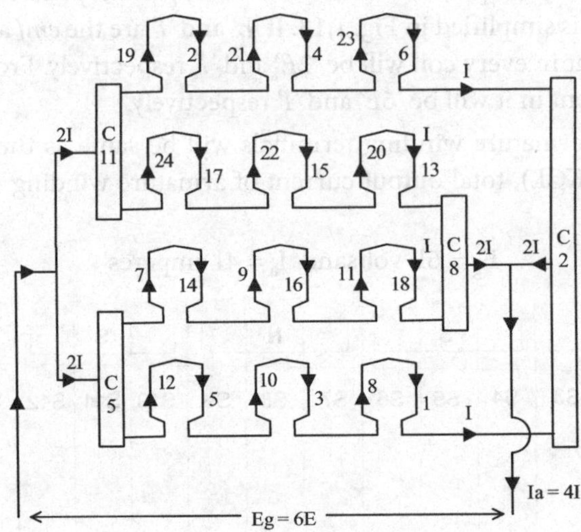

Fig. 1.18: Parallel paths in simplex lap winding

Example 1.2: A 4-pole dc generator having 48 armature coils is lap connected. If the machine generates a voltage of 240V and delivers a current of 80A to the load calculate, (a) the current delivered per brush, (b) the current flowing in each coil and (c) the average voltage induced per coil.

Data:

P = 4, A = 4 (since lap connected), armature coils = 48, Eg = 240V, Ia = 80A

Solution:

Since information about field winding is not provided, assume that, armature current = load current

(a) See Fig. 1.17 for clarity.

The load current of 80A flows out of the (+) ve terminal and back into the (–) ve terminal of the generator. Being a lap connected winding, the number of brushes = number of poles = 4. Two brushes will be (+) ve and 2 will be negative.

∴ Current per brush = 80/2 = 40A

(b) In lap winding, the number of parallel paths (A) = number of poles (P)

Therefore, each brush will gather current from 2 parallel paths.

Consequently, current in each coil will be 80/4 = 20A

(c) Voltage across each parallel path = voltage across winding terminals = 240V

Number of coils per parallel path = 48/4 = 12

∴ Voltage per coil = 240/12 = 20V

In Wave winding:

- Both Y_F and Y_B are odd and have same sign. They may be equal or differ by 2.
- The average pitch is given by,

$$Y_{av} = \frac{Y_F + Y_B}{2} = \frac{Z \pm 2}{P}$$

Where (+) sign is for *Progressive Wave winding* and (–) sign is for *Retrogressive Wave winding. It is necessary that, the average pitch given by above expression should be an 'integer'.*

- By substituting the given values of Z and P in above equation, ($Y_F + Y_B$) can be calculated. Deciding the individual values of Y_F and Y_B is based on whether the average pitch (Y_{av}) is an Even number or an Odd number.
    - If (Y_{av}) is an even number than: $Y_F = (Y_{av} - 1)$ and $Y_B = (Y_{av} + 1)$
    - If (Y_{av}) is an odd number than: $Y_F = Y_B = Y_{av}$
- Resultant pitch $Y_R = (Y_B + Y_F)$.
- Commutator pitch equals to the average pitch (Y_{av}).
- Number of *Parallel paths* (*A*) in the armature winding equals to *A* = *2* regardless of the number of stator poles.
- Number of brushes equals to 2. It can be raised to 4 to distribute the currents.
- Number of commutator segments = number of coils

For more understanding of Simplex wave winding, let us develop it for a *dc* machine of same specifications that were used for Lap winding in Section 1.5, Fig. 1.17. But before doing so, verify whether the mandatory requirement of average pitch as mentioned above will be fulfilled i.e.

$$Y_{av} = \frac{Z \pm 2}{P} \text{ should be an integer} \tag{1.7}$$

In Fig. 1.17, the number of coils were 12, *P* = 4 therefore *Z* = 24. *Yav* by above equation will be either 5.5 or 6.5 which is not an integer. It means that wave winding cannot be developed for *Z* =24, P =4 (though lap winding will be possible). For *Z* = 30, 34, 42 etc., wave winding can be developed with P = 4. Thus there is a restriction on the value of *Z* for given number of poles.

Example:

Let us now develop a simplex wave winding of retrogressive type, double layer, 15 coils and 4 stator poles. Let each coil consist of 1 turn. The winding data is calculated as under. Since every coil has only one turn, conductor can be also be termed as coil side.

- Total number of armature conductors Z = (15 × 2) = 30
- Average pitch $Y_{av} = \frac{Y_F + Y_B}{2} = \frac{Z - 2}{P}$ where (–) sign represents retrogressive type winding.

- Substituting Z = 30 and P = 4 we get, $Y_{av} = 7$.
- Since $Y_{av} = 7$ is an Odd number, ∴ $Y_F = Y_B = Y_{av} = 7$
- Being a double layer winding, number of slots = number of coils = 15
- Number of commutator segments = number of coils = 15
- Commutator pitch = $Y_{av} = 7$.

With above data, develop an unfolded winding diagram using the following steps as shown in Fig. 1.19.

Step 1: Calculate the winding data.

Step 2: Draw 15 commutator segments, C1, C2, C3,..........C15.

Step 3: Draw 15 slots, S1, S2, S3,..........S15 in thin lines.

Step 4: Draw two coil sides in each slot in the form of short parallel lines and number them as 1,2,3,.......30. If all odd numbers (1,3,5,7,9,11,13,15,17,19,21,23, 25, 27, 29) represent coil sides in upper layer, the remaining with even number will represent those in lower layer.

Step 5: Draw two North and two South poles with adjacent poles of opposite polarities. Since the pole pitch is not an integer, the number of slots covered by each pole will not be an integer.

Step 6: Assume and indicate the direction of current which will flow in each coil side by arrows. If current in coil sides under N-Pole is assumed upwards than in coil sides under S-Pole, it will be downwards.

Step 7: Make end connections at commutator side using the value of front pitch $Y_F = 7$. Coil side No.12 is connected to Coil side No. (12+7) = 19 and both are connected to Commutator segment C8. Similarly, Coil side No.10 is connected to Coil side No. (10 + 7) = 17 and both are connected to C7. Thus all other end connections can be made.

Step 8: Make end connections on rear side using the value of back pitch $Y_B = 7$. Coil side No.12 is connected to Coil side No. (12 – 7) = 5. Similarly, Coil side No.19 is connected to Coil side No. (19 + 7) = 26 and so on.

Step 9: Thus, the simplex, retrogressive, double layer, wave type armature winding with 15 coils is developed.

Step 10: Now let us determine the location of brushes on the commutator. For that, observe the direction of current in coil sides that were shown in Step 6. You will notice that at commutator segment C6, the currents from both coil sides 8 and 15 are incoming. Whereas, at commutator segment C10, the currents of both coil sides 16 and 23 are outgoing. At rest all other commutator segments, current of one coil side is incoming and that of other coil side is outgoing.

Therefore, for rotor position as shown in this Fig. 1.19, brushes should be located at C6 and C10. At C6, the brush will receive current from the commutator whereas at C10, the current will leave the brush and enter into the commutator. Therefore 2 brushes will be sufficient. Thus, in Wave winding, only 2 parallel paths are formed regardless of the number of stator poles. By Kirchoff's current law (KCL), total output current of armature winding will be equal to the sum of currents of all parallel paths, therefore $I_a = 2I$ Ampere.

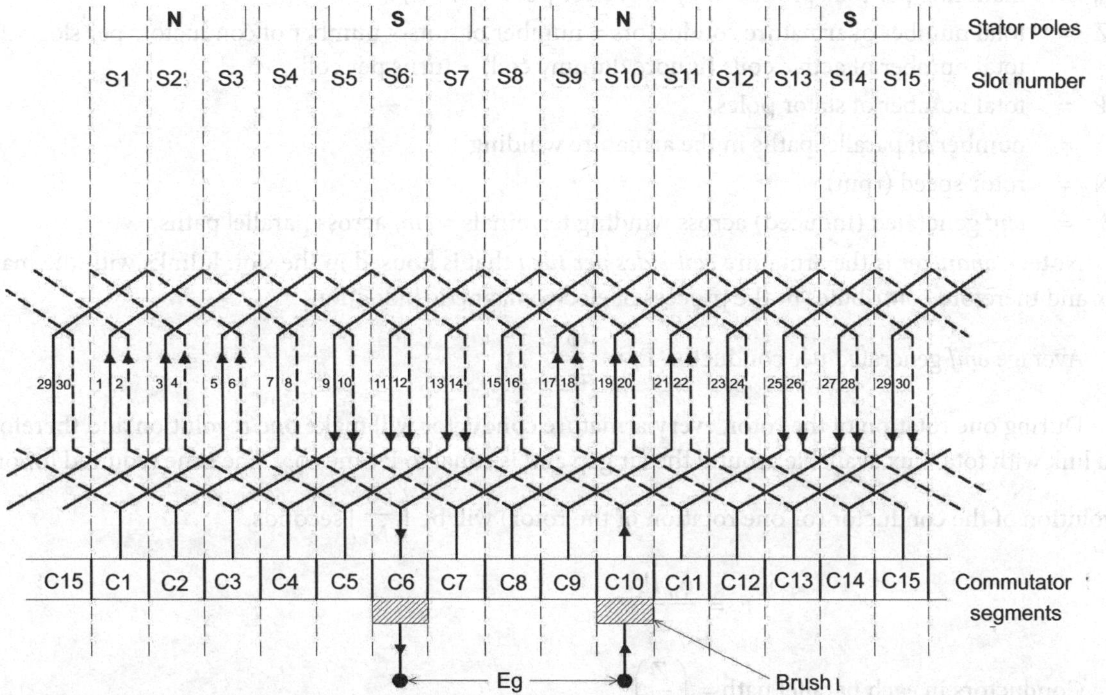

Fig. 1.19: Simplex wave winding for P = 4, 15 armature coils, double layer retrogressive type

1.6.1 Dummy or Idle Coil

In wave winding, there is a restriction on the value of Z for a given number of stator poles. It is, should be an integer.

$$Y_{av} = \frac{Z-2}{P}$$

Therefore, it is not possible to construct a wave winding with any number of armature coils. We have observed this limitation in Section 1.6. For $P = 4$, wave winding can be developed with odd number of armature coils like 15, 17, 21 etc. so that $Z = 30, 34, 42$ respectively. We know that in double layer winding, the number of slots are equal to the number of armature coils. In a *dc* machine, it is necessary that, the number of slots on both sides of the brush axis should be equal which means that the total number of slots should be even and not odd.

When the standard laminations are used in making of armature core for wave winding, the exact symmetry cannot be maintained. Therefore an extra coil is housed in the vacant slot portion to maintain the mechanical balance. It is not connected to the commutator. This coil is called as Dummy coil and used in relatively small machines. It is not preferred in large machines to avoid adverse effect on commutation.

1.7 EMF EQUATION

Let,

ϕ_m = main flux per pole produced by the stator poles (weber)

Z = total number of armature conductors = number of slots × number of conductors per slot = 2 × total number of active coils (ignore dummy coil) × turns per coil

P = total number of stator poles

A = number of parallel paths in the armature winding

N = rotor speed (rpm)

E_g = *emf* generated (induced) across winding terminals = *emf* across parallel paths

Note: *Conductor* is the armature *coil-sides per turn* that is housed in the slot. It links with the main flux and therefore contributes to the process of electromagnetic induction.

Average *emf* generated per conductor, $E_c = \dfrac{d\phi}{dt}$

During one rotation of the rotor, every armature conductor will make one revolution and therefore will link with total flux available around the air gap and is equal to P times ϕ_m. The time required for one revolution of the conductor (or one rotation of the rotor) will be $\left(\dfrac{60}{N}\right)$ seconds.

$$\therefore \quad E_c = \frac{\phi_m PN}{60}$$

Conductors in each parallel path = $\left(\dfrac{Z}{A}\right)$

The *emf* generated in each parallel path will be,

$$E_g = \frac{\phi_m PN}{60}\frac{Z}{A} \text{ volts} \tag{1.8}$$

The rotor speed can also be expressed in radians/sec as,

$$\omega = \frac{2 \neq N}{60} \text{ Rad/sec}$$

$$\therefore \quad E_g = \left(\frac{PZ}{2\pi A}\right)\omega\phi_m$$

Let K_e = Voltage constant,

$$\left(\frac{PZ}{2\pi A}\right) = K_e$$

$\therefore$ $$E_g = K_e \, \omega \phi_m \text{ Volts} \tag{1.9}$$

Since the terminals of parallel paths also appear as terminals of armature winding therefore, Eq. (1.8) and (1.9) represents the total *emf* generated in the armature winding of the *dc* generator. For Lap winding, $A = P$ and for Wave winding, $A = 2$.

Example 1.3: The armature winding of a 4-pole dc generator has 200 turns and runs at 800 rpm. At no-load, the voltage generated by the machine is 240V. Estimate the useful flux per pole when the winding is (a) lap connected and (b) wave connected.

Data:

P = 4, armature turns = 200, N = 800 rpm, Eg = 240V, ϕ_m = ?

Solution:

Number of conductors Z = 200 × 2 = 400

(a) When the winding is Lap connected, number of parallel paths A = P = 4

Since, no-load voltage (i.e. open circuit emf) $E_g = \dfrac{\phi_m PN}{60}\dfrac{Z}{A}$

$$\phi_m = \frac{60AE_g}{PNZ} = 0.045 \text{ Wb}$$

(b) When the winding is Wave connected, number of parallel paths A = 2

$$\phi_m = \frac{60AE_g}{PNZ} = 0.0225\text{Wb}$$

1.8 COMMUTATION

In Section 1.2.1, we have seen that the direction of induced current in all armature conductors reverses after one pole pitch i.e. when they move from the region of North pole to the region of South pole or vice-versa. This results into generation of alternating currents in armature winding though the machine is called as dc generator. The conversion of generated alternating current into direct current is carried out by the commutator and brushes.

In Fig. 1.17 and Fig. 1.19, it is observed that, the directions of armature current on the two sides of brush are always opposite. It means that when the brush shifts from one commutator segment to the next, the corresponding armature coil (referred as coil undergoing commutation) gets short-circuited between the adjacent commutator segments and the direction of current in it gets reversed. As armature rotates, all the armature coils in sequence undergo for commutation process – change of direction of dc current. This process of current reversal in the short-circuited coil is known as commutation. The time required for the reversal is called as commutation period 'tc'.

Refer to Fig.1.20 (a) – Fig.1.20 (e). Three armature coils 1, 2, 3 of a Lap winding are shown. They are connected to three commutator segments C1, C2, C3. A carbon brush is riding on the commutator, as armature rotates. The brush width is equal to the sum of width of one commutator segment and one mica insulation. Assume that due to rotation of armature, all armature coils are moving towards right with respect to the stationary carbon brush. Let the current in each coil be 10A so that the brush current will be 20A.

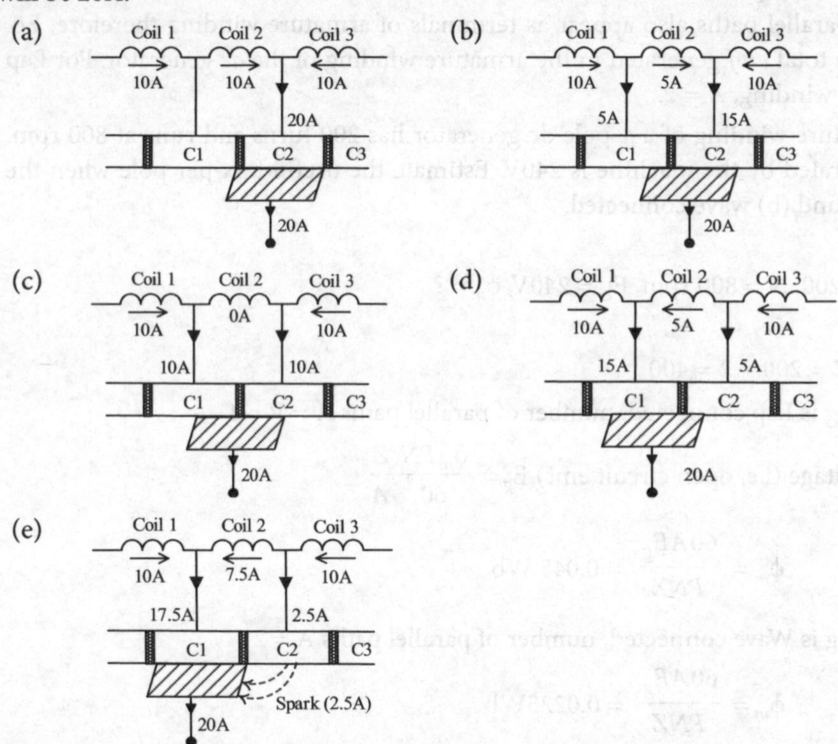

Fig. 1.20: Commutation process

In Fig. 1.20(a), the brush is in contact with commutator segment C2 and one mica insulation therefore, brush current will be the sum of currents in coil-2 and coil-3 (10 + 10 = 20A). This is the beginning of commutation period 'tc'.

In Fig. 1.20(b), the armature has moved towards right therefore, the brush has touched commutator segment C1. Now it is in contact with both C1 and C2 and hence has short circuited coil-2. Since the contact area of brush with C1 is less than that with C2, the contact resistances will be different and therefore current distribution is not equal (5 + 15 = 20A).

In Fig. 1.20(c), coil-2 has reached the middle of short circuit period. The brush contact area with C1 and C2 are equal and currents from coil-1 and coil-3 will directly enter into the brush because, relatively it will be the low impedance path (10 + 10 = 20A). Current in short circuited coil-2 will be zero.

In Fig. 1.20(d), due to further advancement of the coils, the contact area of brush with C1 is greater than that with C2, the contact resistances will be different and therefore current distribution is not equal (15 + 5 = 20A).

In Fig. 1.20(e), the brush is in contact with full commutator segment C1 and one mica insulation therefore, brush current will be the sum of currents in coil-1 and coil-2 (10 + 10 = 20A). This is the end of commutation period 'tc'.

If the current in coil undergoing commutation (coil:2) reverses linearly from +I (+20A) to –I (-20A) within the commutation period 'tc', it is called Linear Commutation. This is shown in Fig. 1.21. Practically, there are factors which try to avoid linear commutation and therefore reversal of current from +I to –I does not complete within the commutation period 'tc'. Some current is left even after the brush has advanced ahead and lost its contact with the previous commutator segment. The remaining current enters into the brush from the previous segment through air resulting into a spark between them.

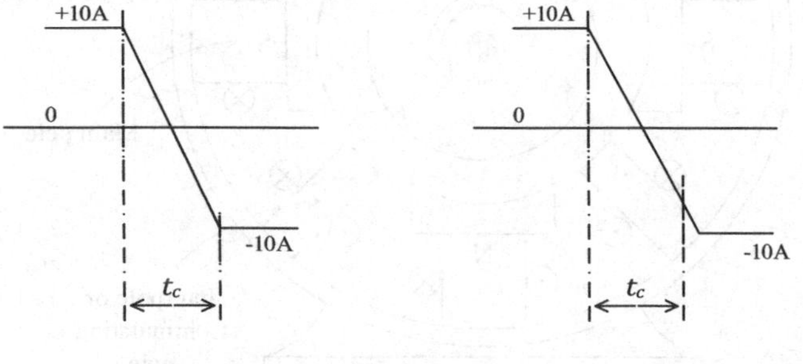

(a) Linear commutation (b) Under commutation

Fig. 1.21: Coil current versus time during commutation

One of the prime factors that prevent linear commutation is the inductance of the coil undergoing commutation. The change in its current results into two voltages in the coil. One due to self-induction and other due to mutual induction. Sum of these two voltages is called *reactance voltage.* It causes the coil current to lag in time compared to that in linear commutation and therefore delays its reversal. Hence the condition is also known as *delayed commutation* or *under-commutation.* It causes sparking between the brush and commutator segment. This condition can be avoided if we ensure coil inductance at minimum value by using a fewest possible number of turns per armature coil.

If the resistive voltage drop in the commutating coil circuit is made more significant compared to reactance voltage than, the delayed commutation effect can be minimized. This can be achieved by increasing the contact voltage drop between commutator and brush which is resistive in nature. Hence carbon brushes are preferred due to their appreciable contact drop. When commutation is improved by using the advantage of resistive voltage drop, the process is known as *resistance commutation.*

Alternately, if we produce a rotational voltage in the commutating armature coil such that it will oppose the reactance voltage than, the commutation process can be improved. This method is called as *voltage commutation. Interpoles* or *commutating poles* are popularly used for voltage commutation.

1.8.1 Interpoles or Commutating Poles

Reactance voltage induced in the armature coil undergoing commutation (i.e. reversal of current) is primarily responsible for the delay in commutation process and results into sparking between the commutator segments and brush. Interpoles are the additional small poles fixed on the inner periphery of the yoke and located on the GNA between the larger main field poles. The number of interpoles are equal to the number of main poles. They do not have pole shoes and do not undergo magnetic saturation. A simple commutating winding of few number of turns but thicker wire is wound over the interpoles. This is shown in Fig. 1.22.

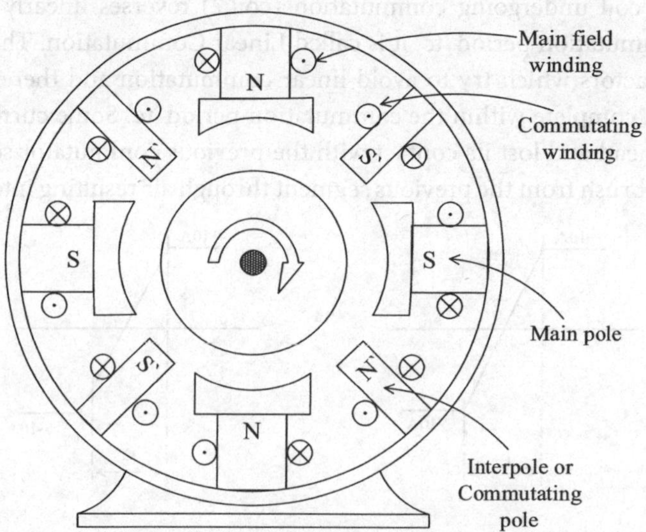

Fig. 1.22: DC generator with main poles and Interpoles

Interpoles are supposed to establish sufficient mmf which will induce a rotational *emf* in the commutating armature coil (i.e the coil undergoing commutation) and cancels the reactance voltage. Also, it should be able to neutralize the cross magnetizing armature mmf in the interpolar region near the tips of main poles. Since both, reactance voltage and cross magnetizing mmf are proportional to the armature current, the commutating winding (also called as interpole winding) on the interpoles is connected in series with the armature winding. The direction of commutating winding must be such that the polarity of interpole will be same as main pole ahead of it (in the direction of rotor rotation) for a generator and that of main pole just behind it for the motor. Interpoles have localized effect in interpolar region only and cannot help in performance of the machine.

1.9 GNA, MNA & BRUSH AXIS

In *dc* machines, the distribution of main flux ϕ_m in airgap is symmetrical about the centre line of field poles i.e. the centre line of North and South poles. This centre line is called as *field-axis, polar axis* or *direct axis* (*d-axis*). It is shown in Fig. 1.23 for a bi-polar *dc* machine. Exactly midway between the field poles but perpendicular to the polar axis is the *Geometrically Neutral Axis* (GNA). It is also called as *interpolar axis*.

When the armature rotates through 360°, the armature conductors will have flux linkage with ϕ_m of different flux densities at different locations. While crossing the field axis, the flux density encountered by them is maximum whereas at GNA, flux density is zero. At field axis, they will cut ϕ_m at right angle whereas when they pass across GNA, every conductor will move almost parallel to ϕ_m. Consequently, maximum *emf* is induced in the armature conductors on field axis and zero on GNA. Thus, when alone is present, GNA acts like a magnetically neutral zone and therefore, is also called as *Magnetic Neutral Axis* (MNA).

The brushes are placed on MNA to improve commutation which reduces sparking between the trailing edge of brush and commutator. At MNA, the short circuited coil undergoing commutation links with no flux density and therefore the rotational voltage induced in the coil is negligibly small. The axis passing through the brushes is called as '*Brush axis*' and since it is electrically at quadrature to the d-axis (i.e. field axis), it is also called as *quadrature axis* or *q-axis.*

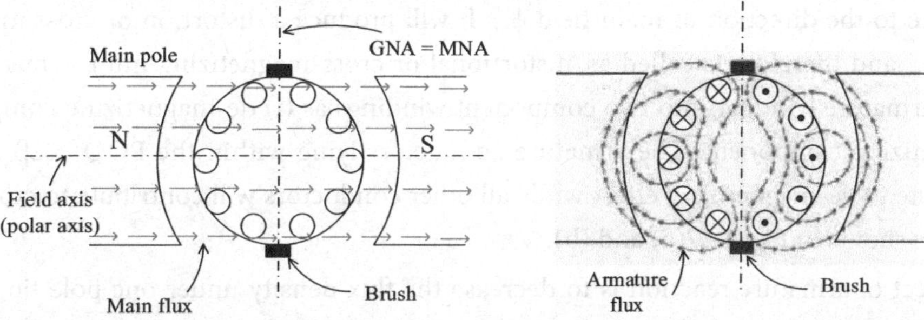

Fig. 1.23: Main flux when **Fig. 1.24:** Armature flux when

1.10 ARMATURE REACTION

Effect of armature flux ϕ_a on main flux ϕ_m is called as Armature Reaction effect. The distribution of main flux ϕ_m in the airgap when $I_a = \phi_a = 0$ are shown Fig. 1.23. At no-load, the armature current will be zero. But when load is connected (electrical load for generator mode and mechanical load for motor mode), $I_a \neq 0$. This armature current I_a will establish its own flux called as armature flux ϕ_a. Fig. 1.24 presents the distribution of armature flux ϕ_a when the main flux is absent ($\phi_m = 0$)

In a practical dc machine, both ϕ_m and ϕ_a exist simultaneously when the load is connected. Resultant of the two is shown in Fig. 1.25. From these figures, you will notice that the effect of armature current (which has produced ϕ_a) is to displace the resultant flux in the direction of rotation of rotor when the machine is acting as a dc generator. However when it operates as a dc motor, the displacement of resultant flux is against the direction of rotation. This can be verified by applying right hand thumb rule as well as the Fleming's rules. The Magnetic Neutral Axis (MNA) that is always at right angle to the resultant field too gets displaced due to armature reaction. Therefore, the brushes should be shifted forward in the direction of rotation of the rotor when the machine is operating as generator and shifted backward opposite to the direction of rotation in order to bring them on the new MNA. Let the angle by which the brushes are shifted from GNA be denoted by β. This is shown in Fig. 1.26 for a dc generator. It is found that the resultant flux density at the leading edges of the pole shoes has weakened while it has increased at the trailing edges.

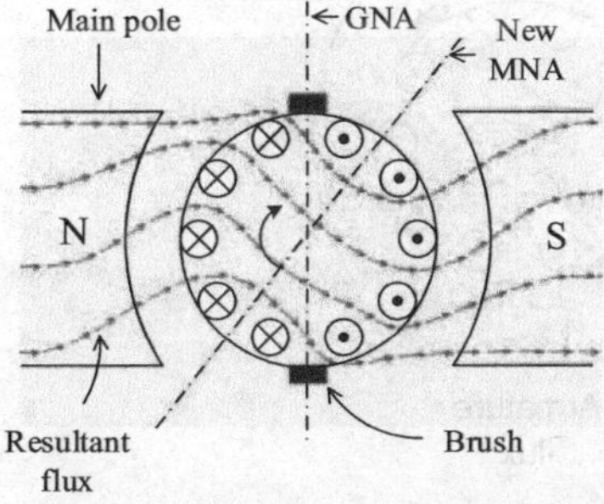

Fig. 1.25: Resultant flux in a *dc* generator due to armature reaction

After the brush shift, armature flux ϕ_a will also get shifted as shown in Fig 1.26. The armature mmf '*Fa*' can be resolved into two components '*Fd*' and '*Fc*'. The direction of '*Fd*' is in direct opposition to the direction of main field ϕ_m and therefore is called as 'de-magnetizing mmf'. The direction of 'Fc' is in quadrature to the direction of main field ϕ_m. It will produce a distortion or cross magnetization of main field ϕ_m and therefore is called as 'distortional or cross-magnetizing mmf'. Thus the effect is to resolve the armature winding into two component windings as (i) de-magnetizing component and (ii) cross magnetizing component. The armature conductors lying within the POQ = 2β and XOY = 2β will contribute to de-magnetizing effect while all other conductors will contribute to cross magnetizing effect. This is shown in Fig. 1.27(a) and (b).

The effect of armature reaction is to decrease the flux density under one pole tip and increase it under the other causing saturation. Consequently, the resultant flux becomes less than what it would had been in the absence of armature reaction when ϕ_m alone is present and therefore, the generated emf also decreases. The increased flux density at some parts of the machine will increase the iron loss. It will also increase the maximum voltage between adjacent segments of the commutator at load than at no-load. If it goes beyond 30V, there is a possibility of a continuous spark between adjacent segments which may culminate into ring of fire. The armature reaction also causes delayed commutation, hence more sparking at the brushes. These reasons lead to faster wear and tear of commutator and brushes.

The effect of armature reaction is to decrease the flux density under one pole tip and increase it under the other causing saturation. Consequently, the resultant flux becomes less than what it would had been in the absence of armature reaction when ϕ_m alone is present and therefore, the generated emf also decreases. The increased flux density at some parts of the machine will increase the iron loss. It will also increase the maximum voltage between adjacent segments of the commutator at load than at no-load. If it goes beyond 30V, there is a possibility of a continuous spark between adjacent segments which may culminate into ring of fire. The armature reaction also causes delayed commutation, hence more sparking at the brushes. These reasons lead to faster wear and tear of commutator and brushes.

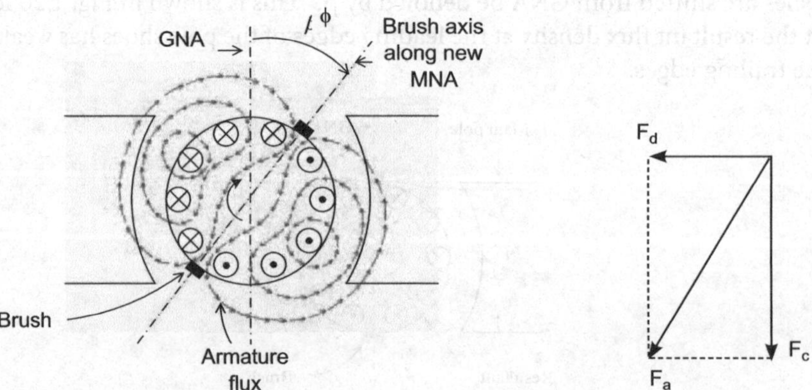

Fig. 1.26: Armature flux alone after brush shift

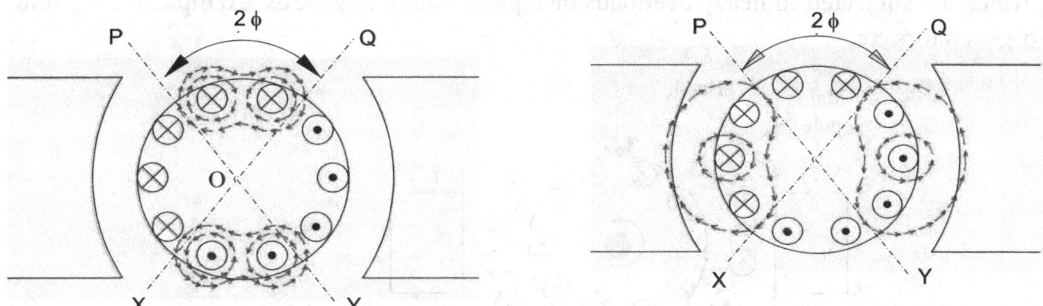

Fig. 1.27(a): Demagnetizing armature conductors **Fig. 1.27(b):** Cross magnetizing armature conductors

Armature reaction effect can be reduced by adopting the following measures:

- Shifting of brush axis from GNA to MNA
- Increasing the length of airgap at the pole tips so as to avoid magnetic saturation at pole tips which is caused by overcrowding of flux due to armature reaction. The length of airgap at pole tip can be 1.5 to 2 times the airgap length at pole centre.
- Increasing reluctance of pole tip
- Use of compensating winding of nullify the armature reaction effect in polar region
- Use of interpoles to nullify the armature reaction effect in interpolar region

1.10.1 Compensating Winding

The cross magnetizing armature reaction is proportional to armature current. In case of overloading or rapid changes in load (both in generator and motor), the armature current responds to the changes in load by making similar changes in itself. This results in the shifting of resultant flux in forward and backward direction with every change in load. In other words, if you refer to Fig.1.25, the position of new MNA will keep changing with every change in load. The shifting of resultant flux will induce a *rotational emf* in the armature coils whose magnitude will depend on the rapidity and magnitude of changes in the load. At the instant when an armature coil is located at the peak of a badly distorted resultant flux wave, the rotational *emf* may be high enough to cause dielectric breakdown of the insulation between adjacent commutator segments to which the coil is connected. This will result in a flashover or arcing between the segments. The maximum allowable voltage between adjacent commutator segments is of the order of 30 to 40V.

To address this problem, slots are made in the face of stator field pole shoes and an additional winding called as *compensating winding* is housed in them. This winding is connected in series with armature winding and made to carry the armature current in such a way that the current in them will be in opposite direction to that flowing in the armature conductors directly below the pole shoes which are responsible for the cross magnetizing effect. This is shown in Fig. 1.28. Thus a compensating flux created by compensating winding will be in opposition to the armature flux directly below the pole shoes. This will minimize the distortion of main flux and secure a sparkless commutation which otherwise would be substantial in the absence of compensating winding. It can be verified by applying the right hand thumb rule. Compensating winding causes increase in machine cost and hence is usually employed in big size *dc* machines which are subjected to heavy overloads or rapidly fluctuating loads, example: rolling mill motors, turbo-generators etc.

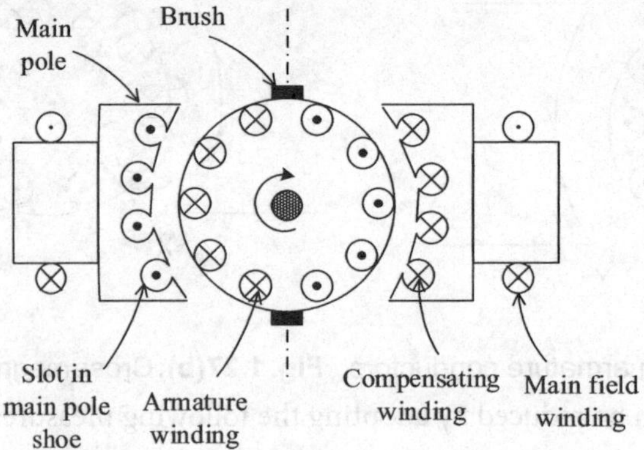

Fig. 1.28: DC generator with compensating winding

1.11 TYPES AND APPLICATIONS OF DC GENERATOR

Dc machines (both, *dc* generator and *dc* motor) are classified depending on the mode of excitation. In other words, they are based on the connection of field winding with main winding and how the same will be energized to establish the main magnetic field. Fig. 1.29 shows the classification.

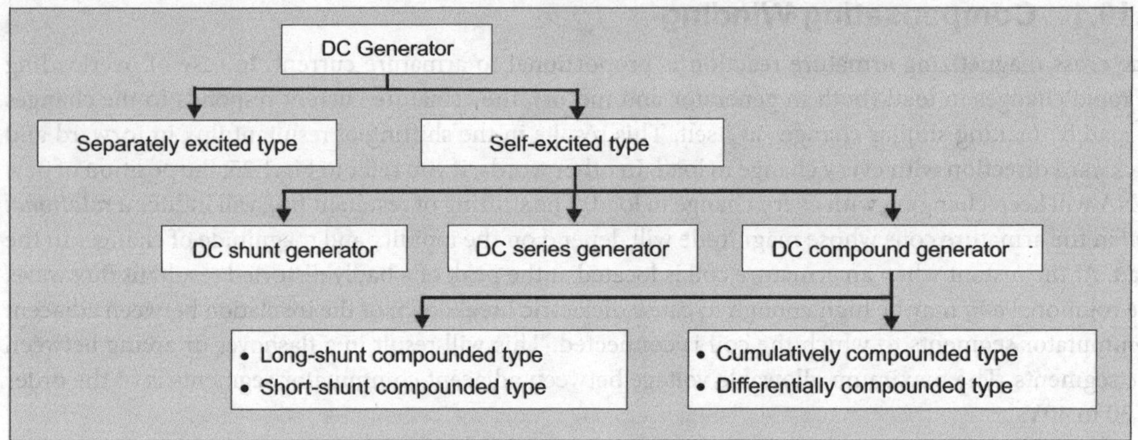

Fig. 1.29: Classification of dc generator

1.11.1 Separately Excited DC Generator

In separately excited *dc* generator, the field and armature windings are not interconnected. The main magnetic field ϕ_m is established either by,

(i) Connecting a separate voltage source across the field winding or by

(ii) Replacing the set of stator poles and field winding by permanent magnets.

Schematic circuit diagram of the former one is shown in Fig. 1.30. The field winding is made of many turns of thin copper wire with effective resistance R_f much higher than the armature winding resistanceand is connected across a *dc* voltage source V_f. If V represents the terminal voltage across armature winding then by applying Kirchoff's voltage law (KVL) in field and armature circuits we have,

$$V_f = I_f R_f \qquad (1.10)$$

The generated emf $E_g = V + I_a R_a$

$$\therefore \qquad V = (Eg - I_a R_a) = K_e \omega\phi_m - I_a R_a \qquad (1.11)$$

In addition to the ohmic drop $I_a R_a$, voltage drop at the brush contacts also take place in the armature circuit. This drop is approximately constant and is of the order of 2 volts with carbon brushes provided that the armature current per cm² contact area of the brushes do not exceed 7A. For simplicity, the brush contact resistance is assumed to be constant and is included in . The performance of a dc generator can be assessed using following characteristics.

- Magnetization characteristic or open circuit characteristic (OCC)
- Internal characteristic
- External characteristic

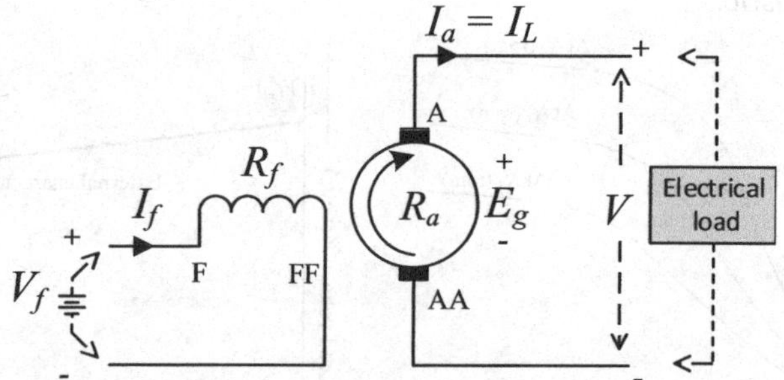

Fig.1.30: Separately excited dc generator

A. *Magnetization characteristic:*

The magnetization characteristic of a *dc* generator is also known as no-load characteristic and is similar to the magnetization curve of a ferromagnetic material. It is a curve of field current (I_f) versus the open circuit voltage across armature winding terminals with the machine rotated at a constant speed. From above voltage equation, during open-circuit, armature current $I_a = 0$, therefore, resistive voltage drop $I_a R_a = 0$ and the armature terminal voltage equals to generated *emf* (V_a=E_g). At $I_f = 0$, the externally provided excitation is absent but still a small flux is available between the poles which generates a small *emf* in the armature winding. This is the residual flux due to residual magnetism already available in the main poles of all types of *dc* generator. It is shown in Fig. 1.31. In absence of residual magnetism, once the field winding need to be connected to *dc* supply voltage.

Initially at low values of I_f, the main flux ϕ_m is small, poles are unsaturated and therefore E_g increases linearly. Later, due to magnetic non-linearity, some reduction in flux takes place (4 to 8%) and therefore a greater increase in I_f is required to produce the same rise in E_g as that in the lower part of the curve. DC generators are usually designed so that the operating point 'P' which corresponds to the rated value of armature terminal voltage lie on knee of the magnetization curve. Since E_g α speed (N), a family of OCC can be obtained for different values of speed. The magnetization curves (OCC) of dc generator at different speeds ($N_1 > N_2 > N_3$) are shown in Fig. 1.31.

B. *External characteristics:*

This characteristic shows the relationship between armature terminal voltage V_a and the load current (same as armature current I_a provided that the field current I_f and rotor speed are held constant at their rated values. At no-load, the armature circuit is open, armature current $I_f = 0$ and therefore the no-load terminal voltage and generated emf are equal ($V_{a(nl)} = E_g$). With rise in electrical load, current I_a will increase in direct proportion to the load and result into the rise in resistive voltage drop $I_a R_a$. This is evident from the following equation,

$$V = (Eg - IaRa) \tag{1.12}$$

where Ra actually includes the resistance of armature winding, resistance of brush and resistance of brush contact. The voltage drop due to brush contact is approximately constant, of the order of 2V per brush provided that the current per cm² contact area does not exceed 7A. Apart from the IaRa drop, there is an additional drop in the terminal voltage from no-load to full load condition and it is due to the demagnetizing armature reaction. Both effects will cause voltage drop in the armature circuit resulting into the fall of terminal voltage as shown in Fig. 1.32. This terminal voltage drop is about 5 – 10% from no-load to full load condition. The resulting curve is called as external characteristic.

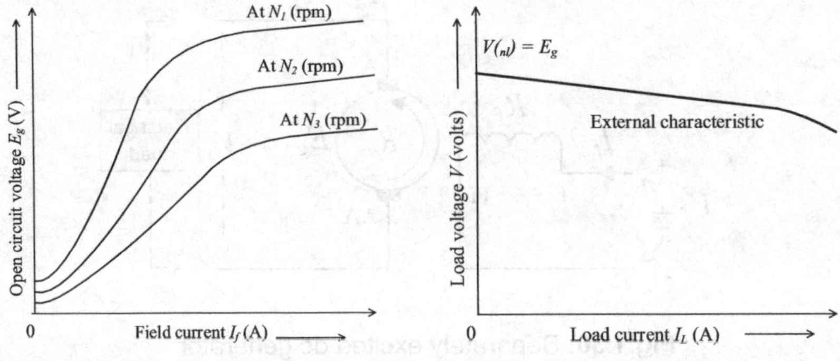

Fig. 1.31: OCC of dc generator **Fig. 1.32:** External characteristic of dc generator

1.11.2 Self Excited *DC* Generator

In self-excited *dc* generator, the field winding is connected either in parallel or in series with armature winding and accordingly, the machine is called as *dc* parallel generator and *dc* series generator. DC compound generator has two field windings, one is connected in series and the other in parallel with armature winding. In all types of self-excited *dc* generators, the machine derives its excitation from the *emf* generated in armature winding and no separate voltage source is connected across the field winding. Initially, when the rotor is at standstill, the generated *emf* $E_g = 0$, therefore the machine cannot excite it's field winding to produce the main field ϕ_m. To help the machine to self-excite and build up the generation of *emf* from this stage, following conditions should be fulfilled.

(a) It is necessary that there should be some residual magnetism in the field poles.
(b) Field winding must be connected in a direction so as to aid the residual magnetism.
(c) The field winding resistance should be less than the critical resistance value.
(d) The rotor speed should be greater than the critical speed value.
(e) The field circuit should be closed.

1.11.3 DC Shunt Generator

Dc shunt generators are the widely used types of *dc* generator. They need no external power for excitation and provides almost steady terminal voltage from no-load to full load. Schematic circuit diagram of a *dc* shunt generator is shown in Fig. 1.33. The voltage building process starts with rotation of the armature by means of prime mover. Due to the presence of a small amount of residual main flux ϕ_m in the airgap, the armature conductor have flux linkages with it and this process induces a small magnitude of *emf* E_g in the armature winding. Since the two windings together make a close circuit, the *emf* E_g causes some current to flow in the field winding. If this shunt field current I_{sh} sets up in the direction so as to aid the residual flux, the resulting airgap field produced by main poles ϕ_m becomes stronger than before and therefore the magnitude of E_g increases. Consequently, I_{sh} increases and builds up E_g further. The mutual cumulative reinforcement between E_g and ϕ_m continues till equilibrium is reached. Thereafter, there is no further rise in E_g. The point of equilibrium depends on the open circuit characteristic, critical resistance and critical speed. Apply KVL in the two close loops of Fig. 1.33.

$$V = (E_g - I_a R_a) = K_e \omega\phi_m - I_a R_a \tag{1.13}$$

$$V = I_{sh} R_{sh} \tag{1.14}$$

$$K_e = \left(\frac{PZ}{2\pi A}\right)$$

A. *Critical resistance (R_c)*

Critical resistance R_c of a *dc* shunt generator is the maximum magnitude of its field winding resistance with which the machine will just excite for a given speed N. For any further rise in R_{sh} beyond R_c, the machine fails to self-excite and cannot build up its generated voltage. Refer to Fig. 1.34. It shows straight lines of different slopes representing the shunt field winding resistances R_{sh1}, R_{sh2}, R_{sh3} obtained by the relation $V_a = I_{sh} R_{sh}$. Three open circuit characteristic curves OCC1, OCC2, OCC3 corresponding to different speeds N_1, N_2 and N_3 are also plotted.

It is observed that, R_{sh1} is tangent to OCC1 which corresponds to speed N_1, R_{sh2} is tangent to OCC2 which corresponds to speed N_2 and R_{sh3} is tangent to OCC3 which corresponds to speed N3. Therefore, R_{sh1} will be termed as the critical resistance at speed N_1, R_{sh2} will be the critical resistance at speed N_2 and R_{sh3} will be the critical resistance at speed N_3.

Now observe the points of intersection P_A and P_B between (i) OCC1 and R_{sh3}, (ii) OCC1 and R_{sh2} respectively. In the first case, voltage coordinate of point P_A represents the maximum emf E_{gA} which the generator can build up at speed N_1 when the shunt field winding resistance is R_{sh3}. If the winding resistance is increased to R_{sh2}, the maximum *emf* that could be generated decreases to E_{gB} as given by point P_B. For rise in winding resistance to R_{sh1}, the resistance line will become tangent to OCC1 at which the machine will just self- excite. Hence R_{sh1} will be called the critical resistance R_c at speed N_1. For any further rise in field winding resistance the R_{sh} line will not intersect with OCC1 and the machine will fail to self-excite, giving only a small voltage due to residual magnetism. Thus the critical resistance value depends on operating speed.

B. *Critical speed (N_c)*

Critical speed N_c of a *dc* shunt generator is the minimum value of rotor speed with which the machine will just self-excite for a given value of its shunt field resistance R_{sh}. For any further drop in rotor speed below N_c the machine will fail to self-excite and cannot build up its required generated voltage. Refer to Fig. 1.34 again. Consider OCC_1 and R_{sh1} line. Since R_{sh1} is tangent to OCC_1, R_{sh1} will be the critical resistance at speed N_1. Similarly, N_1 will be termed as critical speed N_c for the given shunt field resistance R_{sh1}. If the winding resistance is maintained constant at R_{sh1}, but the speed is reduced below N_1 (say N_2 corresponding to OCC2), than the OCC will not intersect the resistance line which means, the machine cannot self-excite. Thus, the critical speed value depends on the field winding resistance of *dc* shunt generator.

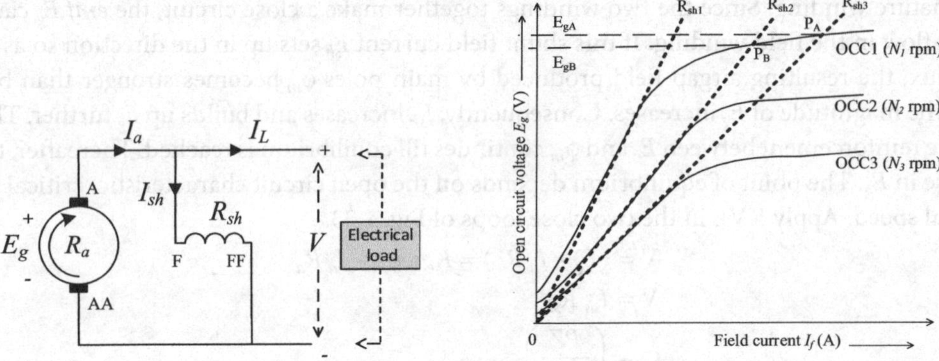

Fig.1.33: DC shunt generator **Fig. 1.34:** Critical resistance and critical speed

C. *No-load characteristic or OCC of a dc shunt generator*

Variable rheostat is connected in series with the field winding of a dc shunt topology and a voltmeter is connected across the armature winding terminals. The rotor is rotated at constant rated speed and the shunt field current I_{sh} is varied in steps by adjusting the rheostat. Corresponding readings of armature terminal voltage are noted along with the values of I_{sh}. A graph of I_{sh} versus $V_{a(nl)}$ is plotted to give OCC. Practically, there is no difference between the OCC of a *dc* separately excited generator and a shunt generator.

D. *External characteristic of dc shunt generator*

External characteristic is the relationship between armature terminal voltage V_a and the load current I keeping the shunt field resistance R_{sh} and rotor speed N constant. The load connected across armature winding terminals is varied in steps and the corresponding values of V_a and I are noted. When the generator is loaded, voltage across armature winding terminals decreases due to following major reasons:

(a) Ohmic voltage drop $I_a R_a$ in armature winding

(b) Voltage drop due to armature reaction effect

(c) Being a self-excited generator, the voltage drop in V_a due to the above two factors results in the drop in shunt field current I_{sh} which reduces the generated emf E_g. This is because, the voltage V_a appears across shunt field winding also.

The first two factors are common with separately excited generator but the third factor which is the most significant is applicable to *dc* shunt generator only. It is due to the third factor, the external characteristic of a *dc* shunt generator is more drooping than that of a separately excited generator. It is shown in Fig. 1.35.

1.11.4 DC Series Generator

The field winding is connected in series with the armature winding and load. Therefore, the circuit should be closed for self-excitation. Since the field winding carries load current, it is made up of thick wire or strips with few number of turns. Series generators are rarely used except as voltage boosters. The circuit connection is shown in Fig. 1.36. If the field current crosses the permitted limit, the magnetic poles may get saturated and the developed voltage falls to zero. This can be observed from its external characteristics. Apply KVL in the two close loops of Fig. 1.36.

$$V = E_g - I_a(R_a + R_{se}) \tag{1.15}$$

1.11.5 DC Compound Generator

DC compound generators have two field windings. One is connected in series with the armature winding and the other in parallel. Both field windings ϕ_{sh} are wound on the same stator poles and are excited simultaneously. The magnetic field established by shunt field winding is ϕ_{se} and that produced by the series field winding is . Resultant of the two is the main magnetic field ϕ_m. If ϕ_{sh} and ϕ_{se} are in same direction, the machine is called as *cumulatively compounded generator* and if they oppose each other, it is known as *differentially compounded*. This is shown in Fig. 1.37.

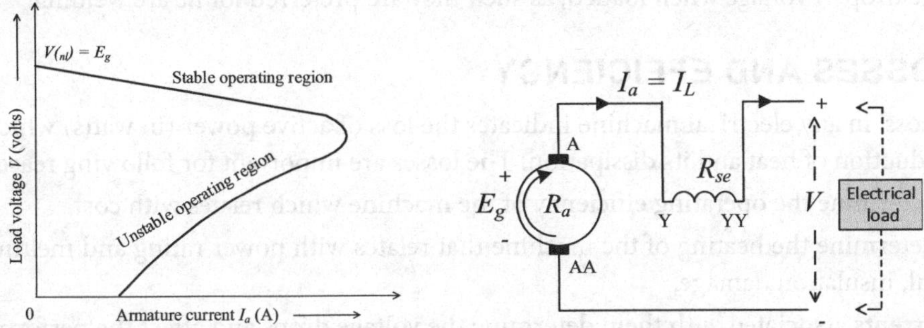

Fig. 1.35: External characteristic of a dc shunt generator **Fig. 1.36:** DC series generator

Compound generator can also be classified as short shunt generator and long shunt generator. It depends on the circuit topology. Circuit connections are shown in Fig. 1.38 (a) & (b). Performance characteristics of a dc compound generator can be attuned to suit the load requirements by adjusting the number of turns of appropriate field winding as over compounded, under compounded and flat compounded.

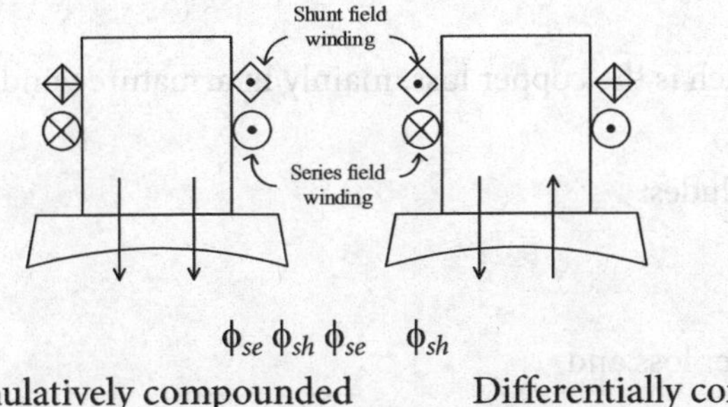

Fig. 1.37: Cumulative and differential type

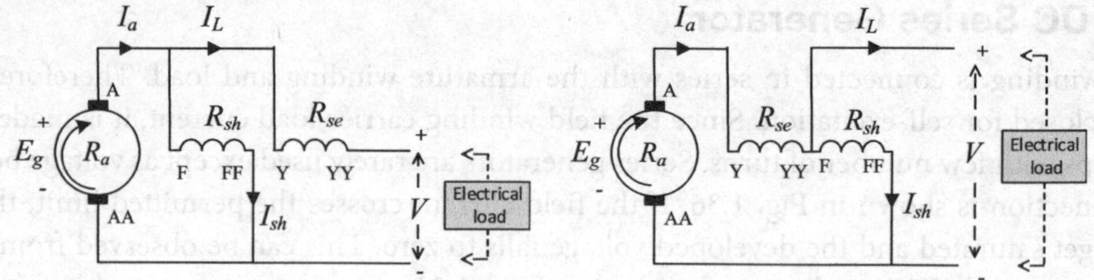

Fig. 1.38 (a): DC short shunt generator **Fig. 1.38 (b):** DC long shunt generator

Amongst the different types of *dc* generator discussed above, separately excited type and cumulatively compounded type generators are commonly preferred. Separately excited generators can provide a wide range of output voltages whereas the self-excited generators may produce unstable voltages in lower ranges particularly when the field resistance becomes equal to critical resistance or more than that. The external characteristics of shunt and separately excited generators are drooping in nature with respect to rise in load. On the other hand, a cumulatively compounded generator produces a substantially flat voltage characteristic or the one which rises with load. Cumulatively compound generator can be used to provide almost constant *dc* voltage irrespective to connected load. Differentially compound generator exhibits large drop in voltage when loaded, as such they are preferred for *dc* arc welding.

1.12 LOSSES AND EFFICIENCY

The word 'Loss' in any electrical machine indicates the loss of active power (in watts) which culminates into the production of heat and its dissipation. The losses are important for following reasons:

- They determine the operating efficiency of the machine which relates with cost.
- They determine the heating of the machine that relates with power rating and melting of winding material, insulation damage.
- The currents associated with them determine the voltage drops and affect the performance.

In general, power loss have three categories:

- loss in electrical circuit,
- loss in magnetic circuit
- loss in mechanical system.
- A high voltage machine may have dielectric loss also.

For a *dc* machine (generator and motor), let us discuss them in detail. Loss can be broadly classified as:

- *Variable loss* which is the copper loss, mainly in armature winding
- *Constant loss* (*Pc*)

The constant loss includes:

- core loss,
- mechanical loss,
- shunt field copper loss and
- stray loss.

1.12.1 Copper Loss

The copper loss is also called as *ohmic loss* or *resistive loss*. It takes place in all the windings where current in flowing against the winding resistance and is given by $I^2 R$ watts. The armature current varies with the magnitude of load and therefore its corresponding $I_a^2 R_a$ loss also changes. Hence, it is also called as *variable loss*. In *dc* machines, the copper loss takes place in the following windings.

(a) Copper loss in armature winding, $I_a^2 R_a$ watts
(b) Copper loss in compensating winding (it carries armature current)
(c) Resistive loss in brushes and brush contacts
(d) Copper loss in series field winding (only in series type and compound type machines)
(e) Copper loss in shunt field winding (only in shunt type and compound type machines)

The current in shunt field winding is normally constant and do not change considerably with any variation in load therefore, the associated copper loss $I_{sh}^2 R_{sh}$ remains same at all load conditions and is treated as one of the constant loss.

1.12.2 Core Loss

The core loss is also called as *open-circuit loss*, *no-load loss* or *iron loss* as it takes place in the iron core of the machine. The armature core of *dc* machine is subjected to rotational magnetization. The losses result due to the cyclic changing flux densities in the magnetic core as armature rotates. It is further classified as:

- Hysteresis loss
- Eddy current loss

Hysteresis loss

It originates from the repeatedly changing orientation of dipoles when subjected to changes in the magnetization of the core due to rotation. In *dc* machines, the armature core and slot teeth thereon experience these changes when moving from North pole to South pole and vice-versa. Also, the stator pole shoes also experience the change in magnetization due to armature reaction. Hysteresis loss can be expressed as:

$$P_h = K_h f B_{\max}^n \frac{W}{kg} \tag{1.16}$$

where, K_h is the proportionality constant dependent on the material characteristics and volume of iron core, f represents the frequency of reversal of magnetization and the exponent n ranges between 1.5 and 2.5 with a value 2.0 being usually used, it is referred as Steinmetz constant. Hysteresis loss increases noticeably when the machine is loaded because the armature *mmf* due to load current changes the space distribution of flux densities significantly. However, this rise in core loss is accounted as a part of stray loss. Hysteresis loss can be minimized by selecting a high permeability material for the magnetic core.

Eddy current loss

This loss takes place due to the induction of eddy currents in the magnetic core when it is subjected to changing flux densities. It is dependent on the squares of maximum flux density (B_{max}) in Wb/m^2, frequency (f) in cycles per second and thickness of single lamination (τ) in meter as,

$$P_e = K_e(B_{max} f\tau)^2 \frac{W}{kg} \tag{1.17}$$

where K_e is the proportionality constant dependent on resistivity and volume of iron core. Eddy current loss is minimized by fabricating the magnetic core from a stack of laminations where every lamination is a thin stamping of required shape and insulated on both sides.

Mechanical loss

This loss consists of friction at the bearings and brushes. Rotating structures of machines cuts air and certain power is lost in that. It is jointly referred as friction and windage loss.

Stray loss

When the machine is loaded, due to leakage and fringing, the flux links with other metallic parts of the machine like outer frame, brush holder, shaft, bearings etc and causes magnetic losses. Secondly, due to the presence of armature slots, the airgap flux density distribution is never uniform. The resulting losses are called as stray load loss. Practically, it is difficult to determine this loss with accuracy and usually 1.0 percent of the output power is considered as stray loss.

1.12.3 Efficiency (η)

Efficiency of any system is the measure of its quality which tells us about the level of output that can be derived from the system for given input. For an electric generator, it is expressed as,

$$\eta = \frac{\text{output power in watts}}{\text{input power in watts}}$$

$$\eta = \frac{\text{output power}}{\text{output power + variable loss + constant loss}}$$

$$= \frac{VI_L}{VI_L + I_a^2 R_a + P_c} \tag{1.18}$$

Example 1.4: A four pole dc shunt generator having lap connected armature winding has armature resistance of 0.15 Ω, shunt field resistance of 100 Ω and 64 armature conductors. The effective area of cross section per pole is 2.5 × 10^{-2} m^2. It generates 25 kW at a voltage of 250 V. If the armature is rotated at 600 rpm, calculate the airgap flux.

Data:

P = 4, A = 4 (since lap connected), R_a = 0.15 Ω, R_{sh} = 100 Ω, Z = 64, Per pole area of cross section = m^2, Po = 25 kW, V = 250 volts, N = 600 rpm,

Solution:

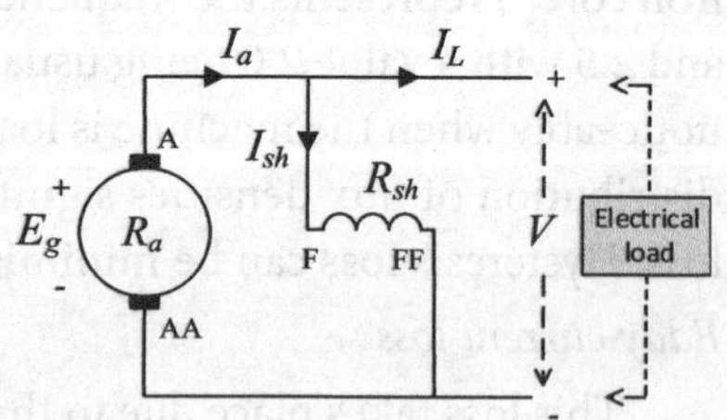

$$\text{Load current } I_L = \frac{P_0}{V} = 100 \text{ A}$$

$$\text{Shunt field current } I_{sh} = \frac{V}{R_{sh}} = 2.5 \text{ A}$$

$$\therefore \quad \text{Armature current } I_a = I_L + I_{sh}$$

$$= 102.5 \text{ A}$$

Generated emf $E_g = V + I_a R_a = 265.375$ volts

But, $$E_g = \frac{\phi_m PN}{60}\frac{Z}{A}$$

∴ $$\phi_m = \frac{60AE_g}{PNZ} = 0.4146 \text{ Wb}$$

Example 1.5: The armature of a four pole, lap wound shunt generator has 120 slots with 4 conductors in each slot. The flux per pole is 0.05 Wb. The armature resistance and shunt field resistance are 0.05 Ω and 50 Ω. Find the speed of the machine when supplying 45 A at a terminal voltage of 250V.

Data:

P = 4, A = 4, slots = 120, 4 conductors/slot, Ra = 0.05 Ω, Rsh = 50 Ω, , IL = 45A, V = 250 volts, N = ?

Solution:

$$Z = 120 \times 4 = 480$$

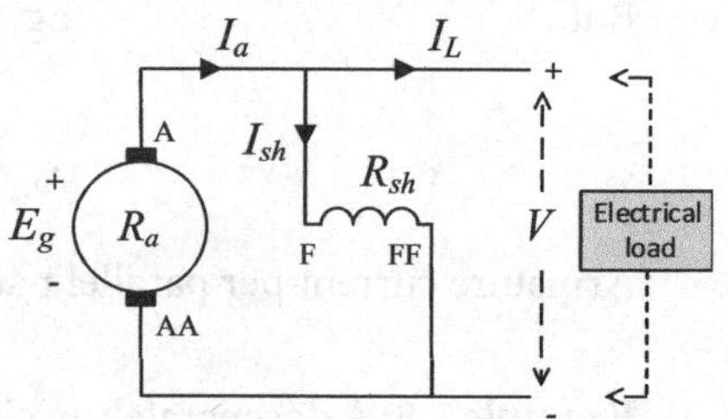

Shunt field current $I_{sh} = \dfrac{V}{R_{sh}} = 5\text{A}$

∴ Armature current $I_a = I_L + I_{sh} = 50$ A

Generated emf $E_g = V + I_a R_a = 252.5$ volts

But, $$E_g = \frac{\phi_m PN}{60}\frac{Z}{A}$$

∴ $$N = \frac{60AE_g}{PZ\phi_m} = 631 \text{ rpm}$$

Example 1.6: A lap wound dc shunt generator having 80 slots with 10 conductors per slot generates at no-load an emf of 400V when running at 1000 rpm. At what speed it should be rotated to generate a voltage of 220V on open circuit.

Data:

Slots = 80, 10 conductors/slot, E_{g1} = 400V at N_1 = 1000 rpm. For E_{g2} = 220V, N_2 = ?

Solution:

Since, $$E_g = \frac{\phi_m PN}{60}\frac{Z}{A};$$

∴ $$\frac{E_{g1}}{E_{g2}} = \frac{N_1}{N_2}$$

Substituting the values

∴ $$N = 550 \text{ rpm}$$

Example 1.7: The lap connected armature of a 4 pole, dc shunt generator is required to supply the following loads connected in parallel at 250V: (i) 5 kW geyser, (ii) 2.5 kW lighting load. The generator has an armature resistance of 0.2 Ω and a field resistance of 250 Ω. The armature has 120 conductors in the slots and runs at 1000 rpm. Allowing 2V for brush contact drops and neglecting friction, calculate flux per pole and armature current per parallel path.

Data:

P = 4, A = 4, Z = 120, $R_a = 0.2\ \Omega$, $R_{sh} = 250\ \Omega$, V = 250 volts, Po = (5 + 2.5) = 7.5kW, contact drop = 2V, N = 1000 rpm, ϕ_m = ?, I_a/A = ?

Solution:

$$\text{Load current } I_L = \frac{P_0}{V} = \frac{7500}{250} = 30\text{A}$$

$$\text{Shunt field current } I_{sh} = \frac{V}{R_{sh}} = 1\text{A}$$

Armature current

Generated emf

But,

$$Eg = \frac{\phi_m PN}{60}\frac{Z}{A}$$

$$\phi_m = \frac{60AE_g}{PNZ} = 0.129\text{Wb}$$

Armature current per parallel path

$$= 31/4 = 7.75\text{A}$$

Example 1.8: A *dc* separately excited generator delivers 80A at 240V to a resistive load when running at 800 rpm. Calculate the load current and voltage if the prime mover speed is reduced to 600 rpm. The armature resistance is 0.1 Ω and contact drop due to brushes = 2V. Neglect armature reaction.

Data:

$R_a = 0.1\ \Omega$, contact drop = 2V, V_1 = 240 volts, I_{a1} = 80A, N_1 = 800 rpm, For N_2 = 600 rpm, V_2 =? I_{a2} = ?

Solution:

$$\text{Load resistance } R_L = \frac{240V}{80A} = 3\Omega$$

When N_1 = 800 rpm, the generated *emf* is,

$$E_{g1} = V_1 + I_{a1} R_a + \text{brush drop}$$

$$= 240 + (80 \times 0.1) + 2 = 250\text{V}$$

Since the field excitation (I_f) is not changed, when N_2 = 600 rpm,

The new terminal voltage $V_2 = I_{a2} R_L = 3I_{a2}$

$$E_{g2} = V_2 + I_{a2} R_a + \text{brush drop}$$

$$= 3I_{a2} + (I_{a2} \times 0.1) + 2$$

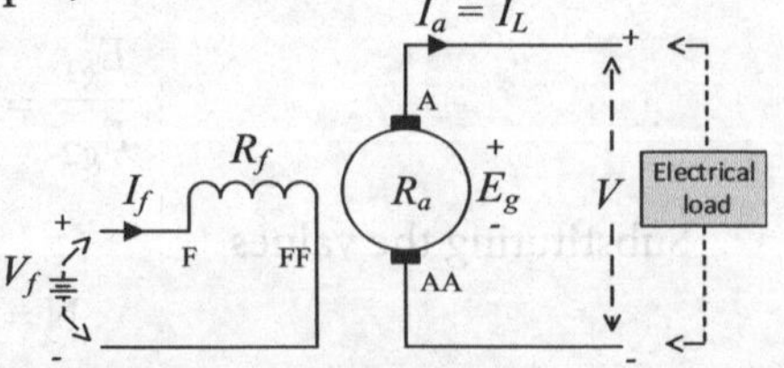

$$\therefore \quad I_{a2} = \frac{(187.5-2)}{3.1} = 59.838\text{A}$$

$$V_2 = 3I_{a2} = 179.5 \text{ volts}$$

Example 1.9: A dc separately excited generator is connected to a dc bus of 440V and supplies 500 kW to the load. If the load is reduced to 250kW, what change will be required in the rotor speed so as to maintain the terminal voltage at same value? Explain this change. Consider 0.02 Ω as resistance between the armature winding terminals and ignore armature reaction effects.

Data:

$R_a = 0.02\ \Omega$, $V_1 = V_2 = 440$ volts, $P_{01} = 500kW$, $P_{02} = 250kW$, $(N_2 - N_1) = ?$

Solution:

Case-1:

$$\text{Output power } P_{o1} = V_1 I_{a1} \therefore I_{a1} = 1136.363A$$

$$\text{Emf generated } E_{g1} = V_1 + I_{a1} R_a$$

$$\therefore \quad E_{g1} = 462.727$$

Case-2:

$$\text{Output power Po2} = V_2 I_{a2}$$

$$\therefore \quad I_{a2} = 568.181A$$

$$\text{Emf generated } E_{g2} = V_2 + I_{a2} R_a$$

$$\therefore \quad E_{g2} = 451.363V$$

Since, $$E_g = \frac{\phi_m PN}{60}\frac{Z}{A};$$

$$\therefore \quad Eg\ \alpha \phi_m\ N$$

Assuming constant field excitation, $\therefore \phi_m = \text{constant}$

$$\therefore \quad \frac{E_{g2}}{E_{g1}} = \frac{N_2}{N_1} \text{ or } \frac{N_2 - N_1}{N_1} = \frac{E_{g2} - E_{g1}}{E_{g1}}$$

The percentage reduction in rotor speed should be,

$$\therefore \quad = \frac{451.363 - 462.727}{462.727} \times 100 = -2.45\%$$

Explanation:

The rotating mechanism of every electrical generator is acted upon by two different torques. One is the driving torque (T_d) provided by prime mover and the other one is counter-acting torque (T_c) (produced due to armature current (I_a). When, $T_d = T_c$, the rotor speed is constant.

If the load P_o is reduced, the armature current I_a will also decrease and so will result into drop in T_c. Therefore, $T_d = T_c$ and the rotor speed will rise. This change will tend to increase E_g as well as the terminal voltage V.

To maintain the terminal voltage constant at its previous value, the rotor speed should be reduced from prime mover side.

1.13 ELECTRICAL MEASURING INSTRUMENTS

Any individual who enters the electrical engineering laboratory to conduct basic experiments will be using the ammeter, voltmeter and wattmeter. It is always useful to understand the operation of some of the measuring instruments before actually using the. In this section, we shall discuss the basic principles of most commonly used instruments. Measuring instruments are broadly classified in following two ways.

(a) Absolute type
(b) Secondary type

(a) Indicating type
(b) Recording type
(c) Integrating type

1.13.1 Absolute Instruments

The absolute instruments are those which present the value of measured quantity in terms of deflection and physical constants of the instrument. It is not required to compare this reading with that of any standard instrument i.e. calibration is not required. Example: Rayleigh current bridge, tangent galvanometer.

1.13.2 Secondary Instruments

These instruments can be used for the measurement of unknown quantity only if their calibration is carried out with standard instrument. Example: Moving iron instrument, moving coil instrument electrodynamic instrument etc.

1.13.3 Indicating Instruments

Indicating instruments indicate the root mean square (RMS) value of the measured quantity that is valid only for the instant when measurement was carried out. Example: Ammeter, voltmeter, wattmeter etc. All indicating instruments that indicate the magnitude of measured quantity by means of pointer deflection require the following three torques for their operation.

- Deflecting torque (T_d)
- Restoring torque or controlling torque (T_c)
- Damping torque (T_{dp})

At steady state deflection of the pointer, $T_d = T_c$

A. *Deflecting torque*

Deflecting torque causes the moving system and the pointer to move from zero position and acquire a steady state deflection proportional to the magnitude of measured quantity. The production of this torque depends upon the principle employed in the instrument. Example: Moving iron type, moving coil type, dynamometer type etc.

B. *Controlling torque*

This torque acts in direct opposition to the deflecting torque and limits the movement of the moving system and ensures that the magnitude of deflection of the pointer is unique and is always the same for a given value of electrical quantity to be measured. In the absence of this torque, with deflecting torque acting alone, the magnitude of deflection will be maximum for all magnitudes of measured quantity and will not return back to zero position even after the electric supply is disconnected. Control torque can be provided by following methods: Spring control method, gravity control method etc.

C. *Damping torque*

Damping torque minimizes the oscillations of the pointer about its final steady state deflection and makes it steady. In the absence of this torque, the pointer after reaching to its final position (where $T_d = T_c$) will start oscillating. Therefore, to damp out these oscillations, damping torque is introduced which is in opposition to the deflecting torque and exists only when the pointer is in motion. Depending on its magnitude, the damping levels can be classified as underdamped, over damped and critically damped. It can be provided by following three methods: Air friction or pneumatic damping, fluid friction damping and eddy current or electromagnetic damping.

1.13.4 Recording Instruments

These instruments not only indicate the value of measured quantity valid for the instant when measurement was carried out but also provides a record of variations in it over a predetermined period of time. Example: Cardiogram, X-Y plotter, strip chart recorder, storage Oscilloscope, data logger etc.

1.13.5 Integrating Instruments

The integrating instruments are those which measure, register and integrate the quantity being measured over a predefined period of time. Example: Energy meter, odometer etc.

UNIT SUMMARY

- The electromechanical device that converts mechanical energy into electrical energy is called as electric generator whereas the one which converts electrical energy into mechanical energy is called as electric motor.
- The generator receives mechanical energy in rotational form from a device generally referred as prime mover.
- The motional emf induced in a conductor is given by where is $e = Blv \sin\theta$ volts the angle between velocity vector $\underline{v}$ and magnetic field vector $\underline{B}$. When the conductor moves at right angle to the flux direction, $\sin\theta=1$.
- Fleming's right hand rule is used to determine the direction of motional emf induced in a conductor having flux linkages with a magnetic field.
- The *emf* actually generated by a *dc* generator is of alternating type. It is converted into *dc* by commutator and brushes.
- The armature core and stator pole shoes are laminated to minimize the eddy current loss.
- The stationary part called of a dc machine is called as stator and consists of Yoke, poles and field winding. The armature core, armature winding and commutator together represents the rotational part called as armature.
- The armature coil-sides that lie in the slots of armature core are termed as conductors.
- The number of conductors = twice the number of turns.
- The armature winding is of two types: Lap type and Wave type.
- The number of armature conductors per stator pole is called as pole pitch.
- The electrical and mechanical angles are related by $\theta_e = \frac{P}{2}\theta_m$ where P = number of poles.
- In lap winding, the number of parallel paths A = P whereas in wave winding A = 2.
- The generated emf = no-load voltage = open circuit voltage $E_g = \frac{\phi_m PN}{60}\frac{Z}{A}$ volts.

- When the brush shifts from one commutator segment to the next, the corresponding armature coil gets short-circuited and the direction of current in it gets reversed. This process of current reversal in the short-circuited coil is known as commutation.
- Interpoles are employed to cancel the reactance voltage by inducing a rotational emf in the commutating coil and thus improves commutation.
- The centre line of North and South poles is called as field-axis, polar axis or direct axis (d-axis). An axis exactly midway between the adjacent poles but electrically perpendicular to the polar axis is also called as interpolar axis.
- Effect of armature flux ϕ_a on main flux ϕ_m is called as Armature Reaction.
- Armature reaction effects are of two types: (a) cross-magnetizing and (b) demagnetizing.
- Compensating winding neutralizes the cross magnetizing armature reaction effect.
- Armature voltage equation for a separately excited and shunt generator is $V = (E_g - I_a R_a)$
- Armature voltage equation for a dc series generator is $V = E_g - I_a (R_a + R_{se})$
- In self excited generators, availability of residual magnetism in field poles is necessary.
- Critical resistance R_c of a *dc* shunt generator is the maximum magnitude of its field winding resistance with which the machine will just excite for a given speed N.
- Critical speed N_c of a *dc* shunt generator is the minimum value of rotor speed with which the machine will just excite for a given value of its shunt field resistance R_{sh}.
- DC compound generator has two field windings, one is connected in series with armature winding and the other in parallel.
- In compound generator, if ϕ_{sh} and ϕ_{se} are in same direction, it is cumulatively compounded and if they oppose each other, it is known as differentially compounded.
- Indicating instruments indicate the RMS value of the measured quantity that is valid only for the instant when measurement was carried out.
- Recording instrument provides a record of variations in the measured quantity over a predetermined period of time.
- Integrating instrument integrates the measured quantity over a predefined period of time.

EXERCISES

Multiple Choice Questions

1. In dc shunt generator, the machine builds up its voltage from zero due to,
   (a) permanent magnets (b) residual magnetism in the poles
   (c) external voltage source (d) armature current
2. Dummy coil is used to,
   (a) to maintain mechanical balance (b) to minimize the power loss
   (c) to fulfil the pole pitch (d) to generate emf
3. Carbon brushes are mounted,
   (a) on MNA between adjacent poles (b) at the centre of North pole
   (c) at the centre of South pole (d) in the armature slots

4. The emf induced in the armature of a dc generator is,
   (a) directly proportional to main flux and inversely proportional to rotor speed
   (b) inversely proportional to main flux and directly proportional to rotor speed
   (c) directly proportional to main flux and to rotor speed both
   (d) inversely proportional to main flux and to rotor speed both
5. The electrical angle between adjacent opposite poles of a 4-pole generator is
   (a) 45° (b) 180°
   (c) 90° (d) 360°
6. The armature of a dc generator having lap winding consists of 72 slots with 6 conductors per slot. The number of commutator segments will be,
   (a) 216 (b) 432
   (c) 108 (d) 72
7. A 4-pole dc machine has 12 slot armature with lap wound progressive winding and 2 coil sides per slot. The front pitch will be,
   (a) 3 (b) 6
   (c) 7 (d) 5
8. In a dc machine, the armature mmf is always directed along the,
   (a) polar axis (b) brush axis
   (c) interpolar axis (d) none of these
9. Due to armature reaction, the shift in MNA is,
   (a) in the direction of rotation of rotor for dc generator and against the direction of rotation in dc motor
   (b) in the direction of rotation of rotor for dc generator and dc motor both
   (c) in the direction of rotation of rotor for dc motor and against the direction of rotation in dc generator
   (d) against the direction of rotation of rotor for dc generator and dc motor both
10. In a dc machine, the armature reaction and inductance of commutating armature coil results in,
    (a) linear commutation (b) delayed commutation
    (c) over commutation
11. Under commutation results into,
    (a) sparking at the leading edge of brushes (b) sparking at the trailing edge of brushes
    (c) Sparking at the middle of brushes (d) no sparking at all
12. Interpoles in dc machine are employed to counteract the,
    (a) cross magnetizing effect of armature mmf in commutating zone and the reactance voltage
    (b) demagnetizing effect of armature mmf in commutating zone and the reactance voltage
    (c) reactance voltage alone
    (d) armature mmf

13. Polarity of interpoles is,

(a) same as that of main pole ahead in dc generator and the main pole behind in dc motor
(b) same as that of main pole ahead in dc motor and the main pole behind in dc generator
(c) same as that of main pole ahead in dc generator and in dc motor both
(d) same as that of main pole behind in dc generator and in dc motor both

14. In a dc machine having interpoles as well as compensating winding,

(a) Interpole winding and compensating winding are connected in series with armature winding
(b) Interpole winding is connected in series with armature winding whereas compensating in parallel
(c) Interpole winding and compensating winding are connected in parallel with armature winding
(d) Interpole winding is connected in parallel with armature winding whereas compensating in series

15. The critical resistance in dc generator is the resistance of its,

(a) armature winding (b) field winding
(c) commutating winding (d) compensating winding

16. Effect of armature reaction is to,

(a) weaken the main flux (b) strengthen the main flux
(c) reverse the main flux (d) protect the main flux

17. For voltage build up in a dc series generator, it is necessary that the,

(a) armature circuit should be closed (b) armature circuit should be open

18. For voltage build up in a dc shunt generator, it is necessary that the,

(a) field resistance should be greater than critical resistance
(b) field resistance should be less than critical resistance
(c) field resistance should be equal to the critical resistance

ANSWERS															
1.	(b)	2.	(a)	3.	(a)	4.	(c)	5.	(b)	6.	(a)	7.	(d)	8.	(b)
9.	(a)	10.	(b)	11.	(b)	12.	(a)	13.	(a)	14.	(a)	15.	(b)	16.	(a)
17.	(a)	18.	(b)												

Short Answer Questions

1. Name the different components of dc generator.
2. What are the functions of Yoke?
3. Describe construction of a commutator.
4. What will be the effect of rise in speed on the induced emf of a dc shunt generator?
5. What will be the effect of reduction in excitation current on the emf of a separately excited generator?
6. What is the effect of change in load on the terminal voltage of a dc shunt generator? Explain.
7. Why the brushes should be placed on magnetically neutral axis?
8. What is reactance voltage in dc generator?

9. Explain the difference between shunt, series and compound generators based on construction.
10. Distinguish between commutating winding and compensating winding.
11. How will you adjust the terminal voltage of a dc shunt generator?
12. Compare slip ring, split ring and commutator.
13. What should be properties of the insulation between adjacent commutator segments?
14. How to decide the width of a carbon brush?
15. What is the purpose of key hole in the armature core?
16. How laminated core helps in reducing eddy current loss?
17. Why sparking is produced between brush and commutator?
18. What should be done to minimize this sparking?
19. What is armature reaction?
20. Why brush axis should be displaced from GNA?

Long Answer Questions

1. With the help of labelled diagram, explain the construction of dc machine and functions of its different parts.
2. What is a commutator? Explain its need and operation in dc generator.
3. What is the cause of sparking between the commutator and brushes? How it can be minimized?
4. Explain (i) resistance commutation and (ii) voltage commutation
5. Discuss the function and working of interpoles.
6. Why is it required to shift the brush axis in a dc machine? Compare this shift when operating as a generator and as motor.
7. What is compensating winding? Explain its connection and operation.
8. Classify dc machine into different types.
9. Discuss the working of self-excited generators.
10. Explain the terms (i) critical resistance and (ii) critical speed.
11. Explain armature reaction. Discuss its effect on the machine performance.
12. Why do the terminal voltage of a dc shunt generator drop with rise in load? Write relevant equations.
13. Discuss the different types of losses in dc machine.
14. What are the different types of electrical measuring instruments? Briefly explain each type with examples.
15. Discuss briefly the difference types of torques required in indicating instruments.

Numerical Problems

1. A 6-pole dc generator has 250 armature conductors and runs at 1000 rpm. If the no-load emf generated across the armature winding terminals is 500V, calculate the useful flux per pole when the winding is (i) lap connected and (ii) wave connected.

2. In numerical problem 1.1, for the same value of flux per pole, if 600V is to be generated on open circuit across armature winding, at what speed the rotor should be rotated? Consider lap connection.
3. A dc separately excited generator having 0.04 Ω armature resistance and 2V brush drop supplies 125V and 200A to a load resistance when running at 1000 rpm. If the speed is reduced to 800 rpm, what will be the new current and voltage supplied to the load. Neglect armature reaction.
4. A 6 pole dc generator has a simplex wave wound armature and flux per pole of 0.02Wb. If the prime mover speed is 750 rpm, what should be the number of armature turns to generate an emf of 500V. If the rotor speed is increased to 1000 rpm, calculate the emf generated considering the same number of turns and flux per pole.
5. The lap wound armature of a 4 pole dc generator has 90 slots and 6 conductors per slot. If the machine runs at 1400 rpm, what will be the generated emf. If the current per conductor is 120A, determine the rated power output. Consider flux per pole equal to 50 mWb.
6. A 6 pole dc generator has wave wound armature with 36 slots, 2 coil sides per slot and 8 turns per coil. If the axial length of each pole is 18 cm and the mean diameter of the airgap between stator and rotor is 25 cm, calculate the emf generated across armature winding. Assume the average flux density over one pole pitch equal to 0.8 Wb/m2.
7. The lap connected armature of a 4 pole dc machine is 14.5 cm in radius and 21 cm in active axial length. It has 33 slots and 33 coils each having only 11 turns. The arc of each pole shoe is 70% of the pole pitch. If the average flux density per pole is 0.8 Wb/m^2 and rotor speed is 1200 rpm, calculate the armature induced emf. Also find the current per conductor if the armature current is 240A.
8. A dc short shunt compound generator has wave connected armature winding and $R_a = 0.15\Omega$, $R_{se} = 0.05\Omega$, $R_{sh} = 110\ \Omega$. The number of poles are 4 and brush contact drop is 0.9V per brush. If the machine delivers 230V and 50A to the load, calculate the emf generated and current per armature conductor.

Answers of Numerical Problems

1. 0.12Wb/pole, 0.04 Wb/pole;
2. 1200 rpm;
3. 159. 4A, 99. 6V;
4. 333 turns, 666.6V;
5. 630V, 302 kW
6. 650V
7. 389V, 60A
8. 242V, 26A

PRACTICAL

Aim: To dismantle a DC machine

Different parts of a dc machine are shown in Fig. 1.15. It is redrawn below.

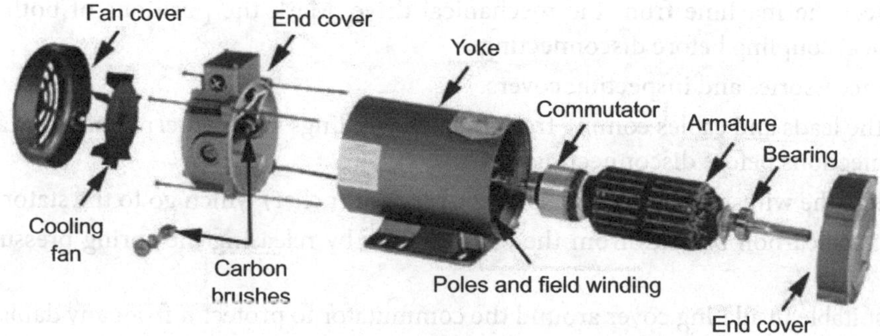

Fig. 1.39: Different parts of *dc* machine

Machine specifications:

Refer to the name plate of the machine and note down following *specifications.*

kW/HP:	Frame:	RPM:
Insulation Class:	Armature Voltage:	Duty:
Type:	Armature Current:	Max. Ambient Temp. (oC):
Cooling:	Field Voltage:	Weight (kg):
Protection:	Field Current:	Enclosure:
Make:		

Precautions:

The following precautions should be taken while dismantling a *dc* machine.

1. Before starting the disassembling or re-assembling of a *dc* machine, refer to the respective manufacturer's manual for specific precautions. Check whether, it is a horizontal assembly machine or a vertical assembly machine.
2. The stator of a *dc* machine has two sides. The side where commutator is located is called 'non-drive end' (N-end) and the other side where the shaft can be coupled with the mechanical load or a prime mover is called as 'drive end' (D-end). The armature should be dismantled or reassembled only from the drive end of the stator.
3. When pulling a ball bearing, force of the bearing puller should be exerted only on the inner race to remove the bearing from the shaft.
4. When using the bearing puller, protect the shaft end.
5. Do not use sharp edged tools.
6. Drain the oil from bearings (if any) before removal.

Procedure:

1. First remove the external electrical connections.
2. Remove air ducts if any present at the place of installation and dismantle the machine from foundation.
3. Disconnect the machine from the mechanical drive. Mark the positions of both halves of the mechanical coupling before disconnecting.
4. Remove accessories and inspection covers.
5. Loosen the leads and cables coming from the two windings into the terminal box. Label the cables and connections before disconnecting them.
6. Disconnect the wires present at the brush gear (brush rocker) which go to the stator.
7. Take out the carbon brushes from the brush holder by releasing the spring pressure present on them.
8. Wrap a suitable insulating cover around the commutator to protect it from any damage during the dismantling.
9. Place wooden blocks under the stator so that both end covers (end shields) become unsupported.
10. Remove the terminal box.
11. Remove the outer bearing cover at non-drive end.
12. Place a suitable insulating material between the stator and rotor to avoid any damage.
13. Unscrew the fixing bolts that tie the end cover to the stator and remove the end cover at non-drive end.
14. Take due care to ensure that the main gasket between end cover and the stator is not damaged.
15. Now proceed to the drive end and remove the outer bearing cover at that side.
16. Repeat steps 12 and 13 at drive end.
17. Take out the armature out of the stator from the D-end side without knocking on commutator.

Observation:

Draw diagram or capture a photograph of every dismantled part of the machine by a camera and paste its print copy in the journal. Discuss briefly the construction, material and functions of each part. Refer Section 1.3 for details.

Conclusion:

The given *dc* machine is of ____________ type with ratings ______kW, ______volts and ______ rpm. It is successfully dismantled and the different components are studied.

APPENDIX

Ampere's Right Hand Thumb Rule

This rule is applied to determine the direction of magnetic field or the flux produced by a current carrying conductor. Imagine that the current carrying conductor is held in the right hand as shown such that the thumb indicates the direction of current. Then, the rest of the fingers will indicate the direction of flux established by this current.

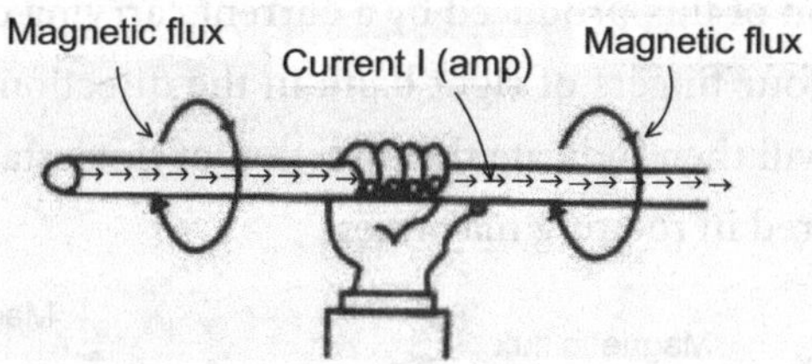

Ampere's Right Hand Thumb Rule

Maxwell's Cork Screw Rule

The direction of magnetic field or flux due to a current carrying conductor can also be determined by applying Maxwell's Cork Screw Rule. Imagine that you are holding a right threaded screw in your hand. Let the screw be rotated such that the screw advances in the direction of current as shown. Then, the direction of rotation of the screw will be the direction of magnetic flux produced by the current.

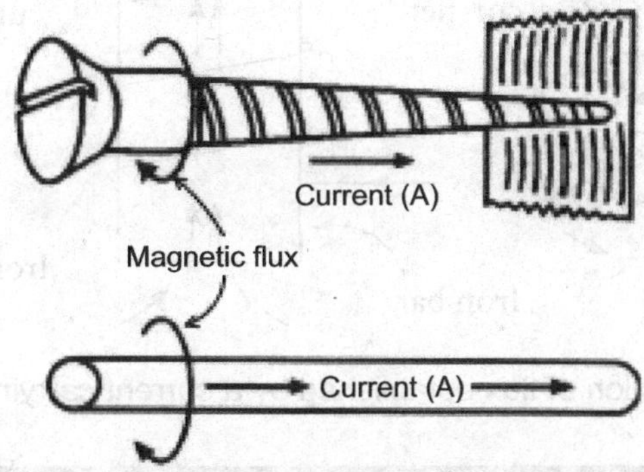

Maxwell's Cork Screw Rule

Dot and Cross Marking

Imagine a current carrying conductor that is perpendicular to the plane of this paper. The direction of current coming out of the paper is represented by a 'Dot' while that entering into the paper is represented by a 'Cross'.

The above discussed Right Hand Thumb Rule or the Cork Screw Rule can be employed to determine the direction of flux produced by these currents. This is as shown in figure.

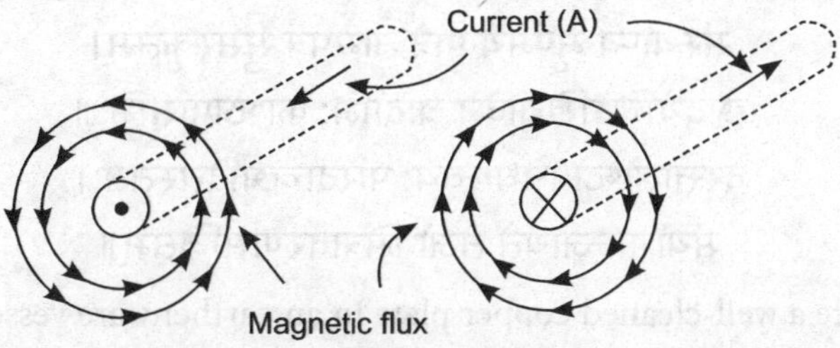

Dot and Cross Marking

Direction of flux established by a current carrying coil

The direction of magnetic flux established by a current carrying coil can be determined by applying the above discussed Right Hand Thumb Rule or the Cork Screw Rule as shown. If the current in the coil is reversed than, the flux will also get reversed.

To determine the direction of flux produced by a current carrying coil (not by a single conductor), it is convenient to orient the four fingers of right hand in the direction of current flowing in different turns of the coil. The thumb will then indicate the direction of flux established on the inner side of the coil. This will be usually required in rotating machines.

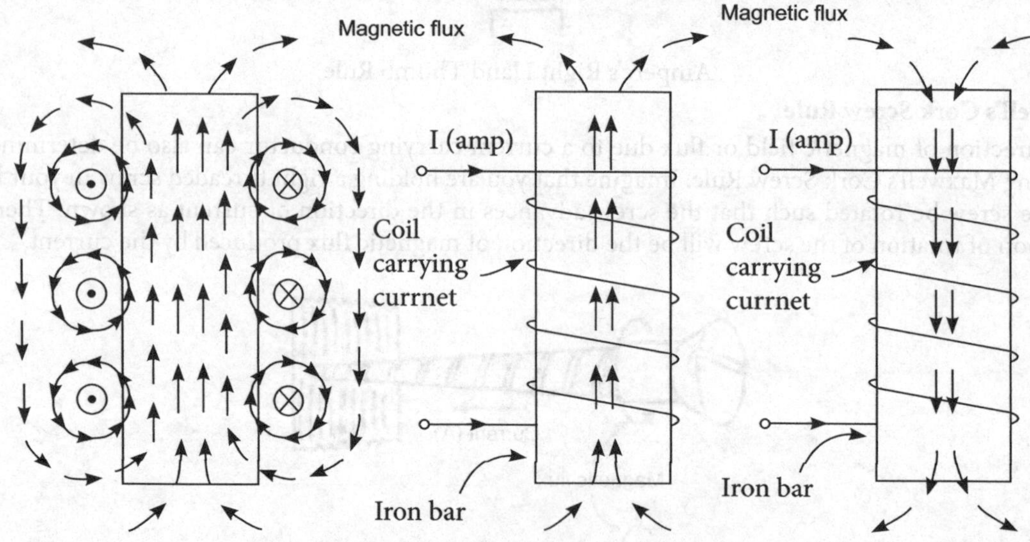

Direction of flux established by a current carrying coil

KNOW MORE

Energy in electrical and magnetic forms exist since the birth of universe. Development of systems for the conversion of energy into electrical form took place since 18th century. However, it is believed that in ancient India, the 'Agastya Samhita' written around 4000 BC by sage Agatsya describes the method to make dry cell with 1.138 volts and 23mA, the electroplating process and hydrogen balloons. It also mentions that water can be split into oxygen and hydrogen. The text in 'Agastya Samhita' says,

संस्थाप्य मृण्मये पात्रे ताम्रपत्रं सुसंस्कृतम्।
छादयेच्छिखिग्रीवेन चार्दाभि: काष्ठापांसुभि:॥
दस्तालोष्टो निधात्वय: पारदाच्छादितस्तत:।
संयोगाज्जायते तेजो मित्रावरुणसंज्ञितम्॥

Which means, "Place a well-cleaned copper plate in an earthenware vessel. Cover it first by copper sulfate and then by moist sawdust. After that, put a mercury-amalgamated zinc sheet on top of the sawdust to avoid polarization. The contact will produce an energy known by the twin name of *Mitra-Varuna*. Water will be split by this current into *Pranavayu* and *Udanavayu*. A chain of one hundred jars is said to give a very effective force.

When a cell was prepared according to Agastya Samhita and measured, it gave open circuit voltage of 1.138 volts, and a short circuit current of 23 mA. The modern electrochemical battery cell resembles Agastya's method of generating electricity [6].

In 1831, the Michael Faraday's law of induction made it clear that an inter relationship exists between electric and magnetic circuits. He also discovered the operating principle of electromagnetic generators in 1831-1832. The first kind of electric generator was a homopolar generator which used a copper disc rotating between a pair of poles of a horseshoe magnet. This is named as Faraday's disc. When the disc was rotated by external means in the field of permanent magnet, it induced an electric current radially outward from the centre of the disc towards its edge. A stationary sliding spring contact 'm' which was connected to external terminal " was used to carry the current from periphery of the disc. The current flowed out through the spring contact 'm', to the external load and back into the centre of the disc through the axle and contact 'B'. This design was found inefficient due to the limited power output. It could produce a small *dc* voltage due to single current path through the magnetic field. The output *dc* current of a homopolar generator is high of the order of 1.0 kA when rotated at very high speeds (of the order of 20,000 rpm) but the output voltage is small of about 5V.

Later, other researchers found that if multiple turns of wire in a coil are used in place of the disc, it could produce more voltage because the output voltage induced in the coil is directly proportional to the number of turns in the coil. So it became possible to design generators to produce any desired voltage by varying the number of turns. Thus wire winding became an essential component of all subsequently designed generators. Thereafter, came the era of hetropolar generators where the conductors and coil sides are made to face the North and South poles alternatively due to which they experience repeated reversals in the direction of magnetic flux. This was not the case in homopolar generators.

Michael Faraday

Faraday's disc

REFERENCES AND SUGGESTED READINGS

1. A. E. Clayton, N. N. Hancock, 'Performance And Design Of-Direct Current Machines,' 3rd Ed., 1958, Oxford & IBH Publishing Co. Pvt. Ltd., New Delhi,
2. A. E. Fitzgerald, C. Kingsley Jr., A. Kusko, 'Electric Machinery,' 3rd Ed., 1971, McGraw-Hill Book Company, Tokyo.
3. P. K. Mukherjee, S. Chakravorti, 'Electrical Machines,' 2nd Ed., 2010, Dhanpat Rai Publications, New Delhi.
4. R. K. Srivastava, 'Electrical Machines,' 2nd Ed., 2013, Cengage Learning India Pvt. Ltd., Delhi.
5. D. P. Kothari, I. J. Nagrath, 'Basic Electrical Engineering,' 3rd Ed., 2012, Tata McGraw Hill Education Pvt. Ltd., New Delhi.
6. Magazine 'Sanskriti', https://www.sanskritimagazine.com/vedic_science/electrical-batteries-in-ancient-india/

QR Code for Further Reading

Unit 1: DC Generator

2 Direct Current Motor

UNIT SPECIFICS

Through this unit the following aspects are discussed:

- Principle of operation of Direct Current (DC) motor;
- Concept and significance of back emf;
- Application of Fleming's Left Hand Rule;
- Concept of torque, speed and BHP;
- Types of DC (dc) motor and voltage equations;
- Losses and efficiency;
- DC motor starter: Necessity and types;
- Brake test and speed control methods of dc shunt and dc series motor;
- Overview of brushless dc motor.

Detail explanation with sufficient number of illustrations and solved examples are covered in this Unit to enable the readers get an effective understanding of the machine. Multiple-choice questions, questions for short and long answers and numerical problems following the lower and higher order of Bloom's taxonomy are provided for practice. References and a dynamic QR code are given at the end for suggestive self-learning.

Laboratory experiments on dc motor along with the required details are also provided to enable the students perform them in laboratory and write effective reports. This is followed by a "Know More" section. It is designed to offer opportunity to the readers to explore more on the topic.

RATIONALE

For a while, imagine your life without fan, pump, blender, refrigerator, vacuum cleaner, washing machine, air conditioner, dishwasher, printer etc. Imagine that the manufacturing industries do not use electric motor. With such a strange assumption, you will be shocked to visualize that the current standard of living of humans has reverted back by at least a century.

Most of the present-day comforts of our daily life right from household appliances, transportation to industry systems have become possible because electric motors are employed to realize them into fact. Electric motors are extremely important in modern day to day life. They turn the wheels of progress at a new speed and help to accelerate the working of existing systems. Every electrical engineer should be aware of the different types of electric motor and their characteristics so as to be able to select the appropriate one for a given application. Electric motors are broadly classified on the basis of the source for which they are designed, as dc motors and ac motors. The conventional ac motors are further classified as induction motors and synchronous motors. This unit will present and explain the details of dc motor. Note that, the dc motors are excellent in terms of controllability and hence are considered as a role model for conventional ac motors.

PRE-REQUISITES

Basics of Electrical Engineering

UNIT OUTCOMES

List of outcomes of this unit is as follows:

U2-O1: Discuss the different types of dc motor and their operation, formulate the voltage and current equations, draw the characteristics and suggest applications;

U2-O2: Explain the significance of back emf, need of starter and discuss the starter types;

U2-O3: Calculate the unknown quantities of dc motor at different operating conditions;

U2-O4: Apply Fleming's left hand rule to determine the direction of torque;

U2-O5: Explain speed control methods and compare them on the basis of merits and demerits;

U2-O6: Discuss the construction and working of brushless dc motor;

U2-O7: Perform experiments on dc shunt and dc series motors and evaluate the performance.

Unit-2 Outcome	EXPECTED MAPPING WITH COURSE OUTCOMES (1 - Weak Correlation; 2 - Medium Correlation; 3 - Strong Correlation)				
	CO-1	CO-2	CO-3	CO-4	CO-5
U2-O1	-	3	-	-	-
U2-O2	-	3	-	-	-
U3-O3	-	3	-	-	-
U4-O4	-	3	-	-	-
U5-O5	-	3	-	-	-
U6-O6	-	3	-	-	-
U7-O7	-	3	-	-	-

2.1 INTRODUCTION

Now, as we have reached to a good understanding of *dc* generators, we can initiate our study on *dc* motors. DC motors are used to convert the *dc* electric energy into mechanical energy of rotational form. This mechanical energy then can be used to drive the mechanical systems right from everyday household tools like fan, pump to heavy industrial machinery like punch press, hoist, crane, automobile, steel plant, paper plant etc.

DC motors are suitable for a wide range of speed control as well as for fast reversals. They offer fast response to any change in reference speed during the speed control and can also generate high starting torque. However, the use of *dc* motors in ordinary industrial applications is not common due to its high initial cost and high maintenance requirement as compared to Variable Frequency Drive (VFD) controlled *ac* motors. The wear and tear of commutator and brushes is one of the common cause in *dc* machines that attract more maintenance. DC motors are preferred for special industrial applications in steel mills, mines and electric trains.

2.2 DC MOTOR: OPERATING PRINCIPLE

DC motor has the similar construction as that of *dc* generator and consequently, a *dc* machine can be used as a generator or as a motor provided that it also has interpoles on the stator. Refer to Section 1.3 and 1.8.1 of Unit-1 to revise the construction of *dc* machine and interpoles.

Now, consider a dc motor as shown in Fig. 2.1. Let the field winding on stator poles be connected across a *dc* voltage source so that the current flowing in it will magnetize the poles and establish a main magnetic flux ϕ_m between them. Also connect a *dc* voltage source V_t across the terminals A and AA of armature winding so as to establish a current in the armature winding. Note that, when the dc machine is used as a generator, the terminals of armature winding are connected across an electrical load whereas if you want to use it as a motor than, the armature terminals are connected across a dc source.

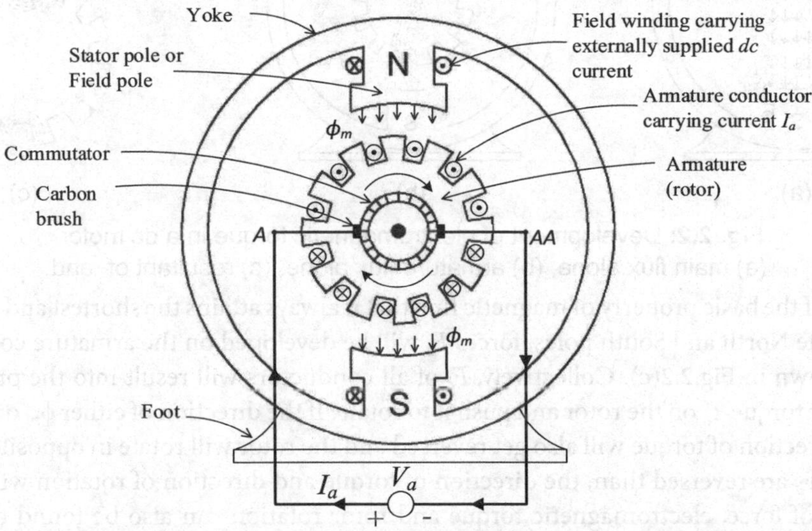

Fig. 2.1: Schematic diagram of a basic dc motor

Since the armature winding comprises of 'Z' number of interconnected conductors that are configured in 'A' number of parallel paths, let the current in each conductor be 'I' such that, I = *Ia* /A. Recall that, for Lap winding, A = P and for Wave winding, A = 2. (See Section 1.2.1, 1.5.1 and 1.6 of Unit-1 for the meaning of conductor and parallel path).

Thus, the armature conductors which are carrying current will be surrounded by the main flux. Interaction between main pole flux ϕ_m and armature current I_a will result into the production of physical force on all armature conductors. If '*l*' metre is the active length of each armature conductor and 'B_{av}'

weber/m² is the average flux density due to field poles than, the force produced on every single conductor will be,

$$F_c = B_{av}\, Il \quad \text{Newtons} \tag{2.1}$$

If the sum of forces produced on all conductors is multiplied by the radius of armature '*r*' than the resulting product will be the electromagnetic torque T_a developed on the armature. This torque will set the rotor in motion. For better understanding of the operation, consider an armature winding having only one coil that is housed in two diametrically opposite slots. When external *dc* currents are passed through both, field winding and armature winding, they will establish the main flux and armature flux respectively. The direction of these currents can be represented by Dot and Cross marking. The direction of fluxes produced by them can be obtained by applying Right Hand Thumb Rule. *(Refer to the Appendix of Unit-1 for details of Dot-Cross marking and Right Hand Thumb Rule).* This is as shown in Fig. 2.2(a) and (b). Depending on the respective directions of and they will aid each other on one side of the armature conductors and oppose each other on the other side. This will increase the net flux density on one side of armature conductors and reduce it on the other side as shown in Fig. 2.2 (c).

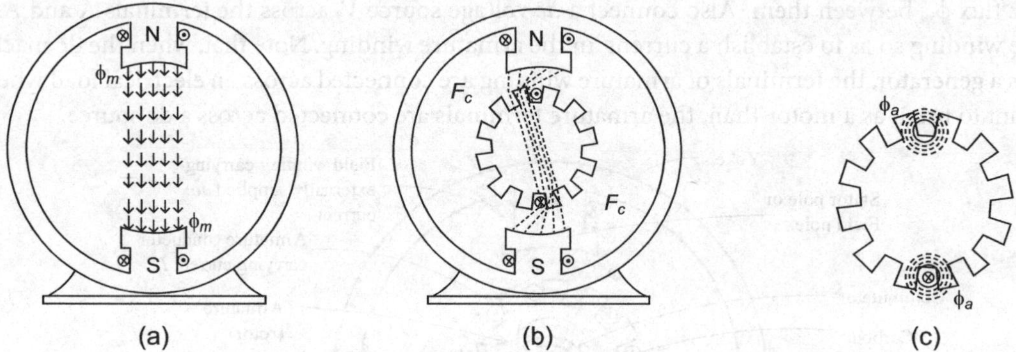

Fig. 2.2: Development of electromagnetic torque in a dc motor, (a) main flux alone, (b) armature flux alone, (c) resultant of and

By virtue of the basic property of magnetic flux that it always attains the shortest and least reluctance path between the North and South poles, forces F_a will be developed on the armature conductors in the direction as shown in Fig.2.2(c). Collectively, F_c of all conductors will result into the production of an electromagnetic torque T_a on the rotor and push it to rotate. If the direction of either ϕ_m or I_a (hence ϕ_a) is reversed, the direction of torque will also get reversed and the rotor will rotate in opposite direction. But if both ϕ_m and ϕ_a are reversed than, the direction of torque and direction of rotation will remain same. The directions of force, electromagnetic torque and rotor rotation can also be found out by applying Fleming's Left Hand Rule. It is explained in the next section 2.2.1.

Once the armature starts rotating, the armature conductors will cut the main flux ϕ_m and have flux linkages with it. By Faraday's law, these flux linkages will induce *motional emf (i.e. speed emf)* in the armature conductors similar to that in generator operation discussed in Section 1.2 of Unit-1. According to Lenz's law, the motionally induced *emf* opposes the *dc* voltage across armature. The polarity of induced *emf* can be determined by applying the *Fleming's Right Hand Rule* and it will be found that this *emf* during motoring is always in opposition to the polarity of source voltage V_t that is connected across the armature winding. Hence it is called as *Back emf* (E_b) or *counter emf.* Thus, at rotor speed N ≠ 0, the armature circuit will have two types of voltages, source voltage V_t and back *emf* E_b. The net voltage in armature winding will be the difference between them i.e. $(V_t - E_b)$. At the time of starting when motor is initially at rest, the motional or speed induced back *emf*, $E_b = 0$. Therefore, to restrict the magnitude of current I_a within safe limits during starting, certain provisions are suggested.

2.2.1 Fleming's Left Hand Rule

This rule is employed to determine the direction of force (*a physical quantity*) developed on a conductor which is carrying current and is placed in a magnetic field. Like *Fleming's Right Hand Rule*, this rule involves three different domains i.e. electrical, magnetic, and physical. To apply this rule, three fingers of left hand i.e., first finger, middle finger and thumb are stretched away from each other such that each of them gets aligned perpendicularly with the remaining two. This is shown in Fig. 2.3 and 2.4.

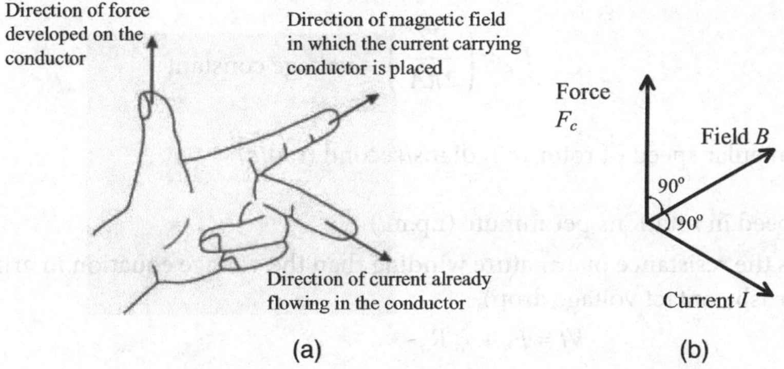

Fig. 2.3: (a) Fleming's Left Hand Rule (b) Vector notation

The alignment of these fingers will be as:

- First finger is oriented in the direction of magnetic field (North to South) in which the conductor under consideration is placed.
- Middle finger is oriented in the direction of current flowing in that conductor.
- Thumb will give you the direction of force developed on the conductor which will set it in motion.

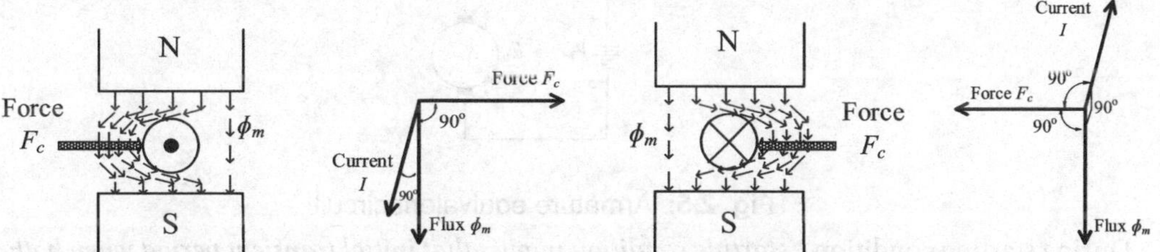

(a) Current I_a coming out of the paper (b) Current I_a entering into the paper

Fig. 2.4: Use of Fleming's Left Hand Rule to determine the direction of Force

2.2.2 Significance of Back emf

It was explained in Section 2.2 above that, after supplying the dc currents both in field winding and armature winding, the interaction between fluxes produced by them results into an initial starting electromagnetic torque on the armature due to which it starts rotating. Due to the rotation of armature, the armature conductors that are housed in its slots will cut the main flux ϕ_m. This cutting action will induce a back emf E_b in the armature conductors whose direction is always opposite to the *dc* source voltage for motoring action.

The process of production of *back emf Eb* in a *dc* motor is identical to the production of *generated emf Eg* in a *dc* generator. Therefore, for the same *dc* machine, $|E_b| = |E_g|$. Hence, from equation (1.8),

$$\therefore E_b = \frac{\phi_m PN}{60}\frac{Z}{A} \text{ Volts} \tag{2.2}$$

It can be simplified as,

$$E_b = K_e \omega \phi_m \tag{2.3}$$

Where,

$$K_e = \left(\frac{PZ}{2\pi A}\right) = \text{voltage constant}$$

$\omega = \frac{2\pi N}{60}$ angular speed of rotor in radians/second (rad/s)

N = rotor speed in rotations per minute (r.p.m.)

If R_a denotes the resistance of armature winding than the voltage equation in armature loop will be (neglecting the brush contact voltage drop),

$$Vt = E_b + I_a R_a \tag{2.4}$$

$$I_a = \frac{V_t - E_b}{R_a} \tag{2.5}$$

∴ Armature current will be,

To ensure low value of armature copper loss, high output power, and low voltage drop, the armature winding resistance R_a should be less. Ideally, it should be zero.

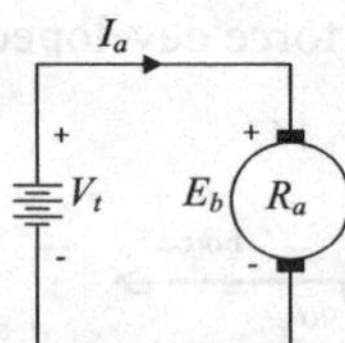

Fig. 2.5: Armature equivalent circuit

During starting condition, *(starting condition implies that initial transient period when both windings are connected across the dc source but the rotor has not yet started rotating)*, speed *N = 0* and therefore from equation (2.2), the back *emf* is zero. Consequently, the armature current during starting condition will be very high. This can be verified from equation (2.5).

Consider a *dc* separately excited motor having rated input voltage and rated armature current of 220V, 11.36A respectively (*Rated current is also referred as full load current*). Let $R_a = 1\Omega$. During starting condition when *N = 0*, back *emf* $E_b = 0$. The armature current from equation (2.5) will be,

$$I_{a(N=0)} = \frac{220-0}{1} = 220A$$

This initial transient starting current is significantly higher than the rated value of current and therefore it produces high copper loss in the armature winding. Consequently, more heat will be dissipated and may result into insulation damage. But once the rotor starts rotating, N ≠ 0. and the magnitude of back *emf* starts building. When the rotor speed reaches close to rated value, E_b reaches close to the magnitude of supply voltage V_t. Obviously, the net voltage in armature winding will now reduce from V_t to $(V_t - E_b)$ and therefore, the magnitude of transient starting current also comes down to a lower steady state motoring value of armature current.

Thus, the presence of back *emf* E_b regulates the magnitude of armature current I_a within safe limits. In the absence of E_b, the current magnitude becomes dangerously high.

2.3 TORQUE AND SPEED

The turning effect of a force is called as torque or moment of the force. It is a measure of the rotational force that can cause an object to rotate about an axis. The unit of torque is Newton.meter (Nm). Sections 2.2 and 2.2.1 have provided a detail explanation about how a force is developed on all armature conductors when they carry current and are placed in the magnetic field of stator poles. This force on each conductor was given by equation (2.1) and is reproduced below.

$$F_c = B_{av} I \ell \text{Newtons}$$

Where B_{av} is the average magnetic flux density in the air gap, I is the current per armature conductor and is the length of armature conductor *(The length of armature conductor l = axial length of armature core = axial length of stator field poles)*. B_{av} can be expressed as,

$$B_{av} = \frac{\text{total main flux}}{\text{total airgap area encountered by this flux}} = \frac{\phi_m P}{2\pi r \ell}$$

Where, Φ_m = main flux per stator pole in weber, P = number of stator poles and r = mean radius of air gap (or by approximation, r = radius of armature core) in meter. The current per conductor can be expressed in terms of armature current as,

$$I = \frac{I_a}{A}$$

Where, A = number of parallel paths in armature winding. In Lap type armature winding, $A = P$ whereas in Wave type winding, $A = 2$. *(See Section 1.5.1 & 1.7 for details)*. Let Z = number of armature conductors in the armature winding. The sum of forces developed on all armature conductors can be obtained by substituting the values of B_{av} and I in above equation of F_c and then multiplying by Z as,

$$F = \frac{\phi_m PZ}{2\pi r} \frac{I_a}{A}$$

The armature torque will be,

$$T_a = \text{Sum of all } F_c \times \text{radius} = F \times r$$

$$\therefore T_a = \left(\frac{1}{2\pi}\right) \frac{PZ}{A} \phi_m I_a \qquad (2.6)$$

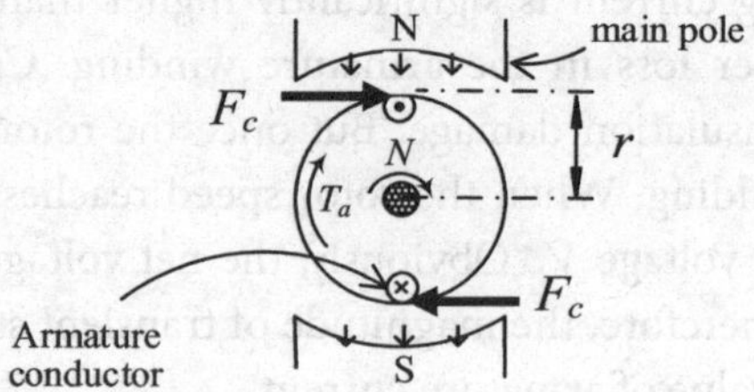

Fig. 2.6: Force and torque

The above torque expression can be simplified by replacing all constant terms as,

$$\therefore \qquad \text{Armature torque } T_a = K_t \phi_m I_a \tag{2.7}$$

$$\text{where, } K_t = \left(\frac{1}{2\pi}\right)\frac{PZ}{A} = \text{Torque constant}$$

Thus, the armature torque is directly proportional to the product of main flux and armature current. Theoretically, torque constant K_t is same as Voltage constant K_e .

2.3.1 Shaft Torque

Whole of the armature torque T_a as calculated above is not available at the motor shaft. Some part of it is lost in overcoming the mechanical power loss. This mechanical power loss is due to the friction at bearings and the opposition of surrounding fluid (air) to rotation. Hence, it is also called as friction and windage loss. Therefore, the shaft torque T_{sh} is,

$$T_{sh} = T_a - \text{torque lost in overcoming mechanical loss}$$

It can also be expressed as,

$$T_{sh} = \frac{\text{Shaft power output in watts}}{\left(2\pi N/60\right)} = \frac{P_{sh}}{\omega}\text{ Nm}$$

where,

$\omega = \dfrac{2\pi N}{60}$ = Angular speed of rotor in radians/second (rad/s)

N = Rotor speed in rotations per minute (r.p.m.)

P_{sh} = Shaft output power (W)

2.3.2 Brake-Horse-power

The mechanical power output available at the shaft P_{sh} can be expressed in watts (W) as well as in brake-horse-power (BHP). It is this power that drives the mechanical load. Hence BHP is the mechanical power available at the shaft. This shaft power P_{sh} can be calculated using equation (2.8). The conversion between two units of measurement is,

$$1 \text{ horse power} = 746 \text{ watts}$$

Example 2.1: A 20 cm long copper conductor is placed in a magnetic field of flux density 1 Wb/m². If a dc current of 1A is passed through the conductor, what will be the force developed on it?

Data: B = 1 Wb/m², l = 0.2 m, I = 1A, Fc = ?

Solution: From equation (2.1),

Force Fc = BIl

$$=1\times 1\times 0.2 = 0.2\text{ N}$$

Example 2.2: A 4 pole, lap wound *dc* motor has 60 slots on the armature and 20 conductors per slot. The flux per pole is 20 mWb. If the armature current is 60A, calculate the torque developed in the armature.

Data: P = 4, A = P = 4 (since lap connected), slots = 60, conductors per slot = 20, $\phi_m = 20\times 10^{-3}$ Wb, I_a = 60A, T_a = ?

Solution:

Total number of conductors Z = 60 × 20 = 1200

From equation (2.6), $T_a = \left(\frac{1}{2\pi}\right)\frac{PZ}{A}\varnothing_m I_a$

$$\therefore T_a = \left(\frac{1}{2\pi}\right)\frac{4\times 1200\times 20\times 10^{-3}\times 60}{4} = 229.3\text{ Nm}$$

Example 2.3: A 6 pole, 480V, lap wound dc shunt motor has 864 armature conductors. The flux per pole is 0.05 Wb and armature current is 100A. If the armature resistance is 0.2 Ω, calculate the rotor speed and torque developed in the armature.

Data: P = 6, A = P = 6, Vt = 480V, Z = 864, ϕ_m = 0.05 Wb, I_a = 100A, R_a = 0.2 Ω, N = ?, Ta = ?

Solution:

From equation (2.4), Back *emf* $E_b = (V_t - I_a R_a) = 480 - 100\times 0.2 = 460\text{V}$

From equation (2.2), Back *emf* $E_b = \frac{\varnothing_m PN}{60}\frac{Z}{A}$

$$\therefore \text{Speed} N = \frac{460\times 60\times 6}{0.05\times 6\times 864} = 639\text{rpm}$$

From equation (2.6), $T_a = \left(\frac{1}{2\pi}\right)\frac{PZ}{A}\phi_m I_a$

Substituting all values,

$$T_a = 687.89\text{ Nm}$$

2.3.3 Ratio of Back Emf to Armature Torque

From equation (2.3), *back emf* is,

$$E_b = K_e\,\omega\phi_m$$

From equation (2.7), armature torque is, $Ta = K_t\,\phi_m\,I_a$

Where $K_t = K_e = \left(\frac{1}{2\pi}\right)\frac{PZ}{A}$ = torque constant = voltage constant (2.10)

$$\therefore \frac{E_b}{T_a} = \frac{\omega}{I_a}$$

$$\therefore E_b I_a = \omega T_a$$

In rotating electrical machines,

Torque (Nm) × Angular speed(rad/s) = Back emf ×Armature current = Power (watts)

Thus in equation (2.10), ωT_a represents power, It also means that, $E_b I_a$ is power. Thus, $E_b I_a$ = electrical equivalent of mechanical power developed in the armature. This can also be verified from the armature voltage equation,

$$V_t = E_b + I_a R_a$$

Multiply both sides by I_a, the power equation will be,

$$V_t I_a = E_b I_a + I_a^2 R_a$$

Where, $V_t I_a$ = Electrical input power drawn by the motor from dc source in watts

$I_a^2 R_a$ = Armature copper loss in watts

$E_b I_a$ = Electrical equivalent of the mechanical power developed in the armature in watts

2.3.4 Speed

The production of electromagnetic torque T_a on the armature results into its acceleration. A drive continues to rotate at a steady state speed when torque T_a equals the opposing load torque T_L. Expression for rotor speed can be derived from the basic voltage equation (2.4), *back emf* equation (2.3) and torque equation (2.7) given above. They are reproduced below.

$$V_t = E_b + I_a R_a$$

$$E_b = K_e \omega \phi_m$$

$$T_a = K_t \phi_m I_a$$

Where, $K_e = K_t = \left(\frac{1}{2\pi}\right)\frac{PZ}{A}$

$\omega = \frac{2\pi N}{60}$ = rotor speed in radians/second, expressed in angular form.

N = rotor speed in rotations per minute (r.p.m.)

Substituting the value of E_b in above voltage equation, the rotor speed in rad/s will be,

$$\omega = \frac{V_t - I_a R_a}{K_e \phi_m} = \frac{V_t}{K_e \phi_m} - \frac{R_a}{K_e \phi_m} I_a \qquad (2.11)$$

From above torque equation,

$$I_a = \frac{T_a}{K_t \phi_m} = \frac{T_a}{K_e \phi_m}$$

Where, $K_e = K_t$. Substituting this value of armature current in above speed equation, the expression for rotor speed in terms of armature torque and input voltage will be,

$$\omega = \left[\frac{V_t}{K_e\phi_m} - \frac{R_a}{(K_e\phi_m)^2}T_a\right] \text{ rad/s} \tag{2.12}$$

From above speed equation, it is observed that the rotor speed is directly proportional to input voltage V_t but it is inversely proportional to the main flux ϕ_m.

Initially after the motor is connected to the source, the electromagnetic armature torque T_a will start building up and the rotor will accelerate from zero speed. At a condition when the magnitude of *Ta* becomes equal to all the opposing torques T_L, the rotor speed becomes constant and such a situation is called as *steady-state condition* or *equilibrium condition*. These opposing torques are also called as load torque T_L and they are primarily due to the opposition offered by mechanical load, friction and windage. Note that during the motoring action, T_L acts in opposite direction to T_a and both act on the same rotating system. Usually, the mechanical loads like hoist, conveyer, winder, lathe machine, fan, pump, blower, propeller, press etc. are mechanically coupled to the motor shaft through a rope-pulley mechanism, gear-train, belt-drum system or by direct coupling.

After the steady state condition is achieved, if the mechanical load on the shaft is increased, the load torque T_L will increase and will become greater than T_a. Obviously, the rotor speed will start drooping. To overcome this rise in T_L, the armature torque T_a should increase till it again becomes equal to T_L. Equation (2.12), indicates that in such a situation, there will be corresponding drop in the rotor speed. This justifies why the rotor speed decreases due to the rise in mechanical load on the shaft.

Example 2.4: A 230V dc shunt motor draws a no-load current of 3A and full-load current of 30A from dc supply. If the armature resistance is 0.4 Ω, what will be the change in back emf from no-load to full-load condition?

Data: $V_t = 230\text{V}$, $I_{(nl)} = 3\text{A}$, $I_{(fl)} = 30\text{A}$, $(E_{b(fl)} - E_{b(nl)}) = ?$, $R_a = 0.4\ \Omega$

Solution:

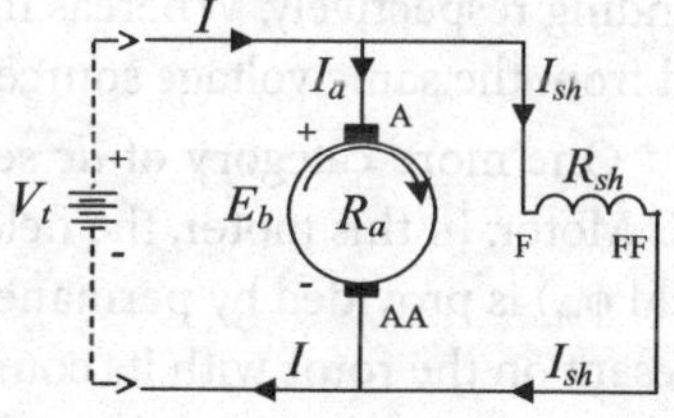

From equation (2.4),

Back emf $E_b = (V_t - I_a R_a)$

Since, R_{sh} is not given and since, normally

$R_{sh} >> R_a$, $I_{sh} << I_a$, $\therefore$ let $I = I_a$

At no-load:

$E_{b(nl)} = (V_t - I_{a(nl)} R_a) = 228.8\text{V}$

At full-load: $E_{b(fl)} = (V_t - I_{a(fl)} R_a) = 218\text{ V}$

There will be a drop in back emf of 10.8V from no-load to full load.

Example 2.5: A 220V, 4 pole, dc shunt motor has 500 armature conductors and two parallel paths. The armature and shunt field resistances are 0.25 Ω and 125 Ω respectively. If the flux per pole is 0.02 Wb and input current is 12A, calculate the speed and torque developed in armature.

Data: Vt = 220V, P = 4, Z = 500, A = 2, $R_a = 0.25\Omega$, $R_{sh} = 125\Omega$, $\phi_m = 0.02$ Wb, $I = 12$A. N, and $T_a = ?$,

Solution:

Shunt field current $I_{sh} = \frac{V_t}{R_{sh}} = 1.76\text{ A}$

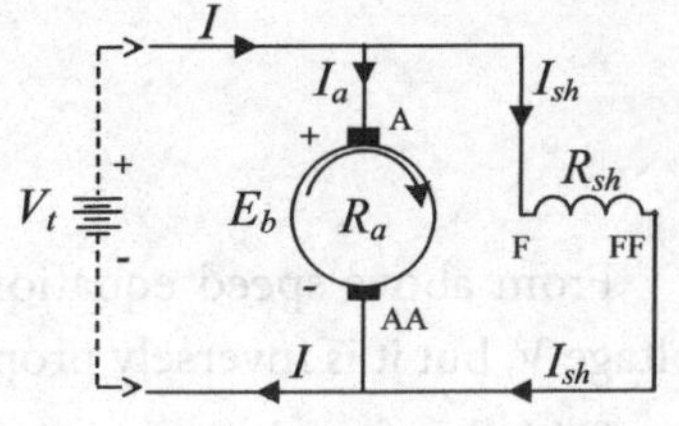

Armature current $I_a = I - I_{sh} = 10.24\text{A}$

Back emf $E_b = (V_t - I_a R_a) = 217.44\text{A}$

Back emf is also given by equation (2.2) as,

$$E_b = \frac{\phi_m PN}{60}\frac{Z}{A}$$

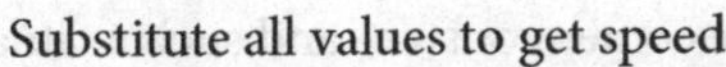

Substitute all values to get speed

$\therefore$ $N = 652\text{ rpm}$

From equation (2.10), the electrical equivalent of mechanical power developed in the armature is, $E_b I_a = \omega T_a$

$\therefore$ $E_b I_a = 2\,226.58\text{W}$

The rotor angular speed $\omega = \frac{2\pi N}{60} = 68.27\text{ rad/s}$

$\therefore$ Armature torque =

$$T_a = \frac{E_b I_a}{\omega} = 32.61\text{Nm}$$

2.4 TYPES OF DC MOTOR

The classification of *dc* motor is based on the mode of excitation of field winding. It is similar to that of *dc* generator as discussed in Section 1.11. Accordingly, *dc* motor is classified into different types as shown in Fig. 2.7. The main difference between a separately excited *dc* motor, *dc* shunt motor and *dc* series motor is that in the first type, two separate voltage sources are connected across the armature winding and field winding respectively. Whereas in shunt and series type *dc* motors, the armature and field windings are fed from the same voltage source.

One more category of *dc* separately excited motor exists. It is called as Permanent Magnet (PM) DC Motor. In this motor, the field winding is absent and instead, separate excitation (i.e. main magnetic field ϕ_m) is provided by permanent magnets fixed in place of poles on the stator. The armature winding present on the rotor with its commutator-brush arrangement is connected across a *dc* voltage source.

DC compound motors have two field windings that are wound on the same poles. Due to the variation in winding connections and variation in flux directions produced by these field windings, the operating characteristics of different types of *dc* motor differ significantly from each other and hence appropriate type of *dc* motor needs to be selected for a given mechanical load. Following are the three important characteristics:

(a) Torque - armature current characteristic

(b) Speed - armature current characteristic

(c) Speed - torque characteristic

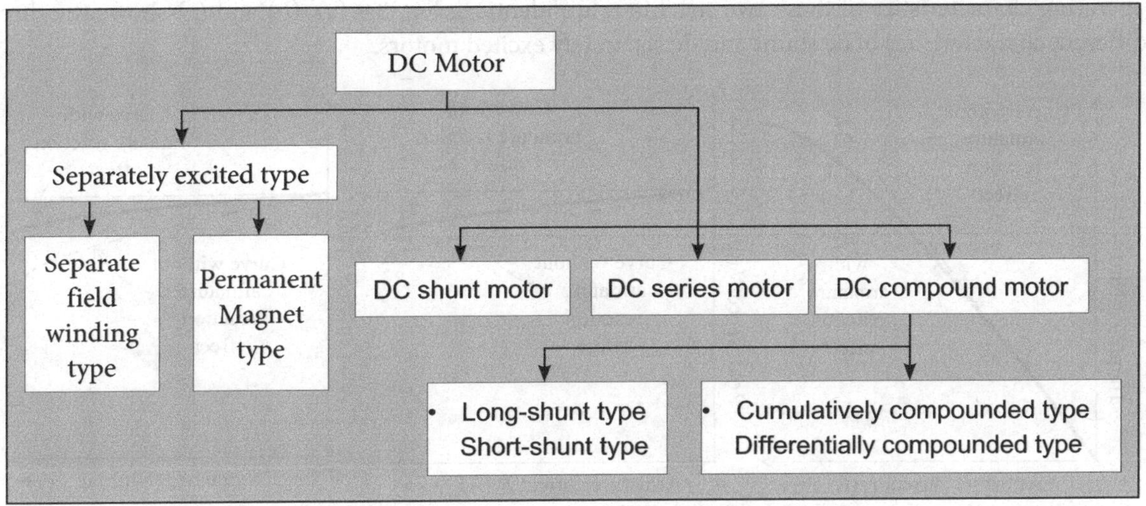

Fig. 2.7: Classification of dc motor

2.4.1 DC Shunt and Separately Excited Motors

In a dc shunt motor, the armature and field windings are connected in parallel with each other across the same input source whereas in *dc* separately excited motor, they are connection-wise isolated from each other and two separate voltage sources are employed for them. The circuit diagrams are shown in Fig. 2.8. Let R_{sh} and R_f represent the resistance of field winding in *dc* shunt and *dc* separately excited motors respectively. R_a is the resistance of armature winding. The corresponding voltage and current equations follow the Kirchhoff's laws and are given below.

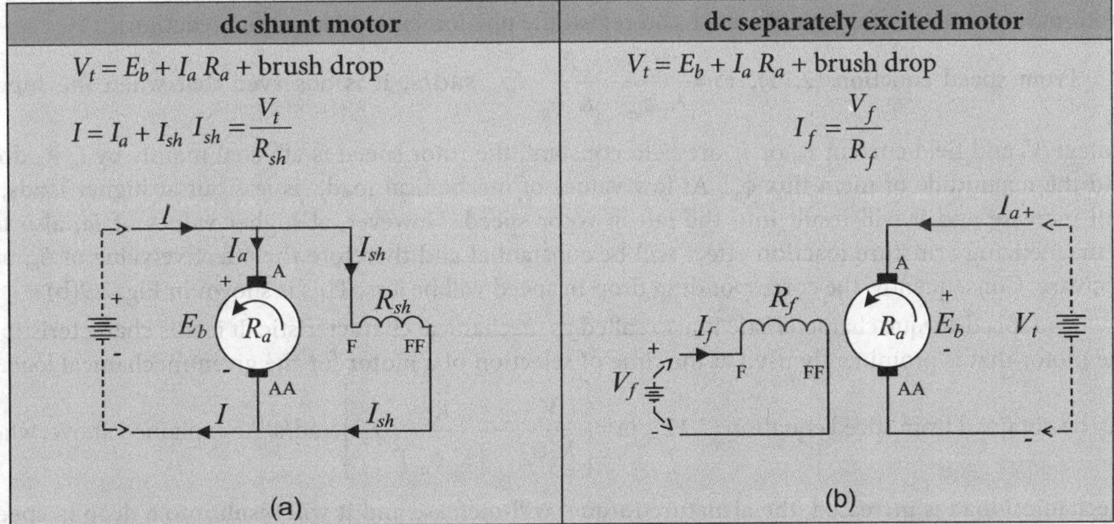

dc shunt motor	dc separately excited motor
$V_t = E_b + I_a R_a$ + brush drop	$V_t = E_b + I_a R_a$ + brush drop
$I = I_a + I_{sh}$ $I_{sh} = \dfrac{V_t}{R_{sh}}$	$I_f = \dfrac{V_f}{R_f}$
(a)	(b)

Fig. 2.8: Circuit diagrams of (a) DC shunt motor and (a) DC separately excited motor

In these motors, the source voltage connected across field winding is usually constant. Therefore the main flux ϕ_m is almost constant (if armature reaction is assumed to be negligible) and as a result, operating characteristics of these two machines are identical. Fig. 2.9 (a), (b) and (c) shows the three different characteristics of dc shunt and dc separately excited motors.

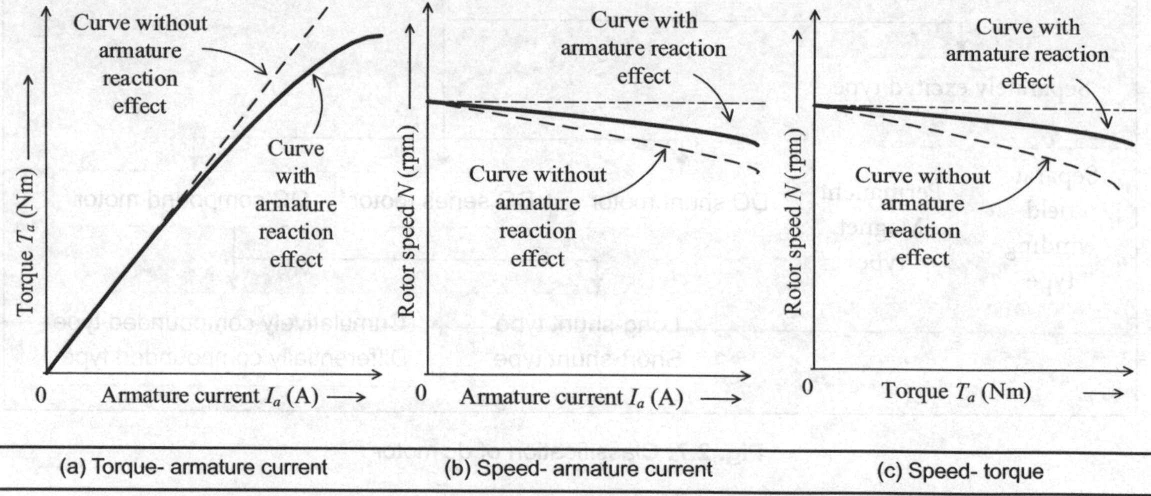

(a) Torque- armature current (b) Speed- armature current (c) Speed- torque

Fig. 2.9: Characteristics of dc shunt motor and dc separately excited motor

The torque equation (2.7) $T_a = K_t\,\phi_m I_a$ reveals that, if ϕ_m is held constant than, $T_a \propto I_a$ thus giving a linear relationship between them. At low magnitudes of mechanical load, less armature torque T_a is sufficient to overcome the load torque and therefore the current I_a is less. Consequently, the armature flux ϕ_a is less and hence it's demagnetizing armature reaction effect on ϕ_m is negligible. Thus ϕ_m remains almost constant. However at higher magnitudes of mechanical load, T_a, I_a and ϕ_a are notably large and hence the demagnetizing armature reaction effect of ϕ_a on ϕ_m is substantial. This weakens the flux ϕ_m and therefore, the torque current characteristic deviates from the straight line as shown in Fig. 2.9(a). (You may refer to Section 1.10 of Unit 1 and revisit the phenomenon of armature reaction).

From speed equation (2.11), $\omega = \dfrac{V_t}{K_e\phi_m} - \dfrac{R_a}{K_e\phi_m} I_a$ rad/s, it is observed that when the supply voltage V_t and field current I_{sh} or I_f are held constant, the rotor speed is affected mainly by $I_a\,R_a$ drop and the magnitude of main flux ϕ_m. At low values of mechanical load, is less but at higher loads, I_a will increase and it will result into the fall in rotor speed. However, at higher values of *Ia*, also the demagnetizing armature reaction effect will be substantial and therefore the effective value of ϕ_m will decrease. Consequently, the corresponding drop in speed will be less. This is shown in Fig. 2.9(b).

The speed-torque characteristic is also called as mechanical characteristic. It is this characteristic of the motor that is prominently used at the time of selection of a motor for the given mechanical load. It can be obtained from speed equation (2.12), $\omega = \left[\dfrac{V_t}{K_e\phi_m} - \dfrac{R_a}{(K_e\phi_m)^2} T_a\right]$ rad/s. As explained above, when mechanical load is increased, the armature torque will increase and it will result into a drop in speed. However, at higher mechanical load, also the effective value of ϕ_m decreases due to armature reaction and hence the speed-torque characteristic will be less drooping. In *dc* shunt and *dc* separately excited motors, the drop in rotor speed is very less. In medium size motors, the speed drop from no-load to full-load condition is of the order of 5%.

The *dc* shunt motor is commonly used in those applications which require approximately constant speed at all values of load from zero to full load condition. Examples are: centrifugal pump, blowers, fans, conveyers, machine tools, printing press etc. The separately excited *dc* motors are employed in applications where adjustable speed is required.

2.4.2 DC Series Motor

In *dc* series motor, the armature and field windings are connected in series with each other as shown in Fig. 2.10. The armature current and field current will be same as the input current. The voltage and current equations will are,

$$V_t = E_b + I(R_a + R_{se}) + \text{brush drop}$$

$$I = I_a = I_{se}$$

Where R_{se} is the resistance of series field winding. The total resistance in the armature circuit will be $(R_a + R_{se})$.

The main flux produced by field winding will be proportional to armature current, $\phi_m \propto I$

∴ Let $\phi_m = K_f I$

Where K_f = field constant

The torque equation (2.7) will get modified as,

$$Ta = K_t \phi_m I = K_t K_f I^2 \tag{2.13}$$

Fig. 2.10: Circuit diagram of DC series motor

The speed equation (2.11) will get modified as,

$$\omega = \frac{V_t}{K_e \phi_m} - \frac{R_a}{K_e \phi_m} I = \frac{V_t}{K_e K_f I} - \frac{(R_a + R_{se})}{K_e K_f} \tag{2.14}$$

Substitute the value of $I = \sqrt{\dfrac{T_a}{K_t K_f}}$ from equation (2.13) into equation (2.14), the speed equation in terms of torque will be,

$$\omega = \frac{V_t}{\sqrt{K_e K_f}} \frac{1}{\sqrt{T_a}} - \frac{(R_a + R_{se})}{K_e K_f} \tag{2.15}$$

Using above equations, the characteristics of dc series motor are constructed as shown in Fig. 2.11. From equation (2.13), at lower values of armature current I, the armature torque is directly proportional to the square of armature current I^2. Therefore, the relationship between them will be parabolic. But at higher values of mechanical load, the armature current I will be high. Since I is flowing through field winding also, the main flux ϕ_m will increase. This will tend to saturate the magnetic circuit. Simultaneously the armature flux ϕ_a will also increase and enhance the demagnetizing armature reaction effect. As a result, the rise in ϕ_m will be compensated by the rise in ϕ_a and therefore, the resultant air gap flux will remain constant at ϕ_m. Hence, from equation (2.13) again, in case of magnetic saturation, torque will remain proportional to I and the characteristic curve between them will be a straight line. This shown in Fig. 2.11(a).

Equation (2.14) represents the relationship between speed and armature current. Since $\frac{(R_a + R_{se})}{K_e K_f}$ is a constant, the speed-current characteristic will be hyperbolic. This shown in Fig. 2.11(b). The speed-torque characteristic of *dc* series motor can be constructed by using equation (2.15). It shows that, if the magnetic saturation and armature reaction effects are neglected, the speed varies inversely as the square root of torque $\sqrt{T_a}$. The relationship between them is hyperbolic as shown in Fig. 2.11(c).

In *dc* series motor, since the torque is proportional to square of armature current I^2, for the same rise in torque, increase in motor current is less as compared to that in a *dc* shunt and *dc* separately excited motor where the torque is proportional to armature current I. Hence during heavy loads or overloads, the current and temperature rise in series motor remains within reasonable values.

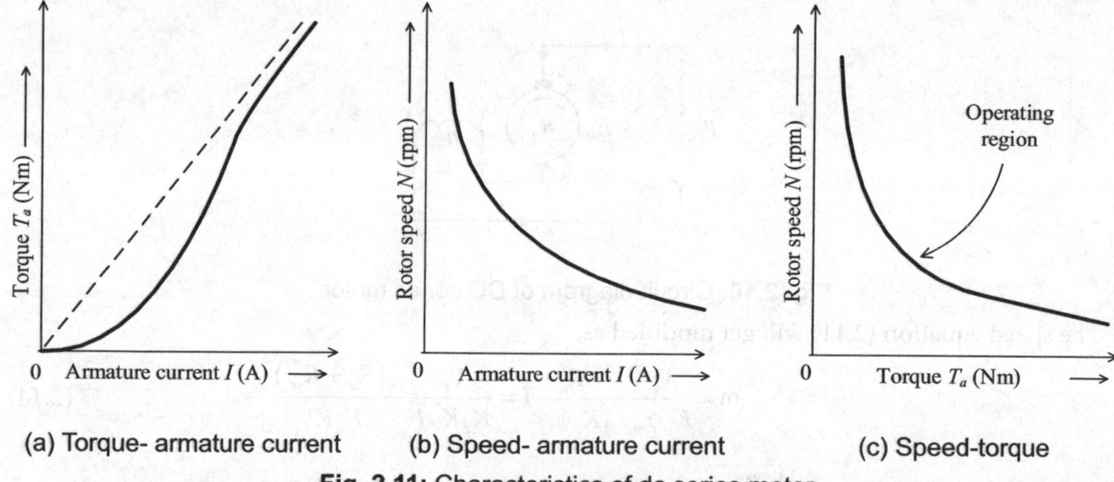

(a) Torque- armature current (b) Speed- armature current (c) Speed-torque

Fig. 2.11: Characteristics of dc series motor

The outstanding feature of *dc* series motor is that; it can generate high starting torque. In other words, at zero speed, the generated torque is very high. This merit allows the use of *dc* series motor in those applications which are loaded right from the beginning and require high starting torque. Example: traction, hoist, crane etc.

As a precautionary measure, it should be ensured that the *dc* series motor is never started at no-load. If the motor is switched ON to the supply at no-load where the torque is very less, it over-speeds and may lead to mechanical damages. The initial starting current at no-load is so large that it causes blowing of fuses. However, blowing of fuses protects the motor from further damage. This is evident from Fig. 2.11(c). Hence, at least 15% of full load should be present right from the starting condition.

DC series motors are referred as Constant Power Motor, because in the operating region, power remains almost constant irrespective of speed of operation. On the other hand, shunt motors are regarded as Constant Torque Motor where the torque is almost constant irrespective of speed. In case of *ac* motor also, shunt characteristics are referred as Constant Torque characteristics and series characteristics are referred as Constant Power characteristics.

When a *dc* series motor is connected across an *ac* source, the direction of torque *Ta* do not change much and rotor continues to rotate in same direction. This is because $T_a = K_t \phi_m I$. With alternating supply connected across series motor, both ϕ_m and I will reverse simultaneously. Series motors designed to operate on both ac and dc supply are called as Universal motor. For limiting the eddy current losses in Universal motor operating on ac supply or rippled dc supply, the magnetic circuits of field poles are made with laminated stampings. Common applications are drilling machine, grinder and kitchen mixer etc.

DC shunt motor has larger field winding inductance compared to that of armature circuit inductance. This causes the field flux to lag behind the armature current with ac excitation. The torque production is ineffective hence, dc shunt motors are not operated on ac supply.

2.4.3 DC Compound Motors

Some applications require a combination of the operating characteristics of *dc* shunt and *dc* series motors. In such cases, *dc* compound motors are employed. As mentioned above, *dc* compound motors have two field windings that are wound on the same stator poles. One of them is connected in series with armature winding and the other one is connected in parallel with armature winding. Series field winding carries large armature current hence, it has few number of turns of thicker wire. The shunt winding is connected in parallel hence, it shares almost full supply voltage but carries low current. These field windings when excited establishes their respective magnetic fields ϕ_{se} and ϕ_{sh}. Resultant of these two fields is the main field ϕ_m. If ϕ_{se} and ϕ_{sh} are in same direction, ϕ_{se} will strengthen ϕ_{sh} and the machine is called as *dc cumulatively compounded motor*. However, if they are in opposite directions, ϕ_{se} will weaken ϕ_{sh} and it is called as *dc differentially compounded motor*.

Based on the winding connections, compound motors are also classified as (1) long shunt type and (2) short shunt type motors. The circuit diagrams are shown in Fig. 2.12 followed by their voltage and current equations.

Different characteristics of *dc* compound motor are shown in Fig. 2.13. A *dc* differentially compounded motor possesses inherent instability during the starting as well as running conditions. It tends to race under heavy overloaded conditions. In such motors under overload condition, the series field flux may become excessively large due to large starting current, therefore, net flux may reverse in direction. This causes the motor to start in opposite direction. Hence, *dc* differentially compounded motors are rarely used.

$$V_t = E_b + I_a (R_a + R_{se}) + \text{brush drop} \tag{2.16}$$

$$I = I_a + I_{sh}$$

$$I_{sh} = \frac{V_t}{R_{sh}}$$

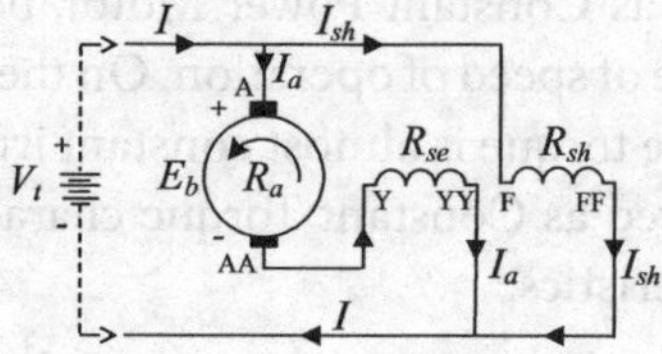

(a) DC long shunt type compound motor

$$V_t = E_b + I_a R_a + IR_{se} + \text{brush drop} \tag{2.17}$$

$$I = I_a + I_{sh}$$

$$I_{sh} = \frac{V_t - IR_{se}}{R_{sh}} = \frac{E_b + I_a R_a}{R_{sh}} \tag{2.18}$$

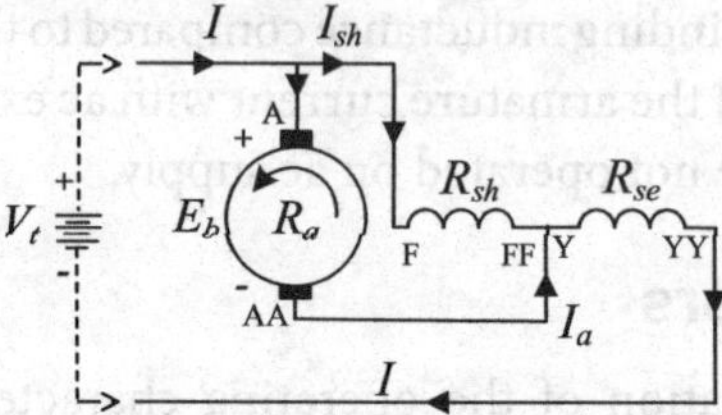

(b) DC short shunt type compound motor

Fig. 2.12: DC compound motors

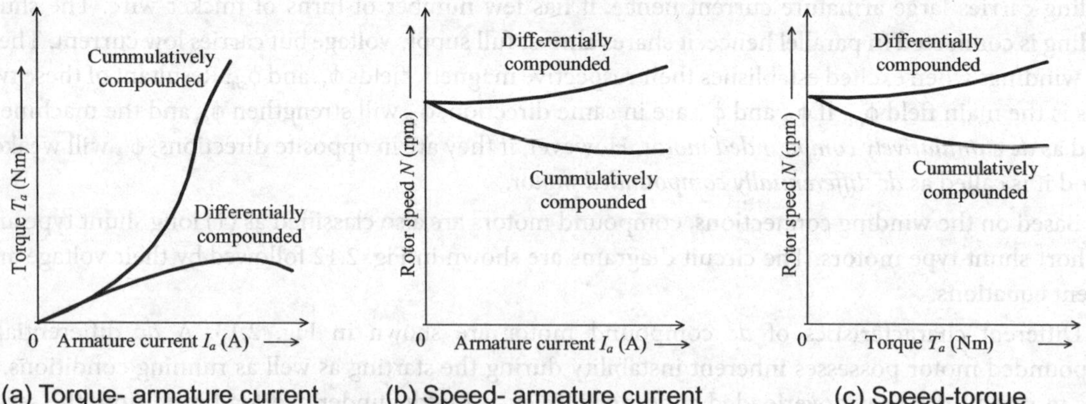

(a) Torque- armature current (b) Speed- armature current (c) Speed-torque

Fig. 2.13: Characteristics of dc compound motor

2.5 STARTING OF DC MOTOR

Usually, *dc* motors below 1 HP (0.746 kW) rating are directly connected to the power supply. This is because the magnitude of armature resistance *Ra* in small size motors is relatively high and therefore they do not allow the starting current to exceed safe limits. Series motors designed to function as Universal Motor for kitchen mixer, grinder and portable drilling machine are a few examples. Similarly, the tiny *dc* motors below 24 V rating used in toys are connected directly to the source. However, such practice cannot be allowed in larger capacity motors.

The voltage equations in different types of *dc* motor were discussed above and are reproduced below.

$V_t = (E_b + I_a R_a)$ in *dc* shunt and *dc* separately excited motors

$V_t = E_b + I_a (R_a + R_{se})$ in *dc* series and *dc* compound motors

From equations (2.2) and (2.3), the back *emf* E_b is given by,

$$\therefore E_b = \frac{\phi_m PN}{60}\frac{Z}{A} = K_e \omega \phi_m$$

Where *N* and are the rotor speeds expressed in r.p.m. and rad/s respectively.

During starting condition when the dc supply is just connected to the motor and the rotor is yet to rotate, speed N = ω = 0 and therefore the back emf $E_b = 0$. From above voltage equations, it will be found that the magnitude of armature current Ia during starting condition will be restricted only by the winding resistances R_a and R_{se}. It may be noted that, in moderate and larger capacity motors, the armature and series field windings are designed to have low magnitude of R_a and R_{se} so as to reduce the copper loss and voltage drops. Consequently, if rated value of input voltage is applied during starting condition than, the motor will draw dangerously high magnitude of transient armature current which is also called as starting current.

For example, consider an 8.8 kW (11.8 hp), 220V *dc* shunt motor. Let $R_a = 0.5\Omega$. The rated value of armature current (i.e. highest current which the machine can carry safely) will be 8800/220 = 40A. From above voltage equation, during starting condition, since $E_b = 0$, the starting value of I_a will be 220/0.5 = 440A). Such heavy starting current may result into:

(i) Excessively large starting torque, causing damage to coupling between motor and load shaft,

(ii) Significant sparking between the commutator and brushes and damage to them,

(iii) Creation of hot spots in the winding and weakening of the insulation,

(iv) Blowing off the fuse,

(v) Damage to the rectifier (AC to DC converter) that is supplying power to the motor, and

(vi) Large voltage dip in the supply lines thereby affecting the performance of other electrical equipment which are connected to the same lines.

Once, the rotor starts rotating, back emf E_b starts building up and consequently the current decreases. In view of this, it is advisable to take appropriate measures and limit the magnitude of armature starting current below rated value that can be handled safely so that the motor develops sufficient starting torque without causing electrical damage to armature windings/commutator segments/ brushes/ supply system and without causing any mechanical damage to rotating parts. This necessity introduces the need of a device called as dc motor starter. These starters can be either of (i) Manual type or (ii) Automatic type. The common types of manual starter that are used for starting of dc shunt motor are:

(i) 3-Point starter

(ii) 4-Point starter

Following types of manual starters are commonly used for *dc* series motor.

(i) 2-Point starter or No-load release starter – drum controller

(ii) No-volt release starter.

Starters are the devices that confines the magnitude of armature current within safe limits during starting condition. The manual starters mentioned above have a provision of adding external resistances in series with armature winding before the input supply is switched ON. After connecting the supply, as the rotor accelerates and E_b starts building, these additional resistances are then removed in steps. In addition, these starters also have motor protection features like No-Volt release and Over-Load release.

2.5.1 Three – Point Starter

Three-point starter is used for the starting of *dc* shunt motor. Externally, it has three terminals *L* (Line), *A* (Armature) and *F* (Field) hence known as 3-point starter. *L*-terminal should be connected to the *dc* source, *A*-terminal to the armature winding and *F*-terminal to the field winding of motor. This is shown in Fig. 2.14. The assembly of a practical 3-Point Starter is shown in Fig. 2.15.

Internally, it has a set of multiple resistors R_1, R_2, R_3, R_4, R_5 that are connected between the copper studs (1, 2, 3, 4, 5 and RUN). One more stud is located before Stud-1 and is named as 'OFF'. It is not connected to any resistor. There is a movable arm with handle which can be moved on different studs starting from 'OFF' to 'RUN' position. This movable arm is attached to a spiral spring which helps to bring the movable arm back to 'OFF' position when supply is disconnected or during the abnormal condition. A conducting strip below the moving arm conducts electricity from supply to studs. There are two protective electromagnets with corresponding coils. One is called as Holding Magnet- Holding Coil (HC) or 'No-volt release and the other one is called as Over Load Release (OLR). HC is connected in series with field winding whereas OR is connected in series with armature winding. The function of HC coil is to hold the movable arm on the 'RUN' stud during the normal running condition and release it whenever, the field current becomes zero or is very low or when the input supply voltage gets disconnected. Function of OLR coil is to disconnect the motor from the supply during overload condition. Under overload condition- the conventional OLR effectively short circuits the HC, causing the movable arm to come back to OFF position due to spring action. When commercial OLR relays are in use, they are directly connected to the armature circuit.

During disconnected state, the movable arm is maintained at 'OFF' stud by the spring. It indicates that the motor is disconnected from supply. To start the motor, the handle is moved to different studs in sequence from '1' to 'RUN'. When it comes in contact with Stud-1, the series combination of HC coil and field winding gets connected across the supply whereas the armature winding gets connected in series with all the starting resistors ($R_1 + R_2 + R_3 + R_4 + R_5$). Since the current begins to flow in both the armature and field windings, the motor starts running. As the speed increases, the back *emf* E_b also increases. Therefore, the external resistors are no more required in the armature circuit. Hence the movable arm is moved ahead from Stud-2 to 'RUN' thereby cutting the external resistors in steps.

There is a soft iron piece attached to the movable arm. When it comes on 'RUN' stud, the soft iron piece touches the holding magnet and remains there for the normal running of the motor till the HC coil is carrying sufficient current. In case of power failure or if the field winding gets open circuited accidentally, HC gets demagnetized and the spiral spring brings the handle back to 'OFF' position. Note that during running condition, the main flux ϕ_m should not be allowed to become zero or become very low otherwise, the rotor speed increases to dangerously high value. This is because, speed *N* is inversely proportional to ϕ_m (See equation 2.2). In other words, the field current should not become zero.

Another protective device is the Over Load Release Magnet (OLR). In case, the armature current exceeds the preset value due to overload, OLR Magnet becomes stronger and attracts the movable soft iron bar towards it. As a result, terminals *a* and *b* of the HC coil gets short circuited. The HC Magnet gets demagnetized and releases the handle. It gets pulled back to 'OFF' position by the spiral spring and thus the motor is disconnected from supply.

In case of 3-point starter, HC is connected in series with the field winding. When 3-point starter is used with *dc* shunt motor having provision of speed control using field weakening, the force of attraction produced by HC may even become so weak that spring force can bring the movable armature back to OFF stud.

With DC shunt motor having provision to raise the speed above base speed using field weakening, 4-point starter is used. In case of 4-point starter, the HC is connected directly across supply voltage. In case of voltage failure, the HC is demagnetized causing the movable arm to come back to OFF stud due to spring action. However, it does not provide field failure protection.

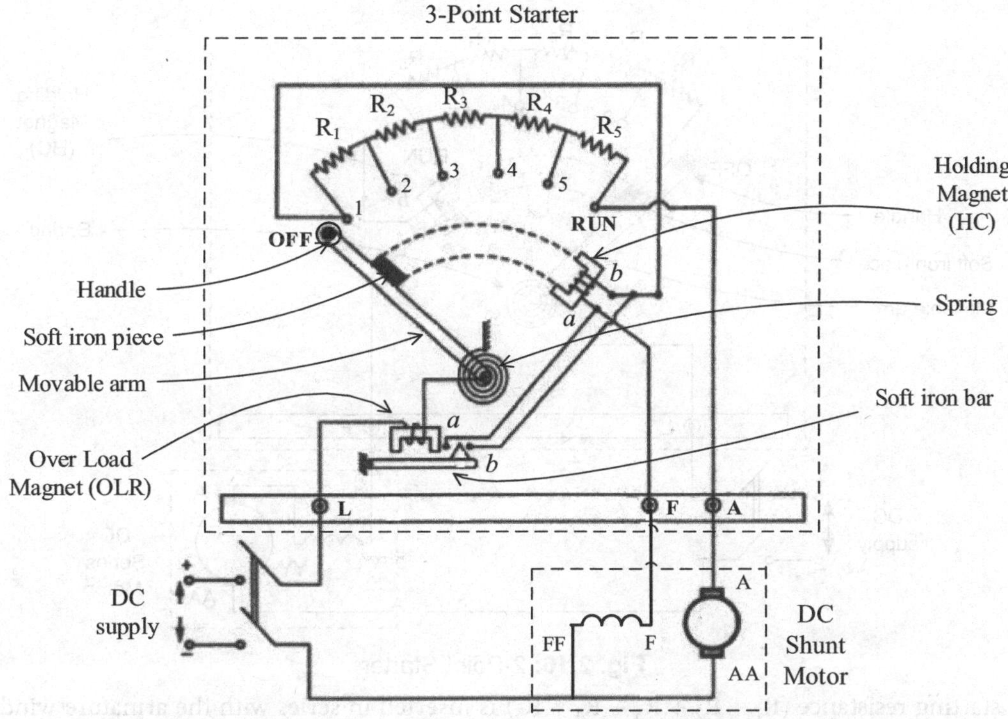

Fig. 2.14: 3-Point Starter circuit

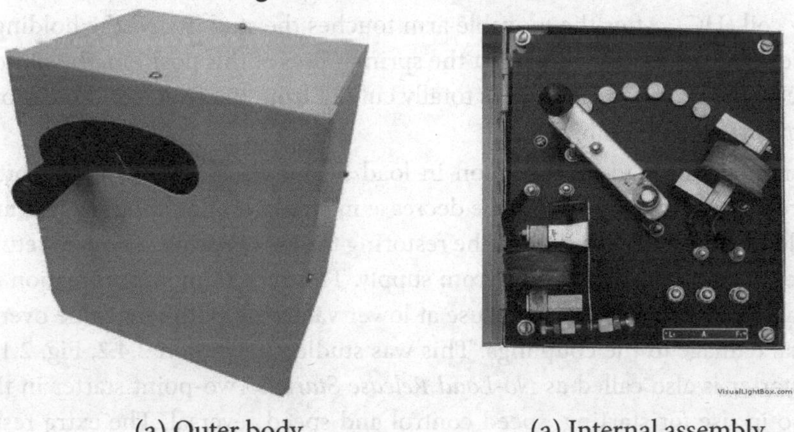

(a) Outer body (a) Internal assembly

Fig. 2.15: 3-Point Starter glimpses

2.5.2 Two – Point Starter

Two-point starter is used for the starting of *dc* series motors. Externally, it has two terminals *L* (Line) and *F* (Field). *L*-terminal is to be connected to the *dc* source and *F*-terminal to the field winding of motor. This is shown in Fig. 2.16.

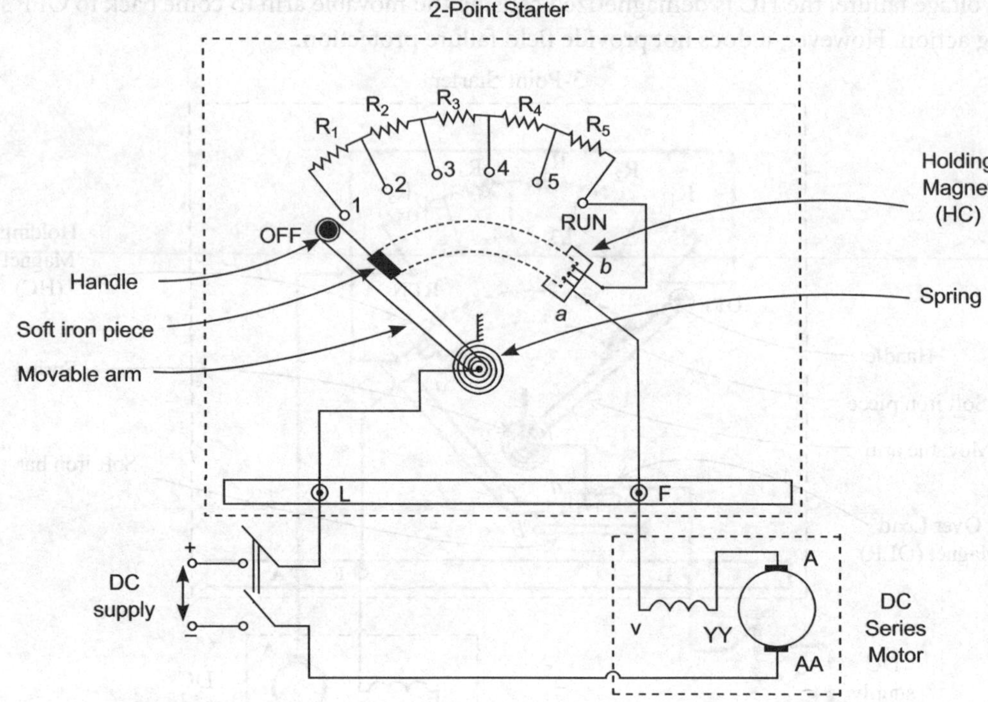

Fig. 2.16: 2-Point Starter

A starting resistance ($R_1 + R_2 + R_3 + R_4 + R_5$) is inserted in series with the armature winding for restricting the armature current within safe limit during starting condition. It is then gradually cut out as the rotor accelerates. The armature current passes through the series field winding, armature winding and the holding coil (HC). After the movable arm touches the stud 'RUN', the holding magnet attracts the soft iron piece and holds the arm against the spring. Thus at this position, the motor has reached its final normal speed, the starting resistance is totally cut out from the circuit and the motor receives rated input voltage for normal operation.

In case of removal of load or reduction in load below a safe value, the magnitude of armature current *Ia* gets reduced. This results into the decrease in strength of holding magnet and hence, it is no more able to hold the movable arm against the restoring torque of spring. The arm returns back to 'OFF' position and the motor gets disconnected from supply. This type of motor protection against very light loads is important in *dc* series motors because at lower values of *Ia*, the machine over-speeds and may cause mechanical damage to the couplings. This was studied in Section 2.4.2, Fig. 2.11(b). Hence, this type of 2-Pole starter is also called as *No-Load Release Starter.* Two-point starter in the form of drum controller is also in use for starting, speed control and speed reversal. The extra resistances in series starter are rated to carry full load current continuously for low speed operation.

The conventional starters discussed above for *dc* shunt and *dc* series motors' are based on resistance control. For limiting the starting current, controlled voltage starting using solid state starters based on solid state *ac-dc* full bridge and half bridge thyristor controlled rectifier / converter operating on *ac* supply and *dc-dc* chopper operating on *dc* supply are quite common nowadays.

2.6 SPEED CONTROL OF DC MOTOR

All electric motors have a specified rated speed as mentioned on their name plate. When the mechanical load on motor shaft changes, a corresponding change in rotor speed takes place naturally. This was discussed in Section 2.3.4. However, in many applications, an intentional adjustment in speed is required to attain high productivity, proper operation and high quality products. Speed control below and above rated speed is needed in many applications.

Some applications require the motor to rotate at constant speed even when the load is changed. The winder used in paper mill needs fast adjustment of motor speed in accordance with the changing diameter of the roll so as to maintain a constant tension in the paper. In metal cutting machine, the tool must have its speed adjusted according to the size of the work-piece, type of cutting tool, metal to be cut and other factors. Other examples include, electric crane, electric traction, electrical equipment used in mines, textile industry, rolling mills etc. DC motors are most suitable for wide range of speed control and therefore are the preferred choice for many adjustable speed drives. From equation (2.11),

Rotor speed, $$\omega = \frac{V_t - I_a R_a}{K_e \phi_m} \text{ rad/s}$$

The rotor speed ω can be converted into rotations per minute (r.p.m.) by using the following relationship where *N* is rotor speed in r.p.m.

$$\omega = \frac{2\pi N}{60}$$

Thus, rotor speed of a *dc* motor can be controlled by any of the following methods:

(i) By varying the resistance of armature circuit
(ii) By varying the main field (ϕ_m) of the motor
(iii) By varying the input voltage V_t applied to the armature terminals.

The above methods can be applied to all types of *dc* motor.

2.6.1 Speed Control of *dc* Shunt Motor

Following methods are used for speed control of *dc* shunt motor:

(i) By varying the resistance of armature circuit
(ii) By varying the armature input voltage V_t
(iii) By varying the main field ϕ_m

(i) *By varying the resistance of armature circuit*

Fig. 2.17(a) shows the connection diagram for speed control of *dc* shunt motor by varying the armature circuit resistance. A speed-controlling variable resistor $R_{a(ext)}$ is connected in series with armature winding to vary the effective resistance of armature circuit. The speed equation (2.11) will modify as,

$$\omega = \frac{V_t - I_a\left(R_a + R_{a(ext)}\right)}{K_e \phi_m} \text{ rad/s} \tag{2.19}$$

If the main flux ϕ_m and input voltage V_t are held constant than, for constant value of load torque T_L, the variation in rotor speed for different values of external speed-controlling resistance is shown in Fig. 2.17(b). At no-load, $I_a = 0$, $\therefore \omega_0 = V_t / K_e \phi_m$. Load torque is the torque offered by mechanical load to the motor. It opposes the armature torque T_a and hence opposes the rotation of rotor.

Note that if T_L is constant, the armature torque T_a and armature current I_a of a *dc* shunt motor also remains constant. This is because, for a *dc* shunt motor, ϕ_m is normally constant. Therefore from torque equation (2.7), *Ia* remains same. Examples of constant torque load are: low speed hoist, conveyer etc.

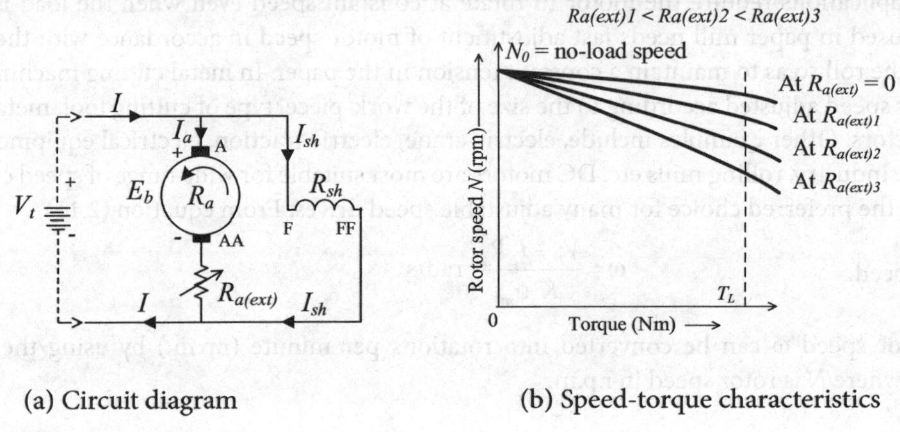

(a) Circuit diagram (b) Speed-torque characteristics

Fig. 2.17: Speed control of dc shunt motor by varying the resistance of armature circuit

When is increased then, from above equation (2.19), it is observed that the rotor speed will decrease. Thus by this method, the speed control is accomplished in downward direction below rated speed. The range of speed control is not constant and depends on the value of load torque T_L as shown in Fig. 2.17(b). The main advantage of this method of speed control is that very low speeds down to creeping speeds of a few r.p.m. can also be obtained.

In this method, the field current I_{sh} and hence the main flux are held constant. Any change in *Ish* results in a variation of speed as shown in Fig. 2.18.

Following are the drawbacks of this method.

Rotor speed above rated value cannot be obtained because at = 0, speed is rated. and cannot be decreased below zero.

At low operating speeds, is high therefore, excessive power loss occurs in . This reduces the motor efficiency resulting in higher operating cost.

At higher values of , the change in rotor speed due to change in the magnitude of load torque TL is high. In other words, for higher values of , the speed regulation is poor.

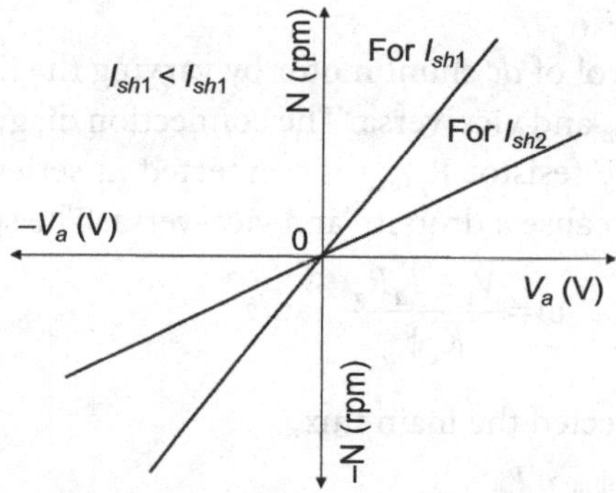

Fig. 2.18: Armature voltage control of *dc* shunt motor

(ii) *By varying the armature input voltage V_t:*

This method of speed control requires a variable voltage source. It is a quite attractive method due to its increased efficiency if solid-state power semiconductor converter circuit is used to obtain the variable *dc* voltage. Following types of devices can be employed to vary the *dc* voltage.

(a) *AC to DC Converter or Rectifier* which converts *ac* voltage into variable *dc* voltage.

(b) *DC to DC Chopper Circuit* which converts fixed *dc* voltage into variable *dc* voltage output.

(c) Autotransformer and uncontrolled Rectifier

(d) Ward Leonard method which requires an additional motor-generator set

Referring back to equation (2.11), it is observed that, the rotor speed is linearly related to the armature voltage when the field current I_{sh} and hence ϕ_m are constant.

$$\omega = \frac{V_t - I_a R_a}{K_e \phi_m} \; \omega \text{ rad/s}$$

For constant load torque *TL*, the variation in rotor speed for different values of armature voltage is shown in Fig. 2.19. The armature voltage control method is preferred for lowering the speed below rated value. This is because at rated input voltage, the speed is also rated. If any attempt is made to obtain the rotor speed above rated value by using this method, it will require a rise in input voltage above rated value. Note that, it is never advisable to apply input voltage higher than the rated value (specified by manufacturer) as that may weaken the insulating material and shorten the life of machine. As shown in Fig. 2.18, the rotor speed and armature voltage are linearly related hence, this is the best method for quick speed reversal.

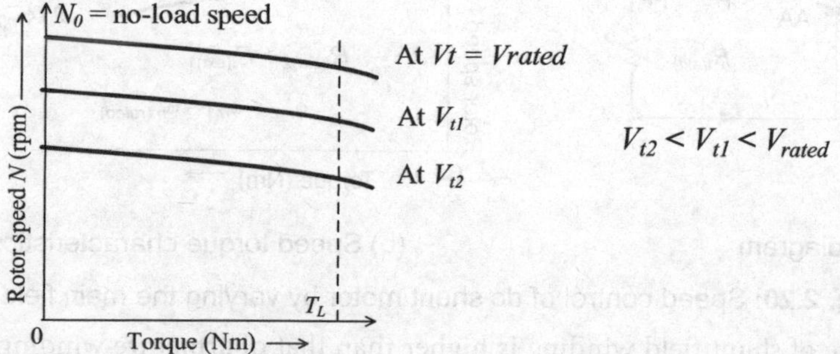

Fig. 2.19: Speed-torque characteristics for speed control of dc shunt motor by varying input voltage

(iii) *By varying the main field ϕ_m:*

In this method of speed control of *dc* shunt motor by varying the field excitation, the speed can be increased by reducing main flux ϕ_m and vice-versa. The connection diagram is shown in Fig. 2.20(a). An external speed-controlling variable resistor $R_{f\,(ext)}$ is connected in series with field winding to vary the field current I_{sh}. Rise in $R_{f\,(ext)}$ will cause a drop in and vice-versa. The speed equation (2.11) is,

$$\omega = \frac{V_t - I_a R_a}{K_e \phi_m} \text{ rad/s}$$

If magnetic saturation is neglected the main flux,

$$\phi_m \propto I_{sh}$$

$$\therefore \omega = \frac{V_t - I_a R_a}{K_e K_f I_{sh}} \text{ rad/s} \tag{2.20}$$

Where, $\phi_m = K_f I_{sh}$ and is the field constant.

Thus, rotor speed is inversely proportional to the field current as well as inversely proportional to .

$$\omega \propto \frac{1}{I_{sh}}$$

At $R_{f(ext)} = 0$, the current and rotor speed, both have rated value. Since the speed is inversely proportional to I_{sh}, for any reduction in speed below rated value by this method, it will require a corresponding rise in I_{sh} above rated value. It means that $R_{f\,(ext)}$ should be reduced below zero which is not possible. However, to obtain speeds above rated value, I_{sh} can be decreased by increasing $R_{f\,(ext)}$. Thus, this method of speed control by varying the field excitation can be employed for getting the speeds above rated value. Hence, it is also called as '*Field Weakening Method*' because rise in speed is attained by reduction in main field. The speed-torque characteristics for different values of main field is shown in Fig. 2.20(b).

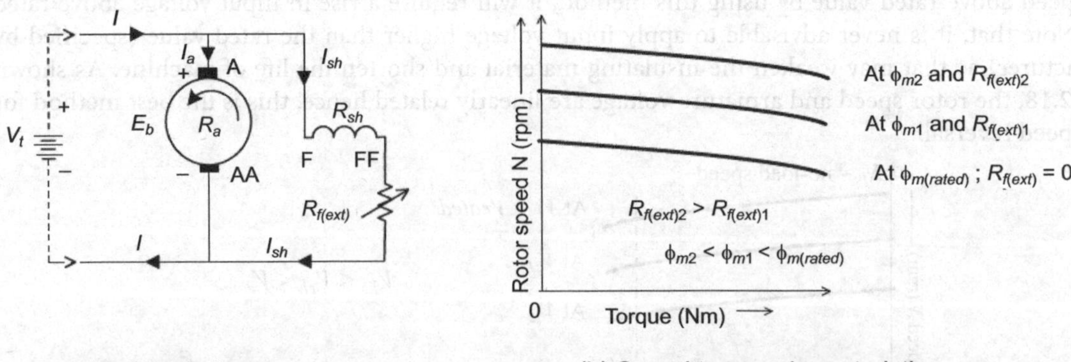

Fig. 2.20: Speed control of dc shunt motor by varying the main field ϕ_m

The inductance of shunt field winding is higher than that of armature winding. We know that, an inductive circuit stores energy in the form of electromagnetic field and due to which it opposes any change in the magnitude or direction of current flowing through it. To reverse the speed by this method, it will demand for the reversal in direction of field current. Any attempt to quickly change the direction of field current, first needs to discharge the already stored energy in field winding due to positive field current and thereafter the new field current in negative direction can be made to flow. However, the winding inductance does not allow this quick change and results in huge spark in the contactor. Hence, this method is not suitable for speed reversal. The shunt field current versus speed characteristics in four quadrants is shown in Fig. 2.21. The advantages of this method of speed control are: Speeds above rated value can be obtained.

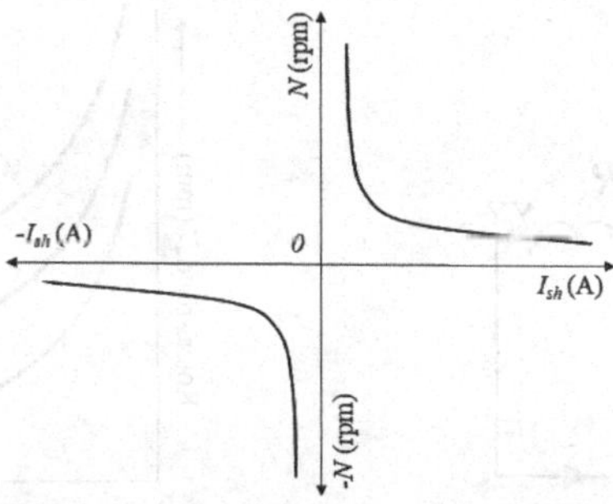

Fig. 2.21: Speed control of dc shunt motor by main field control

(a) Speeds below rated value cannot be obtained.

(b) As the magnitude of shunt field current I_{sh} is normally small, the power loss $I^2_{sh}\ R_{f(ext)}$ is also small, the current rating of external speed-controlling resistor is small and overall efficiency is good.

However, it has the following limitations:

(a) Speeds below rated value cannot be obtained.

(b) There is a limit to higher speed because if the field is weakened too much, there is a loss of stability.

(c) This method is not suitable for speed reversal.

(d) DC Shunt motor should be started using 4-point starter.

2.6.2 Speed Control of *dc* Series Motor

Following methods are used in speed control of *dc* series motor:

(i) *By varying the resistance of armature circuit:*

In this method, the speed is reduced in downwards direction below rated speed. An external variable resistor $R_{a(ext)}$ is connected in series with the armature and field winding. It is also called as speed-controlling resistance. The circuit diagram and resulting speed-torque characteristics of *dc* series motor at different values of $R_{a(ext)}$ are shown in Fig. 2.22 (a) and (b). It can be observed that, how at constant torque load, different speeds are obtained by varying $R_{a(ext)}$ provided that the input voltage *Vt* is held constant. The speed governing equation (2.15) for a *dc* series motor is reproduced below.

$$\omega = \frac{V_t}{\sqrt{K_e K_f}} \frac{1}{\sqrt{T_a}} - \frac{(R_a + R_{se})}{K_e K_f}$$

It will get modified as,

$$\omega = \frac{V_t}{\sqrt{K_e K_f}} \frac{1}{\sqrt{T_a}} - \frac{(R_a + R_{se} + R_{a(ext)})}{K_e K_f} \tag{2.21}$$

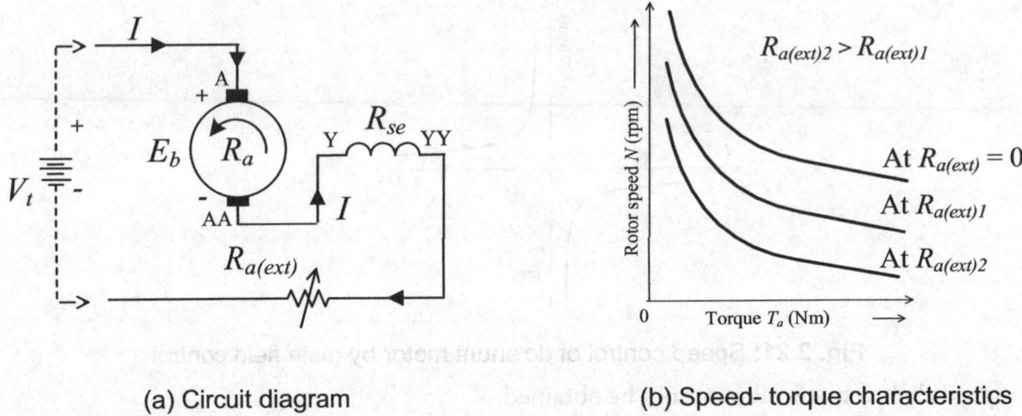

(a) Circuit diagram (b) Speed- torque characteristics

Fig. 2.22: Speed control of dc series motor by varying the resistance of armature circuit

This method of speed control is most economical for constant torque loads like low speed hoist, conveyer belt etc. The major drawback of this method is the high power loss $I^2 R_{a(ext)}$ in the speed-controlling resistor $R_{a(ext)}$ which reduces the overall efficiency. Due to this drawback, it is suitable for intermittent duty loads like crane, hoist because in such loads, the motor is not operated continuously for a longer duration of time. Drum controller is used for starting and speed control of medium capacity *dc* series motor. They have resistances which can carry rated current on continuous basis.

(ii) *By varying the main field* ϕ_m:

In this method, the speed of a *dc* series motor is increased in upward direction by changing the main magnetic flux ϕ_m produced by field winding. It can be varied by any of the following:

(a) Changing the field current 'I' with the help of an external diverter resistance $R_{f(ext)}$ that is connected in shunt with the field winding.

(b) Changing the field *mmf* by using a tapped field winding.

(a) ***By Shunting the Field Winding or Diverter Field Control:***

From equation (2.11), we know that the rotor speed is inversely proportional to the main flux Since this flux is produced by field current, it means that the field current of a *dc* series motor 'I' can be varied to obtain a change in speed. In this method, the field winding is shunted by an external diverter resistance $R_{f(ext)}$ and some part of field current 'I' is bypassed through $R_{f(ext)}$. This is shown in Fig. 2.23(a).

Since $R_{f(ext)}$ is a variable resistor, it can be adjusted to change the field current. The effect of variation in the values of $R_{f(ext)}$ on the speed-torque characteristic of a *dc* series motor is demonstrated in Fig. 2.23(b). This method was useful in traction application for running the trains at higher speed.

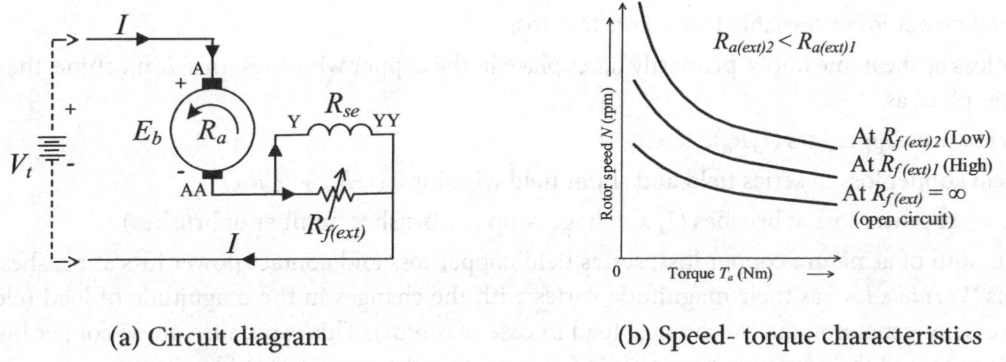

Fig. 2.23: Speed control of *dc* series motor by shunting the field winding

(b) *By using a Tapped Field Winding:*

To employ this method of speed control of a *dc* series motor, it is necessary that the field winding should have tappings as shown in Fig. 2.24. This can be ensured at the design and manufacturing stage of the machine. When the tapping is changed, the effective number of turns of series field winding that will be carrying current, will also change and therefore, it will change the series field *mmf* (ampere-turns).

To attain high speed operation, lesser number of turns should be selected so as to decrease the flux. Thus the main flux ϕ_m produced by series field winding can be varied. This method is preferred in small size *dc* or *ac/dc* universal motors used in domestic appliances.

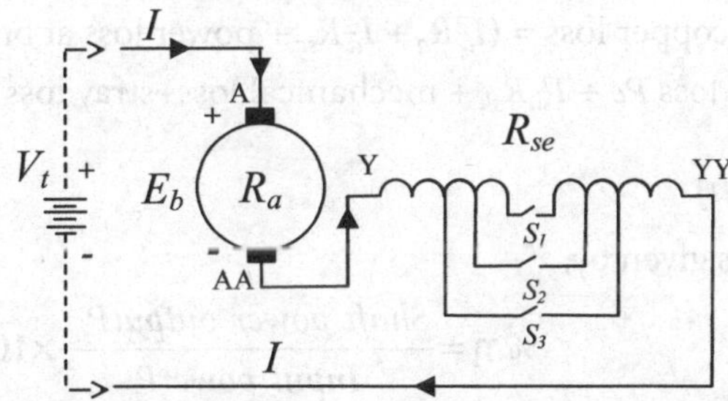

Fig. 2.24: Speed control of dc series motor having tapped field winding

2.7 LOSSES AND EFFICIENCY

In Unit-1, Section 1.12, details about the different types of losses taking place in a *dc* machine (generator as well as motor) and their effect were discussed. Students are advised to go through that section prior to the following explanation. The power equation in all electrical machines is:

$$\text{Output power (W)} = \text{input power (W)} - \text{total power loss (W)}$$

In an ideal machine, the power losses are absent and therefore whatever power is received by the machine as input, it delivers the same amount of power as output thus resulting into 100% efficiency. The total power loss in a rotating machine can be expanded in two ways as:

- Total power loss = total copper loss + total core loss + mechanical loss
- Total power loss = variable loss + constant loss

Copper loss as the name imply, primarily takes place in the copper windings. In a *dc* machine, the copper loss takes place as,

- Armature copper loss ($I_a^2R_a$),
- Field copper loss in series field and shunt field windings, ($I_a^2R_{se} + I_{sh}^2R_{sh}$)
- Contact power loss at brushes (I_a × voltage drop per brush × number of brushes)

The sum of armature copper loss, series field copper loss and contact power loss at brushes is also called as '*Variable loss*' as their magnitude varies with the changes in the magnitude of load (electrical load in case of generator and mechanical load in case of motor). This is because, these copper losses are dependent on and the armature current I_a is dependent on the magnitude of load.

The total core loss P_c, which is also called as Iron loss and No-load loss. It is independent of the changes in the magnitude of load and takes place in the armature core and pole shoes. As explained in Section 1.12, this core loss primarily comprises of hysteresis loss and eddy current loss.

In a *dc* shunt machine and *dc* compound machine, as far as the input voltage *Vt* is held constant, the shunt field current remains same at all values of load. Hence the shunt field copper loss also remains constant. Similarly, the mechanical loss, also called as *friction and windage loss* which represents the loss of power due to friction at the bearings and opposition to rotation by the surrounding air is independent of the magnitude of load. Hence, it too remains constant. The other type of constant loss is the *Stray loss* which represents a small amount of power loss in other metallic parts due to leakage and fringing of the flux. Hence, together, they are called as '*Constant loss.*' Thus,

- Variable loss = total copper loss = ($I_a^2R_a + I_a^2R_{se}$ + power loss at brush contacts)
- Constant loss = core loss *Pc* + $I_{sh}^2R_{sh}$ + mechanical loss+stray loss

2.7.1 Efficiency (η)

Efficiency of a *dc* motor is given by,

$$\%\,\eta = \frac{\text{Shaft power output}P_{sh}}{\text{Input power}P_i} \times 100$$

$$\%\,\eta = \frac{P_{sh}}{P_{sh} + \text{total power loss}} \times 100$$

$$= \frac{P_i - \text{total power loss}}{P_i} \times 100 \qquad (2.22)$$

Where,

Pi = input voltage V_t × input current I

$P_{sh} = T_{sh} \times \omega$

2.8 BRAKE TEST

This is a direct method of determining the efficiency of a *dc* motor in laboratory. It is also called as the *load test* on motor. The motor mechanical output power is calculated using spring balance reading and speed of rotation. The experimental setup consists of a brake drum and belt arrangement as shown in Fig. 2.25(a). The brake drum is mechanically coupled to the shaft of the motor and the belt is wrapped around half periphery of the drum. Two spring balances are attached at the ends of the belt. They are fixed to a stationary steel structure such that, it becomes possible to vary the tension in the belt by tightening or loosening the load adjustment wheels. The circuit connections for a *dc* shunt motor are shown in Fig. 2.25(b).

Initially, the belt is kept loose to ensure no-load condition. The motor is started and made to run at rated speed with the help of a variable field resistor $R_{f(ext)}$. The load-adjustment wheels are tightened gradually till the ammeter reads rated current of the motor. This confirms that rated load is applied. The readings of the two spring balance in kilogram (kg) are noted. Let them be represented by M_1 and M_2. Also, the rotor speed (N), ammeter reading (I) and voltmeter reading (V_t) are noted.

Now, net tension in the belt will be the difference of the two spring balance readings multiplied by acceleration due to gravity.

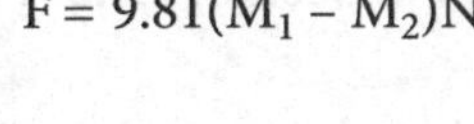

$$F = 9.81(M_1 - M_2)N$$

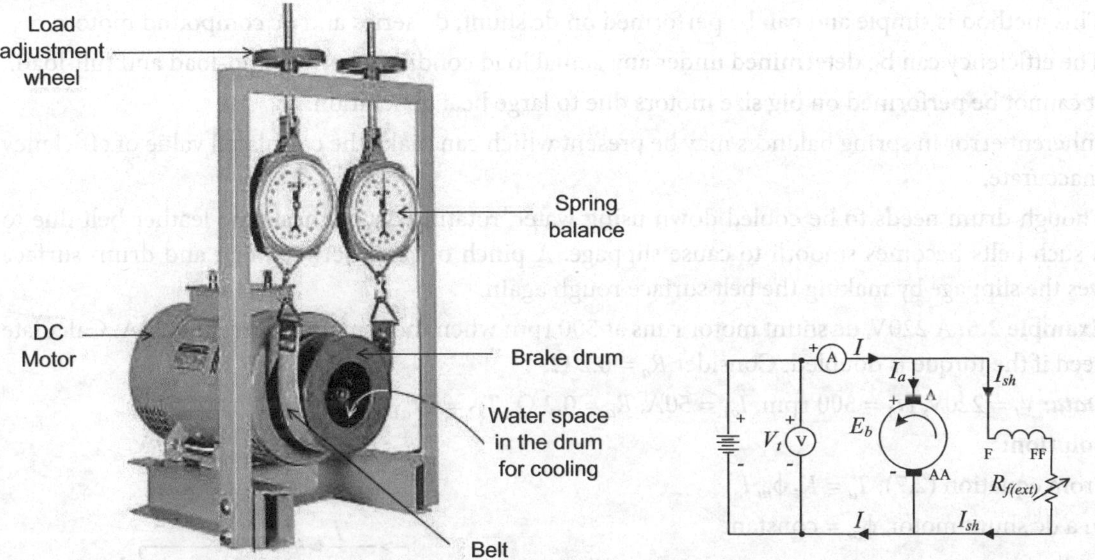

(a) Experimental setup (a) Electrical circuit diagram for dc shunt motor

Fig. 2.25: Brake Test on dc motor,

Shaft torque $T_{sh} = F \times r$, where r = radius of the brake drum in metres + half thickness of belt

$$T_{sh} = 9.81r\,(M_1 - M_2)\text{ Nm} \tag{2.23}$$

Shaft power output will be $P_{sh} = T_{sh} \times \omega$

$$\therefore P_{sh} = T_{sh} \times \frac{2\pi N}{60} \text{ Watts} \tag{2.24}$$

Where, ω = angular speed of the rotor in rad/s $= \frac{2\pi N}{60}$ and N is the rotor speed in rotations per minute (r.p.m). The electrical input power will be the product of input voltage and input current.

$$P_i = V_t I \text{ Watts}$$

$$\therefore \% \text{ Efficiency}\eta = \frac{P_{sh}}{P_i} \times 100 \qquad (2.25)$$

While performing this test, heavy friction is caused between the rotating drum and the stationary belt. Consequently, the resulting friction loss produces heat. Therefore to cool the brake drum, some water should be poured into the space provided for it in the drum before connecting the motor to the supply. This is shown in Fig. 2.25(a).

Brake test is advised only on small size motors because in case of big motors, it is difficult to dissipate the large amount of heat generated between the brake drum and the belt. This test can be performed on all types of *dc* motor. However, while performing it on a *dc* series motor, care should be taken to ensure that the belt on the brake drum is tight enough throughout the experiment even before the motor is connected to the *dc* source. This is because at no-load and at light loads, the *dc* series motor attains dangerously high speed. The reason was discussed in Section 2.4.2. The relative merits and demerits of brake test are:

- This method is simple and can be performed on dc shunt, dc series and dc compound motors.
- The efficiency can be determined under any actual load condition between no-load and full-load.
- It cannot be performed on big size motors due to large heat generation.
- Inherent error in spring balances may be present which can make the calculated value of efficiency inaccurate.

Though drum needs to be cooled down using water, rotating drums heats the leather belt due to which such belts becomes smooth to cause slippage. A pinch of sand between belt and drum surface removes the slippage by making the belt surface rough again.

Example 2.6: A 220V, dc shunt motor runs at 500 rpm when the armature current is 50A. Calculate the speed if the torque is doubled. Consider $R_a = 0.2\ \Omega$.

Data: $V_t = 220\text{V}$, $N_1 = 500$ rpm, $I_{a1} = 50\text{A}$, $R_a = 0.2\ \Omega$, $T_{a2} = 2T_{a1}$, $N_2 = ?$

Solution:

From equation (2.7), $T_a = K_t\, \phi_m\, I_a$

In a dc shunt motor, ϕ_m = constant

$\therefore\ T_a \propto I_a$

Given that: $\frac{T_{a2}}{T_{a1}} = 2$

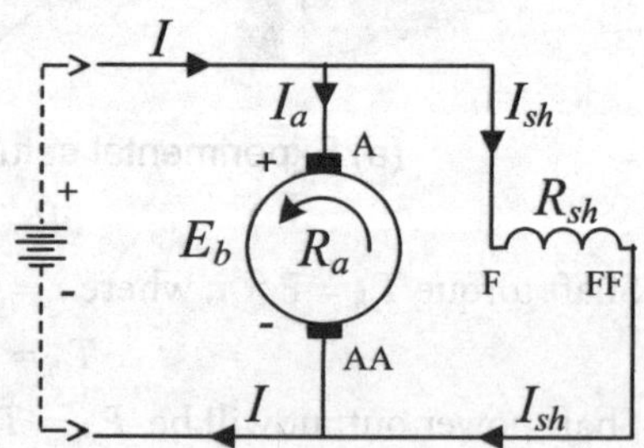

$\therefore \frac{I_{a2}}{I_{a1}} = 2\ \therefore\ I_{a2} = 2I_{a1} = 100\text{A}$

From equation (2.2), $E_b = \frac{\phi_m PN}{60}\frac{Z}{A}$

$\therefore N \propto \frac{E_b}{\phi_m}$

For a *dc* shunt motor, ϕ_m = constant

$\therefore \quad N \propto E_b$

Now, $E_{b1} = (V_t - I_{a1} R_a) = 210V$

$E_{b2} = (V_t - I_{a2} R_a) = 200V$

$\frac{N_2}{N_1} \propto \frac{E_{b2}}{E_{b1}}$

$\therefore \quad N_2 = 476$ rpm

Example 2.7: A *dc* shunt motor supplied at 230V runs at 900 rpm. Calculate the resistance required in series with the armature circuit to reduce the speed to 500 rpm. Assume that the armature current is 25A.

Data: $V_t = 230V$, $I_{a2} = 25A$, $N_1 = 900$, $N_2 = 500$, $R_{a(ext)} = ?$

Solution:

Armature resistance Ra is not given. Normally it is very small and ideally, it should be zero. Hence assuming that $R_a = 0$.

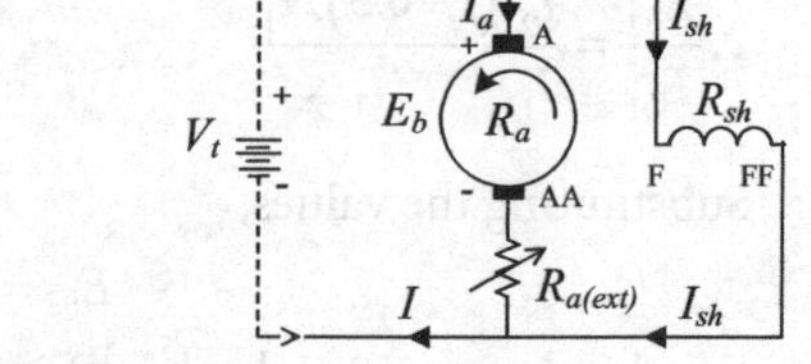

$$E_b = Vt - I_a (R_a + R_{a(ext)})$$

Initially at $R_{a(ext)} = 0$, $N_1 = 900$ rpm

$\therefore \quad E_{b1} = V_t = 230V$

In the second case,

$R_{a(ext)} \neq 0$ and $N_2 = 500$ rpm

From equation (2.2), $E_b = \frac{\varnothing_m PN}{60}\frac{Z}{A}$

$\therefore \frac{E_{b2}}{E_{b1}} = \frac{N_2}{N_1}\frac{\varnothing_{m2}}{\varnothing_{m1}}$

But in a dc shunt motor, flux remains constant, $\varnothing_{m1} = \varnothing_{m2}$

$\therefore E_{b2} = 230 \times \frac{500}{900} = 127.77V$

From voltage equation again,

$$E_{b2} = V_t - I_{a2} (R_a + R_{a(ext)})$$

$$127.77 = 230 - 25(0 + R_{a(ext)})$$

$\therefore \quad R_{a(ext)} = 4.089$

Example 2.8: A *dc* series motor with unsaturated magnetic circuit and negligible resistance when running at a certain speed on a given load takes 40A at 450V. If the torque varies as a cube of speed, find the external resistance that should be connected in series to reduce the speed by 30%.

Data: $V_t = 450V, I_1 = 40A, (R_a + R_{se}) = 0, T_a \propto N^3, N_2 = 0.7 N_1, R_{a(ext)} = ?$

Solution:

Initially, consider $R_{a(ext)} = 0$ and speed = N_1.

Back emf $E_{b1} = V_t - I_1 (R_a + R_{se}) = 450V$

From equation (2.13), $T_a = K_t \phi_m I = K_t K_f I^2$

ie., $T_a \propto I^2$

In this problem it is given that, $T_a \propto N^3$

$\therefore I^2 \propto N^3$

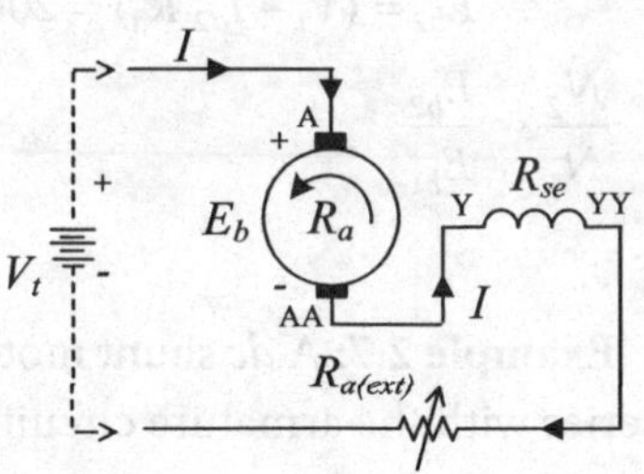

$$\therefore \frac{I_2^2}{I_1^2} = \frac{N_2^3}{N_1^3} = \frac{(0.7N_1)^3}{N_1^3}$$

$I_2^2 = 0.7^3 \times 40^2 = 548.8; \therefore I_2 = 23.42A$

From equation (2.2), $E_b = \frac{\phi_m PN}{60} \frac{Z}{A}$

In a dc series motor, $\phi_m \propto I$

$$\therefore \qquad E_b \propto NI$$

$$\therefore \frac{E_{b2}}{E_{b1}} = \frac{I_2}{I_1} \frac{(1-0.3)N_1}{N_1}$$

Substituting the values,

$$\therefore \qquad E_{b2} = 184.43V$$

For the above circuit, back emf is,

$$E_{b2} = V_t - I_2 (R_a + R_{se} + R_{a(ext)})$$

$$R_{a(ext)} = \frac{V_t - E_{b2}}{I_2} - (R_a + R_{se})$$

Substitute the values,

$$\therefore \qquad R_{a(ext)} = 11.34\ \Omega$$

Example 2.9: A 220V, *dc* series motor is running at a speed of 800 rpm and draws 80A current. If the total resistance of armature and series field windings is 0.15 Ω, calculate the speed when the motor develops half the torque. The magnetic circuit is unsaturated.

Data: $V_t = 220V, N_1 = 800rpm, I_1 = 80A, R_t = (R_a + R_{se}) = 0.15\Omega, T_{a2} = 0.5\ T_{a1}, N_2 = ?$

Solution:

From equation (2.13), $T_a = K_t \phi_m I = K_t K_f I^2$

ie., $T_a \propto I^2$, $\therefore \frac{T_{a2}}{T_{a1}} = \frac{I_2^2}{I_1^2}$

But, it is given that, $\frac{T_{a2}}{T_{a1}} = \frac{1}{2}$

$\therefore \frac{I_2^2}{80^2} = \frac{1}{2}$; $\therefore$ $I_2 = 56.56\text{A}$

Back emf,

$$E_{b1} = (V_t - I_1 R_t) = 208\text{V}$$
$$E_{b2} = (V_t - I_2 R_t) = 211.51 \text{ V}$$

From equation (2.2), $E_b = \frac{\phi_m PN}{60} \frac{Z}{A}$

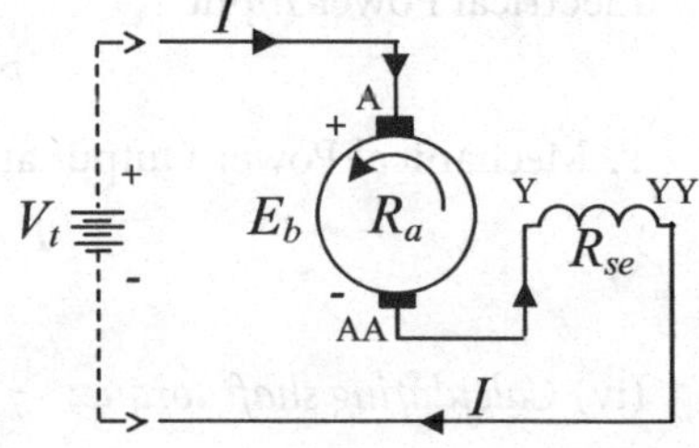

In a dc series motor, $\phi_m \propto I$

$$\therefore E_b \propto NI$$

$\therefore \frac{N_2}{N_1} = \frac{E_{b2}}{E_{b1}} \frac{I_1}{I_2}$

Substitute the values

$\therefore$ $N_2 = 1150$ rpm

Example 2.10: A 4 pole, 220V dc series motor has a wave connected armature winding with 1200 conductors. The flux per pole is 20mWb when the motor is drawing 46A. The iron and friction loss amount to 900W. Armature and series field resistances are 0.25 Ω and 0.15 Ω respectively. Find the (i) speed, (ii) total torque, (iii) shaft power, (iv) shaft torque and (v) efficiency.

Data: $V_t = 220\text{V}$, P = 4, A = 2 (since wave connected), Z = 1200, $\phi_m = 20 \times 10^{-3}$ Wb, $I = 46\text{A}$, $Pc = 900\text{W}$, $R_a = 0.25\Omega$, $R_{se} = 0.15\Omega$. N, T_a, T_{sh}, P_{sh} and η = ?

Solution:

Back emf $E_b = V_t - I(R_a + R_{se}) = 201.6\text{V}$

(i) *Calculating speed:*

From equation (2.2), $E_b = \frac{\phi_m PN}{60} \frac{Z}{A}$

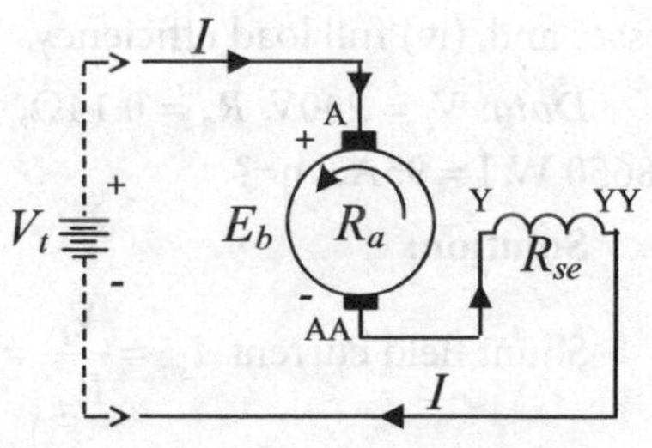

Substitute the values,

$\therefore$ N = 252 rpm

(ii) *Calculating total (armature)torque:*

From equation (2.10), $E_b I_a = \omega T_a$

Where, for a dc series motor $I_a = I$

$$\therefore T_a = \frac{E_b I_a}{\omega} = E_b I_a \frac{1}{\left(2\pi N/_{60}\right)}$$

Or you can use equation (2.6) directly,

$$T_a = \left(\frac{1}{2\pi}\right)\frac{PZ}{A}\varnothing_m I_a$$

Substituting the values

$\therefore$ $T_a = 351.59$ Nm

(iii) *Calculating shaft power:*

$$\text{Total copper loss} = I^2 (R_a + R_{se}) = 846.4\text{W}$$

$$\text{Total loss} = \text{total copper loss} + (\text{iron loss} + \text{friction loss})$$

$$\text{Total loss} = 846 + 900 = 1746\text{W}$$

Electrical Power Input,

$$P_i = V_t I = 10120\text{W}$$

$\therefore$ Mechanical Power Output at the shaft will be,

$$P_{sh} = (\text{Power Input} - \text{Total loss}) = 8374\text{W}$$

(iv) *Calculating shaft torque:*

From equation (2.8),

$$T_{sh} = \frac{P_{sh}}{\omega} = \frac{P_{sh}}{2\pi N/60} = 317.48\text{W}$$

(v) Calculating efficiency:

$$\%\eta = \frac{P_{sh}}{P_i} \times 100 = 82.74\%$$

Example 2.11: A 240V, 25 HP, dc shunt motor has armature and field resistances of 0.14 Ω and 80 Ω respectively. Contact voltage drop is 1V per brush. At full load, the motor takes 95A current from supply. Calculate: (i) Armature and field copper loss, (ii) power loss in the brush contact, (iii) Core + mechanical losses and, (iv) full load efficiency.

Data: V_t = 240V, R_a = 0.14Ω, R_{sh} = 80Ω, Contact drop per brush = 1V, P_{sh} = 25HP = 25 × 746 = 18650 W, I = 95A, , η=?

Solution:

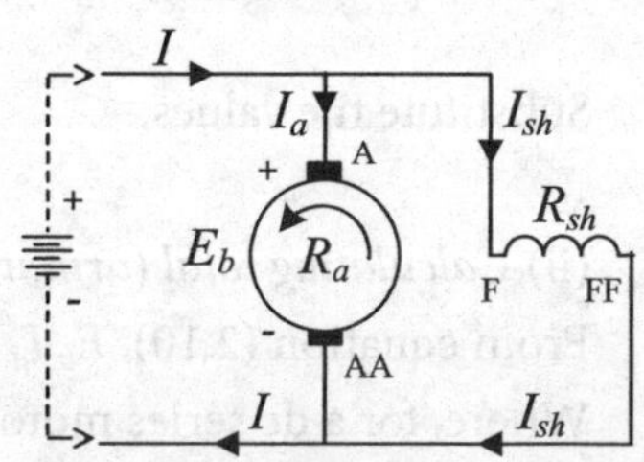

Shunt field current $I_{sh} = \dfrac{V_t}{R_{sh}} = 3\text{A}$

Armature current $I_a = (I - I_{sh}) = 92\text{A}$

(i) *Calculating copper loss:*

Armature copper loss = $I_a^2 R_a$ = 1184.96W

Field copper loss = $I_{sh}^2 R_{sh}$ = 720W

(ii) *Calculating contact power loss at brushes:*

At the contact between commutator and brushes, some power gets lost due to the transfer of armature current between the commutator and brushes. This power loss is treated as a copper loss and is included in the armature copper loss when not specified separately. The machine should have minimum two brushes to provide the incoming and outgoing paths for I_a.

∴ Power loss at brush contact will be,

$$= \text{number of brushes} \times \text{voltage drop per brush} \times I_a$$

$$= 2 \times 1 \times 92 = 184\text{W}$$

$$\therefore \quad \text{Total copper loss} = 1184.96 + 720 + 184 = 2088.96\text{W}$$

(iii) *Calculating Core + Mechanical losses:*

Shaft power output at full load,

$$P_{sh} = 25\text{HP} = 25 \times 746 = 18650 \text{ W},$$

Power input at full load,

$$P_i = V_t I = 22800\text{W}$$

$$\text{Total power loss} = P_i - P_{sh} = 4150\text{W}$$

$$\text{Core loss + Mechanical loss} = \text{total loss} - \text{copper loss}$$

$$\text{Core loss + Mechanical loss} = 4150 - 2088.96 = 2061.04\text{W}$$

(iv) *Calculating efficiency:*

$$\%\eta = \frac{P_{sh}}{P_i} \times 100 = \quad 81.7$$

Example 2.12: A 220V dc motor generates a back emf of 210V while rotating at 1480 rpm. If it draws an armature current of 50A, calculate the torque developed in the armature.

Data: Vt = 220V, E_b = 210 V, N = 1480 rpm, Ia = 50A, Ta = ?

Solution:

From equation (2.10), $E_b I_a = \omega T_a$

Where, the rotor angular speed $\omega = \frac{2\pi N}{60} = 154.9$ rad/s

$$\therefore T_a = \frac{E_b I_a}{\omega} = 67.78 \text{ Nm}$$

Example 2.13: A dc shunt motor runs at 1000 rpm on 220V supply. Its armature and field resistances are 0.5 Ω and 110 Ω respectively and the total current taken from supply is 26A. It is desired to reduce the speed to 750 rpm. Keeping the armature and field currents same, what resistance should be inserted in the armature circuit?

Data: V_t = 220V, R_a = 0.5Ω, R_{sh} = 110Ω, $I_1 = I_2$ = 26A, N_1 = 1000, N_2 = 750, $R_{a(ext)}$ = ?

Solution:

Shunt field current $I_{sh} = \frac{V_t}{R_{sh}} = 2\text{A}$

$$\text{Armature current } I_{a1} = I_{a2} = (I_2 - I_{sh}) = 24\text{A}$$

When $R_{a(ext)} = 0,$

$$E_{b1} = (V_t - I_{a1} Ra) = 220 - 24 \times 0.5 = 208V$$

When $R_{a(ext)} \neq 0,$

$$E_{b2} = V_t - I_{a2} (R_a + R_{a(ext)})$$

$$E_{b2} = 220 - 24(0.5 + R_{a(ext)})$$

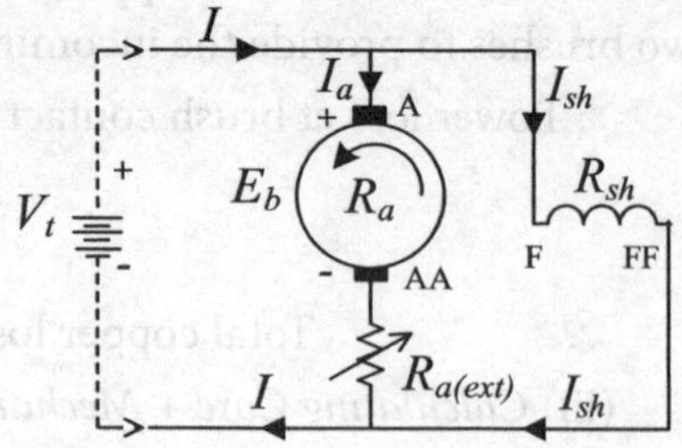

From equation (2.2), $E_b = \dfrac{\varnothing_m PN}{60} \dfrac{Z}{A}$

$$\therefore \frac{E_{b2}}{E_{b1}} = \frac{N_2}{N_1} \frac{\varnothing_{m2}}{\varnothing_{m1}}$$

But in a *dc* shunt motor, flux remains constant

$$\phi_{m1} = \phi_{m2}$$

$$\therefore E_{b2} = 208 \times \frac{750}{1000} = 156V$$

From above equation again,

$$156 = 220 - 24(0.5 + R_{a(ext)})$$

$$\therefore \quad R_{a(ext)} = 2.166\ \Omega$$

Example 2.14: While performing brake test on a 240V, dc shunt motor, it takes 50A from supply and runs at 1400 rpm. The difference between spring balance readings is 10kg and the radius of brake drum is 75 cm. Calculate the shaft torque and efficiency.

Data: $V_t = 240V, I = 50A, N = 1400$ rpm, $(M_1 - M_2) = 10kg, r = 75 \times 10^{-2}$ m, $T_{sh}, \eta = ?$

Solution:

From equation (2.23)

$$T_{sh} = 9.81r(M_1 \sim M_2) = 9.81 \times 75 \times 10^{-2} \times 10 = 73.575 \text{ Nm}$$

Shaft power output will be $P_{sh} = T_{sh} \times \omega$

$$\therefore P_{sh} = T_{sh} \times \frac{2\pi N}{60} = 10.781\text{kW}$$

Input power $P_i = V_t I = 240 \times 50 = 12\text{kW}$

% Efficiency $\eta = \dfrac{P_{sh}}{P_i} \times 100 = 89.84\%$

Example 2.15: A 250V, dc shunt motor runs at 1000 rpm and takes a current of 25A. Calculate the speed when the motor is loaded and takes a current of 50A and the armature reaction weakens the field by 3%. Consider armature and shunt field resistances equal to 0.2 Ω and 250 Ω respectively.

Data: $V_t = 250V, N_1 = 1000$ rpm, $I_1 = 25A, I_2 = 50A, \phi_2 = 0.97\ \phi_1, R_a = 0.2\Omega, R_{sh} = 250\Omega, N_2 = ?$

Solution:

Shunt field current $I_{sh} = \dfrac{V_t}{R_{sh}} = 1A$

In a *dc* shunt motor, the main flux ϕ_m remains constant. It also means that, I_{sh} remains constant.

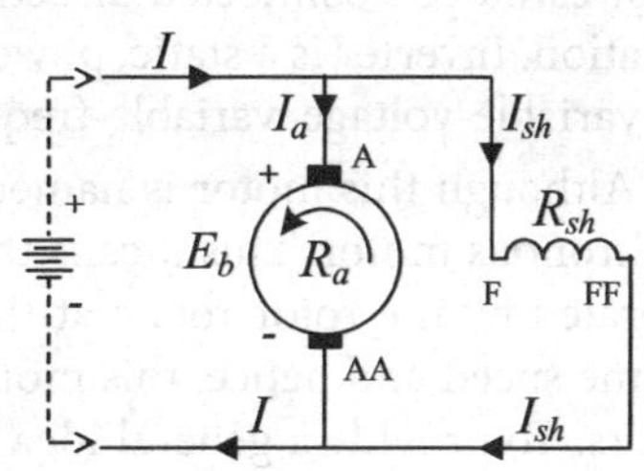

$$\therefore \quad I_{a1} = (I_1 - I_{sh}) = 24\text{A}$$
$$I_{a2} = (I_2 - I_{sh}) = 49\text{ A}$$
$$E_{b1} = (Vt - I_{a1} R_a) = 245.2\text{V}$$
$$E_{b2} = (Vt - I_{a2} R_a) = 240.2\text{ V}$$

From equation (2.2), $E_b = \dfrac{\phi_m PN}{60}\dfrac{Z}{A}$

$$\therefore N \propto \frac{E_b}{\varnothing_m}$$

Due to armature reaction effect, the net airgap flux ϕ will be the resultant of ϕ_m and ϕ_a. Therefore, ϕ will be responsible for the emf and torque generation.

Hence, $\therefore N \propto \dfrac{E_b}{\phi_m}$

$$\therefore \frac{N_2}{N_1} = \frac{E_{b2}}{E_{b1}}\frac{\phi_1}{\phi_2}$$

In this problem,

$$\frac{\phi_1}{\phi_2} = \frac{\phi_1}{(1-0.3)\phi_1} = \frac{1}{0.97}$$

Substituting the values,

$$\therefore \quad N_2 = 994\text{ rpm.}$$

2.9 BRUSHLESS DC (BLDC) MOTOR

The conventional brushed *dc* motor discussed so far has commutator and carbon brushes which are necessary for commutation. Due to friction and transfer of current between the stationary brushes and the rotating commutator, sparking is produced between them which results into their fast wear and tear. Due to the presence of windings both on stator and rotor, the initial cost of a brushed *dc* motor is high. Only because of the excellent performance, high efficiency and large power to volume ratio, they are preferred in industries.

Brushless *dc* (BLDC) motors are the modern electric motors that are becoming increasingly popular in many applications such as domestic appliances, automotives, aerospace, consumer products, medical equipment, industrial automation and instrumentation. The cooling fan used in Switch Mode Power Supply (SMPS), computers and laptop uses BLDC motors. Most of the e-bikes, 3-wheelers and a few 4-wheelers running on road employ BLDC motor for their propulsion.

As the name implies, BLDC motors do not use brushes for commutation; instead, they are electronically commutated. The rotor do not have winding; instead, permanent magnets are fixed on the rotor. BLDC motors have many advantages over brushed *dc* motors and induction motors like better speed versus torque characteristics, high dynamic response, high efficiency, long operating life, noiseless operation, and higher speed ranges etc. In addition, the ratio of torque delivered to the size of motor is higher making it useful in applications where space and weight are critical factors. However, this motor cannot be connected directly to the source; an inverter and position sensors are necessary for its operation. Inverter is a static, power electronic circuit that is used to convert the fixed-voltage *dc* supply into variable-voltage, variable-frequency *ac* supply.

Although this motor is named as brushless *dc* motor but actually, it falls under the category of *ac* synchronous motor. This means that, the magnetic field generated by the stator and the magnetic field generated by the rotor rotate at the same frequency. In other words, these two magnetic fields rotate at same speed and hence, this motor do not experience the "slip" that is normally present in induction motors. To provide a general idea about how it is energized, a block diagram of 3-phase BLDC motor drive is shown in Fig. 2.26.

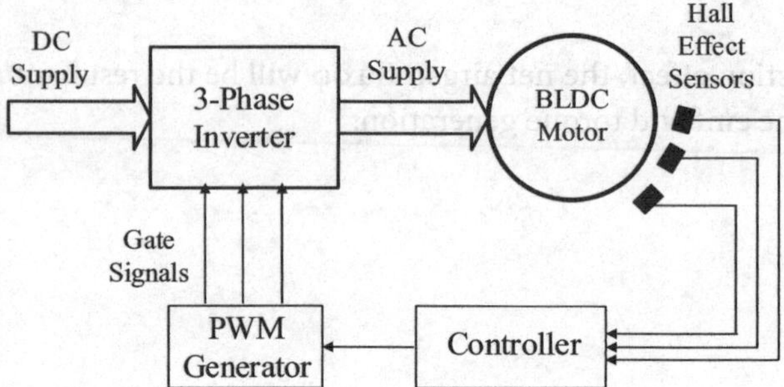

Fig. 2.26: 3-phase BLDC motor drive

2.10 TYPES OF BLDC MOTOR

BLDC motor is a member of permanent magnet synchronous motor family. These motors necessarily have permanent magnets (PM) on the rotor. The stator winding has two types of variants. In one variant, the stator winding is distributed sinusoidally in different slots of the stator and therefore, the shape of induced back *emf* in the stator winding is also sinusoidal. The second variant of stator winding is that of concentrated type and hence it results into trapezoidal shape of induced back *emf*. Thus, PMSMs are classified on the basis of the wave shape of their induced *emf* i.e., sinusoidal and trapezoidal. They are known as:

- Permanent Magnet Synchronous Motor (i.e. with sinusoidal back *emf*) and
- Brushless DC Motor (i.e. with trapezoidal back *emf*).

In addition to the back *emf*, the phase currents in stator winding also have sinusoidal and trapezoidal variations in the respective types of motor. Since, the torque developed is directly proportional to the phase current, it makes the torque output of a PMSM (i.e. with sinusoidal back *emf*) smoother (i.e. lower torque ripple) than that of a BLDC motor which has trapezoidal back *emf*. However, this comes with an extra cost, as the sinusoidal motors take extra winding interconnections because of the coils distribution on the stator periphery, thereby increasing the copper intake by the stator windings. Whereas, the concentrated type stator winding of BLDC motors is relatively simple and also this motor has simplicity of control. Hence, BLDC motors are more popularly used compared to PMSM.

Depending upon the control power supply capability, the motor with correct stator voltage rating is chosen for a given application. Normally, 48V or less rated voltage BLDC motors are used in automotive, robotics, etc. Motors with 100V or higher ratings are used in domestic appliances, automation and in industrial applications. BLDC motors are available in 1-phase, 2-phase and 3-phase configurations. Out of these, 3-phase motors are most popular and widely used.

2.10.1 Classification of BLDC Motor based on Flux Paths

The permanent magnet motors can also be classified based on the path of flux flowing between the stator and rotor. These paths can be either in radial direction or in axial direction. Accordingly, they are classified as:

- Radial flux motors and
- Axial flux motors or Pancake motors.

All conventional motors (including the *dc* machines discussed so far) have flux flowing from stator to rotor and vice versa through the air gap in radial direction. Torque is produced when the radially travelling main magnetic field interacts with the axially flowing armature currents. These machines are commonly used in practice.

The axial flux motors are constructed such that, the flux flow along the axial direction of the shaft. When this axially travelling magnetic field interacts with the radially travelling currents, torque is produced. The axial flux motors are relatively compact in size and may have one or more number of stators, rotors and air gaps cascaded alternately in a sandwich pattern. Such a machine with one stator and one rotor and two stators and one rotor are shown in Fig. 2.27 (a) and (b) respectively. Because of two stators, the power density of the later one is higher than that having only one stator. The air gap in axial flux motors is adjustable during and even after assembly.

In radial flux motors, retention of the magnets against centrifugal force on the rotor is required but there is no such requirement in axial flux motors. In high performance applications, high power density and good acceleration capacity are the highly desirable features. Due to the inherent advantages like compact size, high power density and acceleration capacity, the axial flux motors are being considered in high performance applications like aerospace, aviation, marine and some traction applications including that of the hybrid electric vehicles. The main reason due to which the axial flux PM motor has not become a common choice is its complex manufacturing and assembly requirements.

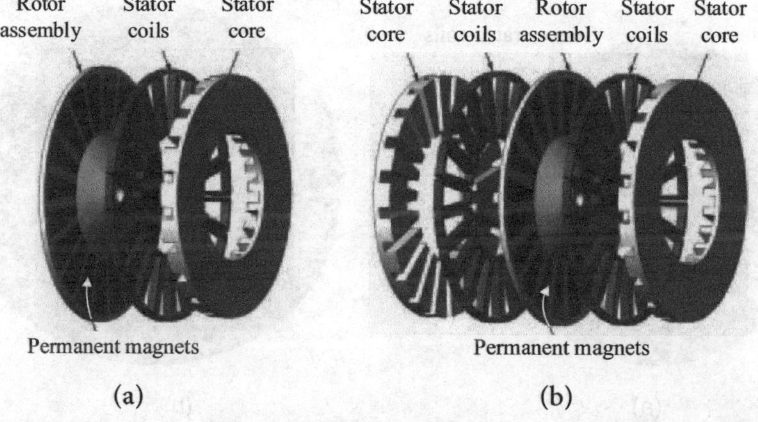

Fig. 2.27: Axial flux PM motor (a) having one stator and one rotor, (b) having two stators and one rotor, Source: [9]

2.10.2 Classification Based on Rotor Position

The brushed *dc* motors discussed earlier in this chapter had electromagnets (comprising of main poles and field winding) on the stationary stator and an armature winding (lap type or wave type) on the rotor. Such type of construction was necessary due to the need of mechanical commutation that is provided by commutator and carbon brushes. However after 1950, the availability of high energy density permanent magnets have made it possible to replace the electromagnets by permanent magnets. Also, the advent of static, semiconductor based high power switching devices like silicon controlled rectifier (SCR), insulated-gate bipolar transistor (IGBT) etc. and the electronic inverter circuits that make use of these switching devices have made it possible to replace the mechanical commutator-brush pair by electronic commutation in the form of inverter circuit. These technological developments led to the birth of PMSM and BLDC machines.

In BLDC motors, permanent magnets are mounted on the rotor whereas the armature winding which is supplied with high input current through inverter is now wound on the stationary stator. This topology simplifies the construction as well as maintenance. Depending upon, whether the rotor is inside the stator or outside the stator, BLDC motor can also be classified as:

- Inner rotor type and
- Outer rotor type

The outer rotor type BLDC motors are mainly preferred in electric vehicles (EV), drones, water pumping and home electronics. In EVs, both types of BLDC motors can be employed. The inner rotor type motor can be coupled to the wheel via a gear system whereas, the outer rotor type motor can be installed inside the driving wheel and operate directly from the wheel without a gear system. Hence, the outer rotor type BLDC motor used in EVs in also called as *Hub Motor*. Such a system allows to remove the gear system thus increasing the overall efficiency. It also provides high torque and helps increasing the mileage of electric bikes. However, since the stator that carries armature winding is enclosed inside the rotor assembly, the heat produced due to power loss in the windings and the stator core is not easily dissipated. As the motor is housed inside the wheel hub, the overall weight of the wheel increases that may lead to stability issues. Fig 2.28 (a), (b) shows the stator and rotor of a Hub motor.

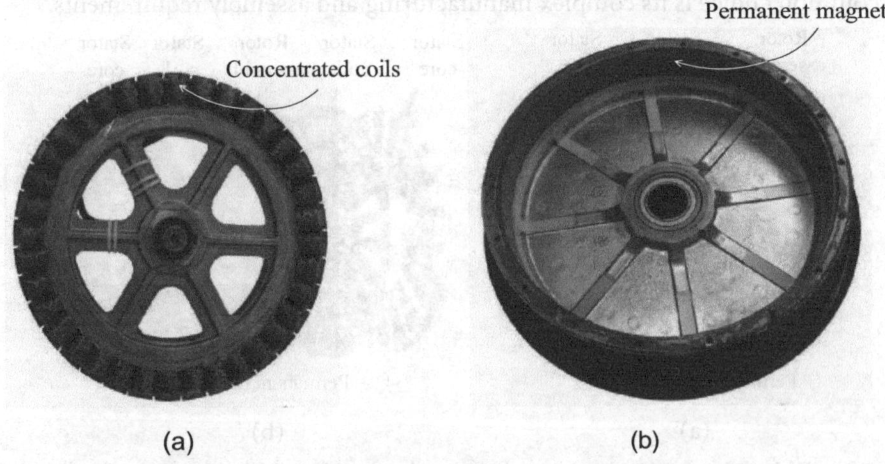

Fig. 2.28: Hub motor (a) stator (inner part), (b) rotor (outer part)

2.10.3 Construction of Inner Rotor Type BLDC Motor

The inner rotor type BLDC motor has two main parts. The stationary part is called as *stator*. It consists of a cylindrical, laminated magnetic core with axially cut slots on the inner periphery and windings housed in these slots. The other one which is free to rotate is called as *rotor*. Both are cylindrical in shape and concentric to each other with a thin airgap between them. Stator consists of an outer frame and salient poles. Concentrated coils are wound over these poles. For a two-pole machine, the coils wound on diametrically opposite poles are connected in series and they constitute one phase of the stator winding. In this way, three phases are formed as shown in Fig. 2.29.

The rotor consists of a cylindrical magnetic core on which, permanent magnets are fixed. Based on the required magnetic flux density in the rotor, appropriate magnets are chosen. Ferrite magnets are used traditionally. They are less expensive but have a demerit of low flux density for a given volume.

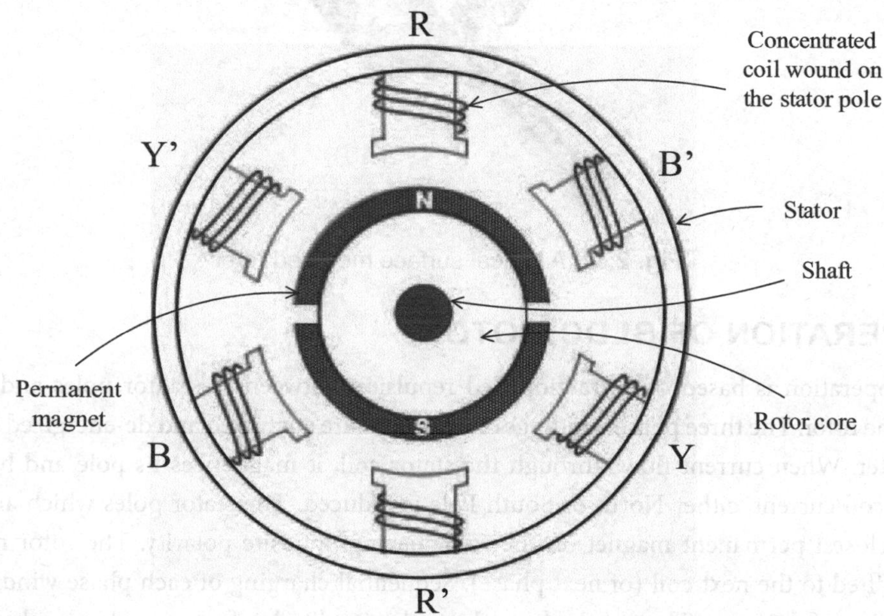

Fig. 2.29: Construction of a BLDC motor

Neodymium iron boron magnets (NdFeB) are the rare earth magnets which are very strong and can provide approximately 20 times the magnetic field per unit volume compared to the regular ferrite magnets. Use of NdFeB magnets allows to generate more torque for the same size of the motor. Based on the different arrangements of permanent magnets on the rotor, it is classified as: (a) Surface mounted type, (b) Embedded type and (c) Inserted type. These three constructions of the rotor are shown in Fig. 2.30. A typical surface mounted rotor is shown in Fig 2.31.

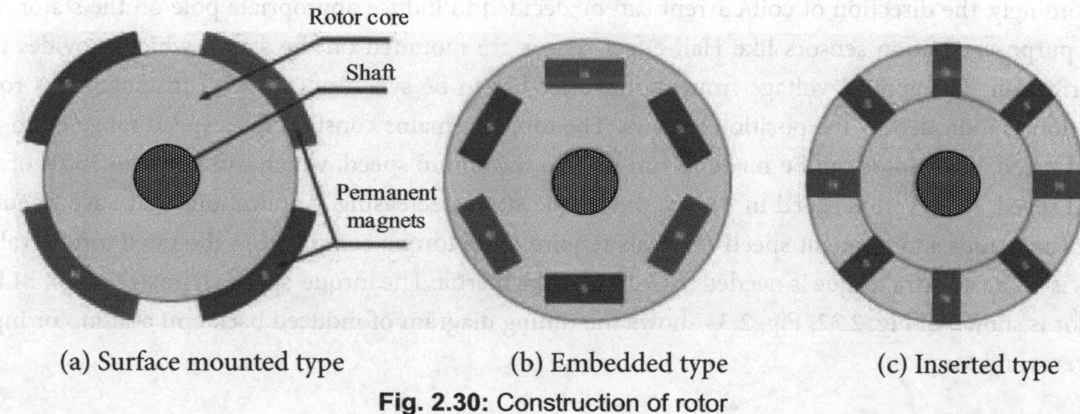

(a) Surface mounted type (b) Embedded type (c) Inserted type

Fig. 2.30: Construction of rotor

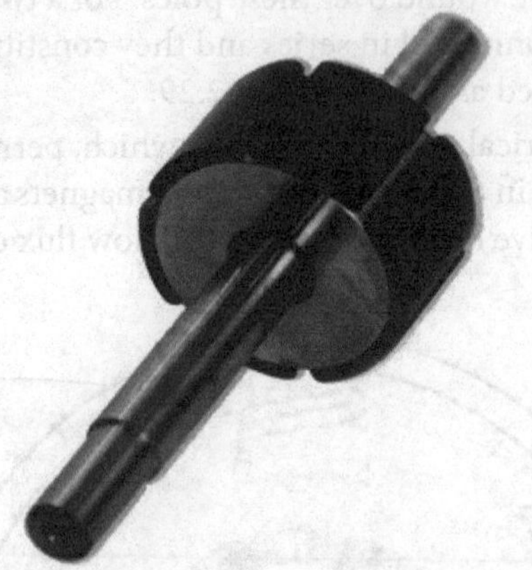

Fig. 2.31: A typical surface mounted rotor

2.11 OPERATION OF BLDC MOTOR

The motor operation is based on attraction and repulsion between the stator poles and permanent magnets of the rotor. The three phase windings on the stator are energized and de-energized sequentially by the inverter. When current flows through the stator coil, it magnetizes its pole and based on the direction of coil current, either North or South Pole is induced. The stator poles which are energized attracts the closest permanent magnet on the rotor having opposite polarity. The rotor moves if the current is shifted to the next coil (of next phase). Sequential charging of each phase winding causes a smooth rotation of the rotor. The torque depends on the amplitude of current, the number of turns of stator winding, the strength and size of permanent magnets, the air gap between the rotor and stator poles, and the radius of rotor.

During the running condition, a trapezoidal shaped back emf is induced in the stator coils. The input current pulses fed to the stator coils are each of 120o duration and should be located in the region where the back emf is constant and maximum. Therefore, before exciting any stator coil, it is necessary to have an advance information about the polarity of rotor magnet which is available close to that coil. Accordingly, the direction of coil current can be decided to induce appropriate pole on the stator. For this purpose, position sensors like Hall-effect sensor are mounted on the stator, which provides this information. The applied voltage from the inverter has to be synchronized with instantaneous rotor position as indicated by the position sensors. The torque remains constant for a speed range up to the rated speed. The motor can be made to run up to a maximum speed, which can be up to 150% of the rated speed. However, as speed increases, the torque starts decreasing. Applications that have frequent starts and stops and frequent speed reversals require more torque compared to the rated torque value. This is because, extra torque is needed to overcome the inertia. The torque-speed characteristic of BLDC motor is shown in Fig. 2.32. Fig. 2.33 shows the timing diagram of induced back emf and motor input currents and torque.

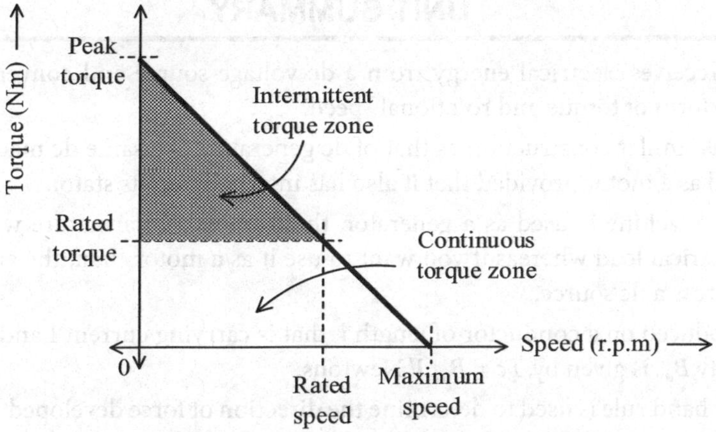

Fig. 2.32: Torque speed characteristic of a BLDC motor

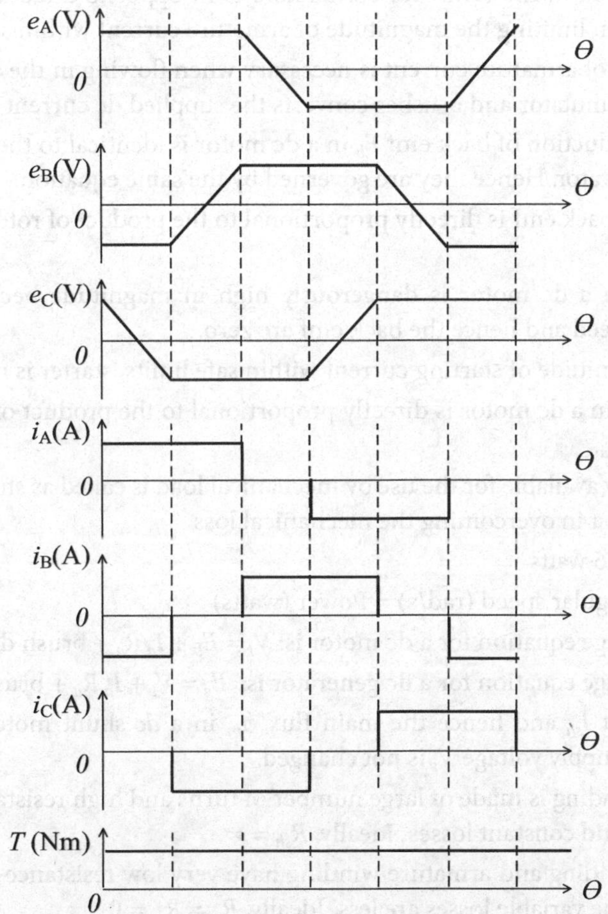

Fig. 2.33: Induced back emf, phase currents and torque waveforms of a BLDC motor

UNIT SUMMARY

- A dc motor receives electrical energy from a dc voltage source and converts it into mechanical energy in the form of torque and rotational speed.
- DC motor has similar construction as that of dc generator. The same dc machine can be used as a generator and as a motor provided that it also has interpoles on its stator.
- When the dc machine is used as a generator, the terminals of armature winding are connected across an electrical load whereas if you want to use it as a motor than, the armature terminals are connected across a dc source.
- The force produced on a conductor of length l, that is carrying current I and placed in a magnetic field of density B_{av} is given by, $Fc = B_{av}\ Il$ Newtons.
- Fleming's left hand rule is used to determine the direction of force developed on a conductor which is carrying current and is placed in a magnetic field.
- Back emf E_b induced in the armature conductors is in opposite direction to supply voltage and therefore is useful in limiting the magnitude of armature current within safe limits.
- Alternating nature of armature current is necessary when flowing in the armature conductors of a dc motor. The commutator and brushes converts the supplied dc current into ac.
- The process of production of back emf E_b in a dc motor is identical to the production of generated emf E_g in a *dc* generator. Hence they are governed by the same equation.
- The magnitude of back emf is directly proportional to the product of rotor speed and main flux $E_b = K_e\ \omega\ \phi_m$.
- Starting current in a dc motor is dangerously high in magnitude because during the starting condition, rotor speed and hence the back emf are zero.
- To restrict the magnitude of starting current within safe limits, starter is used.
- Torque developed in a dc motor is directly proportional to the product of main flux and armature current $T_a = K_t\ \phi_m I_a$.
- The torque actually available for the use by mechanical load is called as shaft torque and is given as, $T_{sh} = T_a$ – torque lost in overcoming the mechanical loss
- 1 horse power = 746 watts
- Torque (Nm) × Angular speed (rad/s) = Power (watts)
- The armature voltage equation for a dc motor is: $V_t = E_b + I_a R_a$ + brush drop
- The armature voltage equation for a dc generator is: $E_g = V_t + I_a R_a$ + brush drop
- Shunt field current I_{sh} and hence the main flux ϕ_m in a *dc* shunt motor are normally constant provided that the supply voltage V_t is not changed.
- The shunt field winding is made of large number of turns and high resistance R_{sh}. This reduces the magnitudes of I_{sh} and constant losses. Ideally, $R_{sh} = \infty$.
- The series field winding and armature winding have very low resistance so that the voltage drops ($I_a R_a$, $I_a R_{se}$) and the variable losses are less. Ideally, $R_a = R_{se} = 0$
- In dc shunt and dc separately excited motors, the drop in rotor speed due to rise in load is very less. Hence, a dc shunt motor is also called as constant speed motor.

- The dc shunt motor is commonly used in those applications which require approximately constant speed at all values of load from zero to full load condition. Examples are: centrifugal pump, blowers, fans, conveyers, machine tools.
- In dc series motor, since the torque is proportional to armature current squared , for the same rise in torque, increase in motor current is less as compared to that in a *dc* shunt and *dc* separately excited motor where the torque is proportional to armature current
- The outstanding feature of dc series motor is that, it can generate high starting torque.
- DC series motor is used in those applications which require high starting torque. Example: traction, hoist, crane, electric vehicle etc.
- For safe starting of dc shunt motor, 3-Point Starter or a 4-Point Starter can be used.
- For safe starting of dc series motor, 2-Point Starter or drum controller can be used.
- Rotor speed of a dc motor can be controlled by any of the following methods:
  - By varying the resistance of armature circuit
  - By varying the main field (ϕ_m) of the motor
  - By varying the input voltage V_t applied to the armature terminals.
- By varying the resistance of armature circuit or by varying the input voltage across armature winding, speed control over a range upto rated speed can be implemented. However, rotor speed above rated value is not possible by this method.
- By varying the shunt field current and hence by changing the main flux, speed control above rated speed can be implemented. However, the rotor speeds below rated value are not possible by this method.
- Brake test is a direct method of determining the efficiency of a dc motor in the laboratory. It is also called as load test on the motor.
- Brushless dc motor consists of permanent magnets on the rotor in place of armature winding. The concentrated coils on the stator are fed a modulated ac supply through an inverter. This is a high speed motors that is popularly being used in domestic applications and electric vehicles.
- The wave shape of emf induced in the stator coils of a BLDC motor is of trapezoidal type.

EXERCISES

Multiple Choice Questions

1. Which of the following can be a normal value of armature resistance of a dc machine?
   (a) 0.002Ω (b) 0.5 Ω
   (c) 15 Ω (d) 150 Ω
2. The electrical equivalent of mechanical power developed in the armature of a dc motor is,
   (a) input power minus all losses
   (b) output power plus all losses
   (c) the product of armature current and back emf
   (d) product of output power and angular speed

3. The wave shape of back emf in a dc motor is,
   (a) trapezoidal (b) unidirectional and constant
   (c) sinusoidal (d) pulsating only in positive direction
4. If a dc generator is made to run as a dc motor in the same direction, the direction of back emf will,
   (a) reverse (b) remain same
   (c) it cannot run as motor (d) back emf will be zero
5. In a dc shunt motor running at rated speed, if the field circuit gets open-circuited, the speed will tend to,
   (a) decrease (b) increase
   (c) remain unchanged (c) cannot be predicted
6. Function of a dc motor starter is to,
   (a) increase the starting torque (b) limit the starting current
   (c) both (a) and (b) (d) none of the above
7. DC motors should not be reversed during running condition when the supply is ON because,
   (a) Back emf will oppose the supply voltage
   (b) high amount of current will flow in the motor and may damage it
   (c) both (a) and (b)
   (d) none of the above
8. A dc shunt motor is fed with a rated supply voltage and rated field current. It also has an external resistance connected in series with armature winding. To obtain rotor speed less than rated value,
   (a) reduce the supply voltage and increase the field excitation
   (b) increase the supply voltage and decrease the armature circuit resistance
   (c) increase the supply voltage and reduce the field excitation
   (d) decrease both, the armature voltage and field excitation
9. A dc shunt motor is supplied with rated voltage and rated excitation. It is driving a mechanical load. If the load torque is doubled than, the speed,
   (a) become half (b) become double
   (c) decrease slightly (d) increase slightly
10. A *dc* series motor has a diverter resistance connected across field winding. If this diverter resistance is increased then the speed will,
    (a) increase (b) decrease
    (c) remain unchanged (d) cannot be predicted
11. A *dc* series motor runs at rated speed on rated excitation current. If an external non-zero resistor is connected in series with field winding, the speed will,
    (a) increase (b) decrease
    (c) remain unchanged (d) cannot be predicted
12. If a *dc* generator is made to run as a *dc* motor in the same direction, the direction of armature current will,
    (a) reverse (b) remain same
    (c) it cannot run as motor (d) armature current will be zero

13. A 120V, *dc* motor having armature resistance of 0.5Ω and back emf equal to 110V will have an armature current equal to,
(a) 20A (b) 460A
(c) 240A (d) 220A

14. A 120V, *dc* motor having armature resistance of 0.5Ω and back emf equal to 110V will have an armature current during starting condition equal to,
(a) 20A (b) 460A
(c) 240A (d) 220A

15. If the current supplied to a *dc* series motor is increased from 10A to 12A then, neglecting magnetic saturation, the percentage rise in its torque will be,
(a) 44% (b) 2%
(c) 20% (d) no change

16. IF the load current and main flux of a dc motor are held constant, and the armature input voltage is increased by 10% than the speed will,
(a) increase by 10% (b) decrease by 10%
(c) remain same

17. If the pole flux ϕ_m of a dc motor tends to change towards zero, the speed will tend to,
(a) become zero (b) approach infinity
(c) no change (d) cannot be predicted

18. For an application requiring high starting torque but approximately constant speed, the most suitable motor will be a *dc*,
(a) shunt motor (b) series motor
(c) compound motor (d) separately excited motor

19. The speed of a dc motor can be controlled by adjusting its,
(a) field current (b) armature input voltage
(c) armature circuit resistance (d) all of the above

20. Brake test for the measurement of efficiency of a dc machine is
(a) an indirect method (b) a direct method
(c) a regenerative method (d) none of the above

ANSWERS															
1.	(b)	2.	(c)	3.	(c)	4.	(b)	5.	(b)	6.	(b)	7.	(b)	8.	(a)
9.	(c)	10.	(b)	11.	(b)	12.	(a)	13.	(a)	14.	(c)	15.	(a)	16.	(a)
17.	(b)	18.	(c)	19.	(d)	20.	(b)								

Short Answer Questions

1. Describe how the direction of rotation of a dc shunt motor can be reversed?
2. Justify why a dc series motor is never run without load.
3. Explain: "The back emf acts as a current regulator"
4. Describe with relevant diagrams the different methods of a dc machine.

5. Four terminals of a dc shunt machine are available in the terminal box but they are un-named. How will you identify the terminals of armature winding and field winding? Note that the two windings are not interconnected.
6. Distinguish between the series field winding and the shunt field winding of a dc compound machine?
7. Explain what would happen if a dc motor is directly switched ON to the supply?
8. Explain the functions of no-volt release in a 3-point starter.
9. Explain why there is a drop in rotor speed due to rise in load?
10. Give examples of electrical and mechanical load.
11. Which losses of a dc compound motor are constant? Justify your answer.
12. State applications of dc shunt and dc series motor.

Long Answer Questions

1. Discuss the different methods of speed control of dc shunt motor. Draw the torque speed characteristics for each of them.
2. "The direction of rotation of a dc series motor remains same even when the input supply is reversed" Justify.
3. Derive the torque speed characteristics of a dc shunt and dc series motor. Compare them.
4. Consider a dc shunt motor connected to a 3-point starter. What will happen if:
   (i) Attempt is made to start the motor with the field circuit open.
   (ii) Field circuit becomes open-circuited when the motor is already running.
   (iii) The motor is overloaded when it is already running.
5. Discuss the advantages and disadvantages of field control method for the speed control of a dc shunt motor.
6. Analyse qualitatively for what will happen when:
   (a) The load torque and field current are held constant but the armature voltage is halved.
   (b) The load torque and armature voltage are held constant but the field current is halved.
7. A dc shunt motor is running at rated load. What will happen when:
   (a) Field winding terminals are interchanged.
   (b) Supply terminals are interchanged.
8. Explain why iron loss is treated as constant loss and the copper loss as variable loss?
9. Discuss the construction and working of a BLDC motor.
10. Explore the various experimental methods for determination of efficiency of a dc motor.
11. What will happen to the direction of torque, if supply voltage is reversed in (i) dc shunt motor and (ii) dc series motor?
12. Why only dc series motor is used as universal motor and not the dc shunt motor?

Numerical Problems

1. A 230V, dc shunt motor has an armature resistance of 0.5Ω and field resistance of 115 Ω. At no-load, it rotates at 1200 rpm and the armature current is 2.5A. On application of rated load, the speed drops to 1120 rpm. Determine the line current and power input when the motor delivers rated load.

2. A 6 pole, lap connected, 230V dc shunt motor has 410 armature conductors. It takes 41A on full load. The flux per pole is 0.05 Wb. Armature and field resistances are 0.1 Ω and 230 Ω respectively. Contact drop per brush is 1V. Determine the rotor speed and total torque developed in the rotor.
3. A *dc* series motor with unsaturated magnetic circuit and negligible resistance when running at a certain speed on a given load takes 50A at 500V. If the load torque varies as cube of speed, find the resistance necessary to reduce the speed by 20%.
4. A 240V, *dc* shunt motor has a field resistance of 400 Ω and armature resistance of 0.1 Ω. The armature current is 50A and the speed is 1000 rpm. Calculate the additional resistance in the field to increase the speed to 1200 rpm. Assume that the armature current remains same and the magnetization curve is a straight line.
5. A 250V separately excited *dc* machine has armature and field resistances of 0.06 Ω and 100 Ω respectively. Determine the total power developed in armature when working as a: (i) generator delivering 25kW output and (ii) motor taking 25kW input from the supply.
6. A 250V, *dc* shunt motor has armature resistance of 0.25 Ω. On-load, it takes armature current of 50A and runs at 750 rpm. If the motor flux is reduced by 10% without changing the load torque, find the new speed of the motor.
7. A 4 pole, 240V, wave connected *dc* shunt motor gives 11.19kW when running at 1000 rpm and drawing the armature and field currents of 50A and 1A respectively. It has 540 conductors. If armature resistance is 0.1 Ω and contact drop per brush is 1V, find: (i) useful flux per pole, (ii) total torque, (iii) useful torque and (iv) efficiency.
8. A dc shunt motor has an armature resistance of 0.1 Ω and it is connected across a 220V supply. The motor runs at 800 rpm and draws an armature current of 20A. Calculate the additional resistance to be connected in series with armature winding to reduce the speed to 520 rpm. Ignore the armature reaction effect.
9. A 460V *dc* series motor runs at 500 rpm taking a current of 40A from supply. The combined resistance of armature and series field winding is 0.8 Ω. If the load is reduced such that the motor takes 30A, calculate the new speed and percentage change in torque. Assume flux as proportional to the field current.
10. A 24V, *dc* series motor takes 40A when giving its rated output at 1500 rpm. The combined resistance of armature and field winding is 0.3 Ω. Determine the value of resistance that should be added in series to obtain rated torque at: (i) starting condition and (ii) 1000 rpm.

Answers of Numerical Problems

1. 33A, 8050W
2. 655 rpm, 130.5 Nm
3. 3.975 Ω
4. 80 Ω
5. P_g = 25600W, P_m = 24400W
6. 828 rpm
7. 13.05m Wb, 112.21Nm, 106.86 Nm, 91.42%
8. 3.815 Ω
9. 679 rpm, 43.75%
10. 5.7 Ω, 1.9 Ω

PRACTICALS

Experiment No: 1

Aim: To reverse the direction of rotation of a *dc* shunt motor.

Machine specifications:

Refer to the name plate of the machine and note down the following specifications.

DC Shunt Motor:		
kW/HP:	Frame:	RPM:
Insulation Class:	Armature Voltage:	Max. Ambient Temp.:
Type:	Armature Current:	Field Current:
Duty:	Make:	

Apparatus:

Device	Make	Range	Quantity
3-Point Starter			01
DC Ammeter			01
DC voltmeter			01
Rheostat			01

Circuit connections:

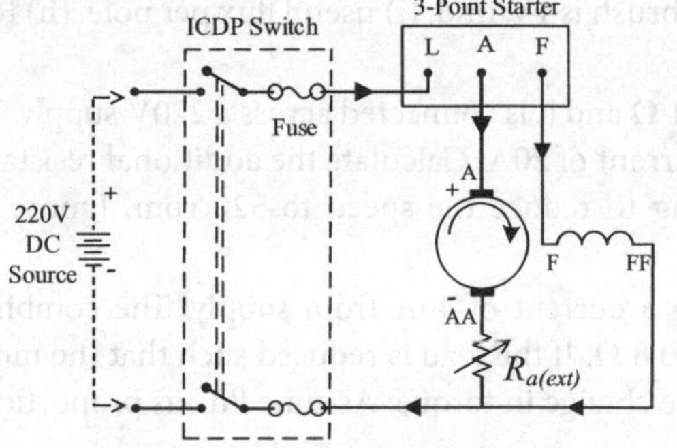

(a) Forward/ normal direction

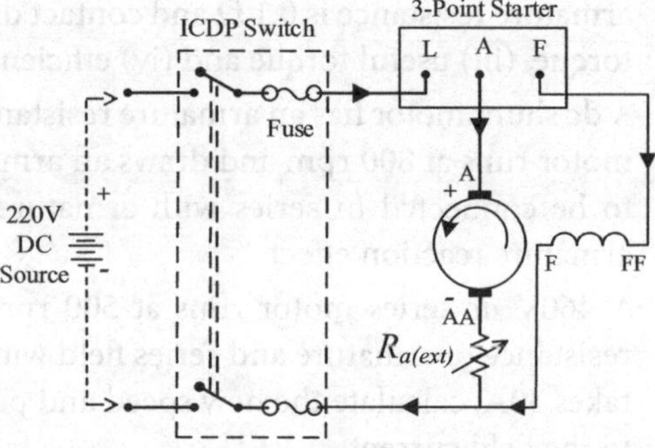

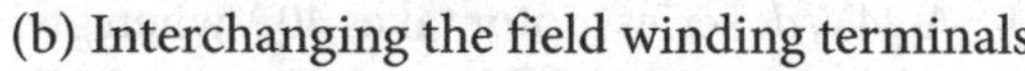

(b) Interchanging the field winding terminals

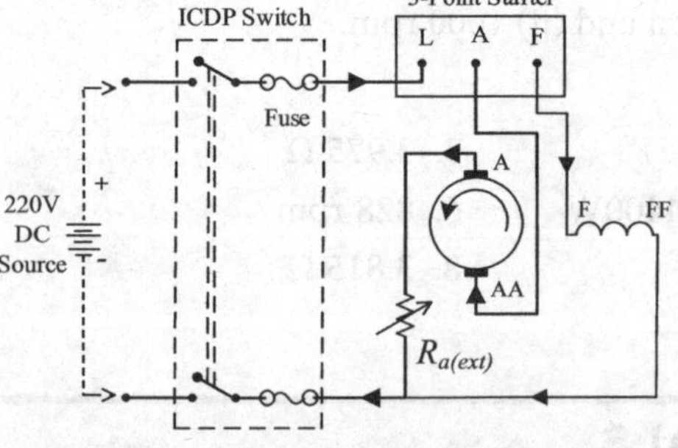

(c) Interchanging the armature winding terminals

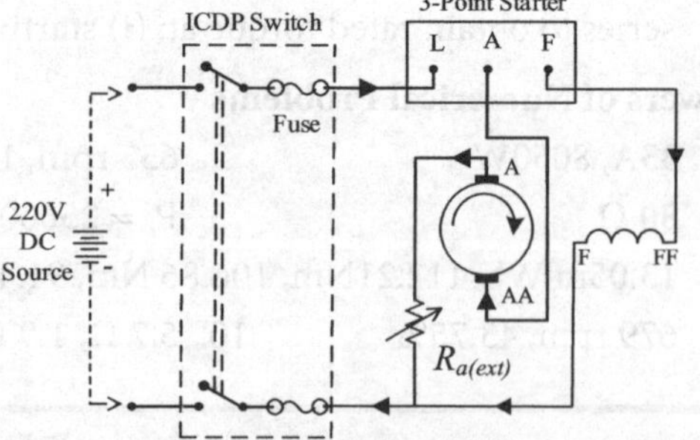

(d) Interchanging the terminals of both, field and armature winding

Fig. 2.34: Connection diagrams for different directions of rotation

Theory:

There are many applications in which the motor is required to rotate in forward and reverse directions. This feature is used to stop, slow down or reverse the direction of motor spinning force. For example: The garage door opener used in garage and automatic sliding door opener used in malls, offices and banks employ a motorized system that opens and closes the gates. The motor used in hoists is required to operate in both directions. These applications require equal torque in both directions.

To reverse the direction of rotation, it is necessary to reverse the direction of developed torque. From equation (2.7) the armature torque is given by, $T_a = K_t \phi_m I_a$ where the main flux is directly proportional to the field current i.e. $\phi_m \propto I_f$.

$$\therefore \qquad Ta \propto I_a I_f$$

Thus to reverse the torque, the direction of either the armature current I_a or the field current *If* should be changed. Simultaneous reversal of both currents will maintain the same direction of torque and the rotor will not be able to rotate in opposite direction. This can be implemented either by reversing the polarities of the *dc* source or by reversing the terminals of the winding. Method of reversal of *dc* motor depends on the type of motor. In different motors, it can be implemented as:

- *DC series motor*: Reversing the direction of current either in field winding or in armature winding.
- *DC shunt motor:* Reversing the direction of current either in field winding or in armature winding.
- *DC separately excited motor:* Changing the polarity of supply voltage either to the field winding or to the armature winding.
- *Permanent magnet DC motor:* Changing the polarity of supply voltage.

Precautions:

1. Make tight connections as shown and switch on the supply after they are verified by the instructor.
2. Ensure that you are standing on a rubber mat while performing the experiment.
3. Stand at a safe distance from the experimental setup when it is running.
4. Do not bend to take readings. Stand straight and avoid parallax error while taking the readings.
5. In this experiment, to reverse the direction of rotation, do not try to open and reverse the winding terminals while the motor is in energised state. First disconnect it from the supply.

Procedure:

1. Make connections as shown in Fig. 2.34 (a)
2. Ensure the starter handle at 'OFF' position before switching on the supply.
3. Ensure the variable rheostat in armature circuit at maximum resistance position before switching on the supply.
4. Turn on the ICDP switch (Iron Clad Double Pole) and move the starter handle gradually towards 'RUN' position.
5. Note the direction of rotation of rotor. Treat it as 'Forward' direction. If it is not possible for you to observe the direction of rotation correctly, vary the speed by adjusting the variable rheostat in armature circuit.
6. Turn off the ICDP switch. The starter handle should return back to 'OFF' position by its own.
7. Reverse the winding terminals as shown in the next diagram (Fig. 2.34 (b))
8. Repeat steps 2 to 8.
9. Repeat the experiment as per Fig. 2.34 (c) and (d).

Observations:

Sr. No	Diagram	Connection	Direction of rotation	Remark
1.	Fig (a)	Normal		
2.	Fig (b)	Interchanging the field winding terminals		
3.	Fig (c)	Interchanging the armature winding terminals		
4.	Fig (d)	Interchanging both, field and armature winding terminals		

Conclusion:

Discussion:

1. *Why starter is used in this experiment?*

   It is known that, starting condition is that short-time interval when the input supply is just fed to the motor but it has not yet started running. In such a condition, the back *emf* induced in the armature winding is zero and therefore the only opposition offered to armature current is due to the armature resistance. Since R_a is normally very small in magnitude, the *armature current* during starting condition is dangerously high. Therefore, to restrict this starting current within permissible limits and to ensure safe starting of the motor, starter is employed.

2. *Which is the preferred method for reversal of rotation of a dc motor when machine is at rest and not connected to supply?*

   Since the field winding carries lesser current than the armature winding, reversing the field current is easier. It requires a low rating, lightweight reversing switchgear and thus reduces the cost. Hence field current reversal is preferred.

   NOTE:

   Field winding of shunt motor has large number of turns and its inductance is large. The stored energy of field winding must be dissipated before it is reconnected in reverse fashion. Strictly no attempt should be made to reverse either the field winding or armature winding when supply is ON and motor is running, as it will cause plugging or counter current operation causing the back *emf* to come in series with supply voltage resulting in very large amount of current to flow in armature circuit. This will eventually damage the DC motor.

   For quick reversal of separately excited *dc* motor, the voltage of armature should be brought gradually to zero and then slowly increased in reverse direction- as is being done in Ward Léonard Control.

3. *Why speed reversal by changing the direction of field current I_f is not possible in PMDC motor?*

   In a permanent magnet *dc* motor, the stator has permanent magnets in place of poles wound with field winding. Therefore, the main flux remains constant and cannot be varied.

4. *If it is not possible to separate the connection between the armature and field winding than, how speed reversal can be done?*

   If it is not possible to separate the armature winding from field winding, it means the armature and field currents cannot be reversed independently. In that case, the speed reversal is not possible.

5. *In a dc compound motor, field current of which field winding should be reversed to reverse the motor speed?*

   A *dc* compound motor consists of two field windings. One is connected in series with armature winding and is called as series field winding whereas, the other one is connected in parallel with armature and is called as shunt field winding. If the current in any one of them is reversed than, the characteristics of the motor will change from differential type to cumulative type and vice-versa. This will create a mismatch with the load characteristics. Hence, currents in both field windings should be reversed to reverse the speed direction. If reversing the terminals of both field windings is not possible than, the armature current alone should be reversed.

6. *Why only dc series motor can be used as a Universal motor and not the dc shunt motor?*

   Universal motor is that motor which can run on *ac* as well as on *dc* supply. From equation (2.7), if the directions of both, the armature current and field current are reversed, the torque remains unidirectional and hence, the motor continues to run in same direction. When the supply voltage to a *dc* series and a *dc* shunt motor is reversed, in both cases, the torque should remain unidirectional regardless of the supply voltage polarity. However, only *dc* series motor is used as *ac* series motor or universal motor. When *ac* source is connected across a series motor, the armature and field currents will change simultaneously in magnitude as well as in direction. They are in-phase because both windings are in series. But when a *dc* shunt motor is connected across an *ac* source, the two winding currents will have a phase difference due to different impedances of the two windings. If the motor runs, it will run poorly.

Experiment No: 2

Aim: To perform brake test on a *dc* shunt motor.

Machine specifications:

Refer to the name plate of the machine and note down the following specifications.

DC Shunt Motor:		
kW/HP:	Frame:	RPM:
Insulation Class:	Armature Voltage:	Max. Ambient Temp.:
Type:	Armature Current:	Field Current:
Duty:	Make:	

Apparatus:

Device	Make	Range	Quantity
3-Point Starter			01
DC Ammeter			01
DC voltmeter			01
Rheostat			01
Tachometer			01
Spring balance loading arrangement			01

Circuit connections:

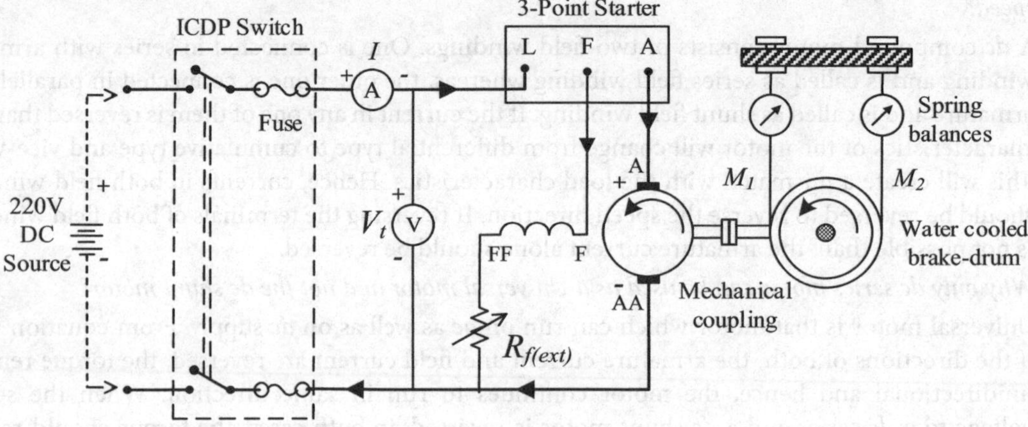

Fig. 2.35: Connection diagram of brake test on a dc shunt motor

Theory:

This test is performed to determine the efficiency of a *dc* motor by direct measurement of the input and output power. For details, refer to Section 2.8.

Precautions:

1. Make tight connections as shown and switch on the supply after they are verified by the instructor.
2. Ensure that you are standing on a rubber mat while performing the experiment.
3. Stand at a safe distance from the experimental setup when it is running.
4. Do not bend to note readings. Stand straight and avoid parallax error while taking the readings.
5. In this experiment, due to friction between the brake-drum and the belt, heat is produced. To dissipate the same, pour permissible amount of water in the space provided in the brake drum when the machine is disconnected from supply.
6. While increasing the load, ensure that the ammeter reading does not exceed the rated current value of the motor.

Procedure:

1. Make connections as shown in circuit diagram.
2. Ensure the starter handle at 'OFF' position before switching on the supply.
3. To ensure the starting of motor at no-load, loosen the belt around the drum.
4. Initially, set the variable rheostat in field circuit at minimum resistance position before switching on the supply.
5. Turn on the ICDP switch (Iron Clad Double Pole) and move the starter handle gradually towards 'RUN' position.
6. Adjust the speed at rated value by varying the rheostat. Now, do not disturb the rheostat throughout the test.
7. Note the readings of input voltage, input current, speed and the two spring balance tensions.
8. Increase the mechanical load on the motor by tightening the belt with the help of load adjustment wheels and note the corresponding readings.

9. Take 8 to 10 sets of reading upto rated load.
10. Loosen the belt and bring the rheostat to initial position.
11. Turn off the ICDP switch. The starter handle should return back to 'OFF' position by its own.
12. Remove the connections and calculate efficiency at different load conditions.

Observations and Calculated Results:

Sr. No.	I (A)	V_t (V)	N (rpm)	M_1 (kg)	M_2 (kg)	($M_1 \sim M_2$)	T_{sh}	P_{sh}	P_i	η

Calculations:

1. Measure the circumference of the brake drum and calculate its radius (r) in meters.
2. Calculate the torque as, $T_{sh} = rg(M_1 - M_2)$ where, $(M_1 - M_2)$ is the difference between spring tensions, g = acceleration due to gravity.
3. Calculate the power output as, $P_{sh} = T_{sh} \times \frac{2\pi N}{60}$ where, N = rotor speed in rpm.
4. Calculate the input power, $P_i = V_t I$
5. Calculate the percentage efficiency as, $\eta = \frac{P_{sh}}{P_i} \times 100$
6. Draw the graph of output power versus and N on the same graph paper

Result and Conclusion:

____

____

Model Graph:

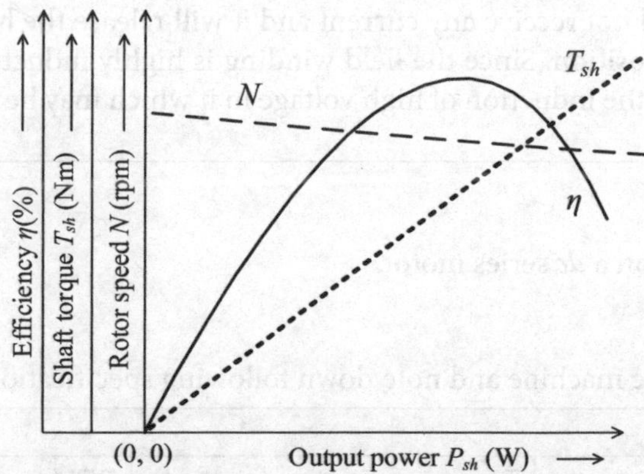

Fig. 2.36: Graph of versus and N

Discussion:

1. *During no-load operation, what is the utility of input power? Also comment on speed and efficiency.*
   During no-load condition, the power output of the motor is zero. A small amount of input power is drawn by the motor from the *dc* source. It is utilized to overcome the no-load losses, namely the mechanical loss and the field copper loss. Since the power output is zero, the efficiency is also zero. The no-load speed is slightly greater than the full load-speed. As the machine is loaded, the speed falls slightly.
2. *Why the speed reduces slightly when the dc shunt motor is loaded?*
   With rise in load, the armature current I_a increases and therefore the voltage drop in armature winding I_aR_a also increases. Consequently, the back *emf* ($E_b = V_t - I_a R_a$) decreases. From equation (2.2), speed N $\propto E_b$. Hence, the rotor speed decreases. Refer to A physical explanation of this phenomenon is given in Section 2.3.
3. *When the motor is loaded, how does it adjusts itself to the new load condition?*
   When the mechanical load on the motor is increased, it means that the load torque *TL* is increased. Note that, *TL* always acts in opposition to the torque *Ta* developed by the motor. Due to rise in load, speed and back *emf* decreases. Consequently, the armature current given by $I_a = \frac{(V_t - E_b)}{R_a}$ increases. Since the armature torque is directly proportional to armature current ($T_a \propto \phi_m I_a$), T_a also increases till it becomes equal to T_L.
4. *Why the starting torque in a dc shunt motor is less than that in a dc series motor?*
   The starting torque of a *dc* shunt motor is directly proportional to armature current ($T_a \propto I_a$) (whereas in a *dc* series motor, it is directly proportional to square of armature current ($T_a \propto I_a^2$). This is because, in a *dc* shunt motor, the main flux ϕ_m is constant whereas in a *dc* series motor, it is directly proportional to armature current I_a. During the starting condition, armature current is high in magnitude and is also called as starting current. Due to these reasons, the starting torque of a *dc* shunt motor is less than that of a *dc* series motor.
5. *What will happen if the field winding of a dc shunt motor gets open-circuited during running condition?*
   With the use of 3-Point starter, in the event of field winding getting open-circuited, the no-volt coil (HC) of the starter will not receive any current and it will release the handle. Thus the starter will return back to 'OFF' position. Since the field winding is highly inductive, sudden interruption of current in it will cause the induction of high voltage in it which may be dangerous to its insulation.

Experiment No: 3

Aim: To perform brake test on a *dc* series motor.

Machine specifications:

Refer to the name plate of the machine and note down following specifications.

DC Series Motor:		
kW/HP:	Frame:	RPM:
Insulation Class:	Armature Voltage:	Max. Ambient Temp. (°C):
Duty:	Armature Current:	Make:

Apparatus:

Device	Make	Range	Quantity
2-Point Starter			01
DC Ammeter			01
DC voltmeter			01
Rheostat			01
Tachometer			01
Spring balance loading arrangement			01

Circuit connections:

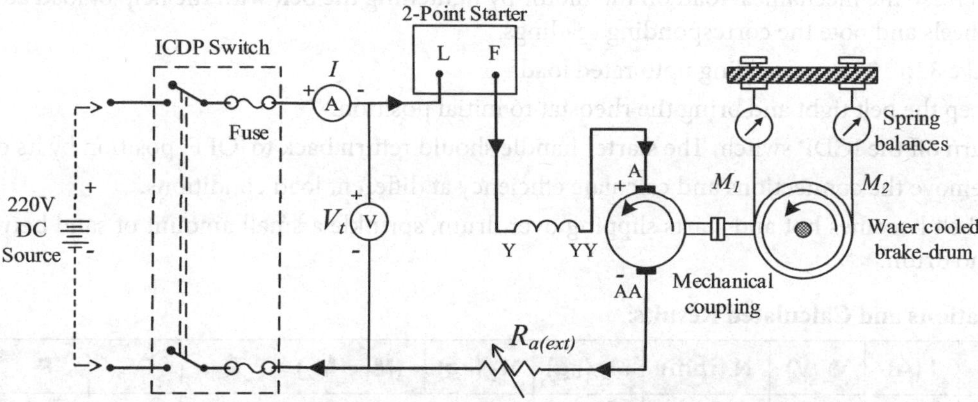

Fig. 2.37: Connection diagram of brake test on a *dc* series motor

Theory:

This test is performed to determine the efficiency of a *dc* motor by direct measurement of the input and output power. For details, refer to Section 2.8.

Precautions:

1. Make tight connections as shown and switch on the supply after they are verified by the instructor.
2. Ensure that you are standing on a rubber mat while performing the experiment.
3. Stand at a safe distance from the experimental setup when it is running.
4. Do not bend to note readings. Stand straight and avoid parallax error while taking the readings.
5. In this experiment, due to friction between the brake-drum and the belt, heat is produced. To dissipate the same, pour limited amount of water in the space provided in the brake drum when the machine is disconnected from supply.
6. While increasing the load, ensure that the ammeter reading does not exceed the rated current value of the motor.
7. Do not apply full voltage across the *dc* series motor terminals at no-load.

Procedure:

1. Make connections as shown in circuit diagram.
2. Ensure the starter handle at 'OFF' position before switching on the supply.
3. Initially, to ensure the starting of motor on-load, tighten the belt around the drum.
4. Pour water in the drum for cooling.
5. Initially, set the variable rheostat in armature circuit at maximum resistance position before switching on the supply.
6. Turn on the ICDP switch (Iron Clad Double Pole) and move the starter handle gradually towards 'RUN' position.
7. Adjust the speed at rated value by varying the rheostat. Do not disturb it throughout the test.
8. Note the readings of input voltage, input current, speed and the two spring balance tensions.
9. Increase the mechanical load on the motor by tightening the belt with the help of load adjustment wheels and note the corresponding readings.
10. Take 8 to 10 sets of reading upto rated load.
11. Keep the belt tight and bring the rheostat to initial position.
12. Turn off the ICDP switch. The starter handle should return back to 'OFF' position by its own.
13. Remove the connections and calculate efficiency at different load conditions.
14. If belt becomes hot and starts slipping over drum, sprinkle a small amount of sand between belt and drum.

Observations and Calculated Results:

Sr. No.	I (A)	V_t (V)	N (rpm)	M_1 (kg)	M_2 (kg)	($M_1 \sim M_2$)	T_{sh}	P_{sh}	P_i	η
1.										
2.										
3.										
4.										

Calculations:

1. Measure the circumference of the brake drum and calculate its radius (r) in meters.
2. Calculate the torque as, $T_{sh} = rg(M_1 - M_2)$ where, $(M_1 - M_2)$ is the difference between spring tensions, g = acceleration due to gravity.
3. Calculate the power output as, $P_{sh} = T_{sh} \times \frac{2\pi N}{60}$ where, N = rotor speed in rpm.
4. Calculate the input power, $P_i = V_t I$
5. Calculate the percentage efficiency as, $\eta = \frac{P_{sh}}{P_i} \times 100$
6. Draw the graphs of output power P_{sh} versus η, T_{sh} and N on the same graph paper

Model Graphs:

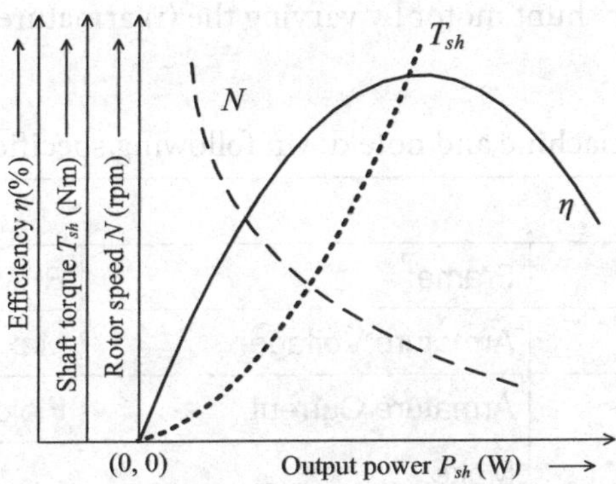

Fig. 2.38: Graph of P_{sh} versus η, T_{sh}, I and N Result and

Conclusion:

Discussion:

1. *In brake test, how is the mechanical power developed by the motor utilized?*

   In brake test, the mechanical power developed by the motor is absorbed (actually wasted) in the form of heat by the brake drum due to its friction with the belt. Hence, the drum is water-cooled.

2. *Why a dc series motor should not be operated at no-load and light loads?*

   From equation (2.15), speed of a *dc* series motor is inversely proportional to the square root of torque $\left(N \propto \frac{1}{\sqrt{T_a}}\right)$. At no-load, the torque is zero and at light loads, it is very small. Running a *dc* series motor at light loads with full applied voltage is known as 'racing'. This is not desirable from the mechanical strength point of view of the machine.

3. *What are the advantages and applications of a dc series motor?*

   A *dc* series motor is capable to generate high starting torque and meet fluctuations in load. It is a constant power motor. Due to these advantages, its preferred applications are traction, lift, crane, hoist etc.

4. *What type of starter is used for starting a dc series motor?*

   2-Point starter and drum-controller are used.

5. *What is the effect of magnetic saturation on the speed and torque of a dc series motor?*

   After magnetic saturation, the speed becomes almost constant. Torque becomes proportional to armature current I_a instead of $(I_a)^2$. After saturation, the characteristics are like those of a dc shunt motor.

Experiment No: 4

Aim: To control the speed of *dc* shunt motor by varying the (i) armature circuit resistance and (ii) field.

Machine specifications:

Refer to the name plate of the machine and note down following specifications.

DC Shunt Motor:		
kW/HP:	Frame:	RPM:
Insulation Class:	Armature Voltage:	Max. Ambient Temp. (°C):
Type:	Armature Current:	Field Current:
Duty:	Make:	

Apparatus:

Device	Make	Range	Quantity
3-Point Starter			01
DC Ammeter			01
DC voltmeter			01
Rheostat			01
Rheostat			01
Tachometer			01

Circuit connections:

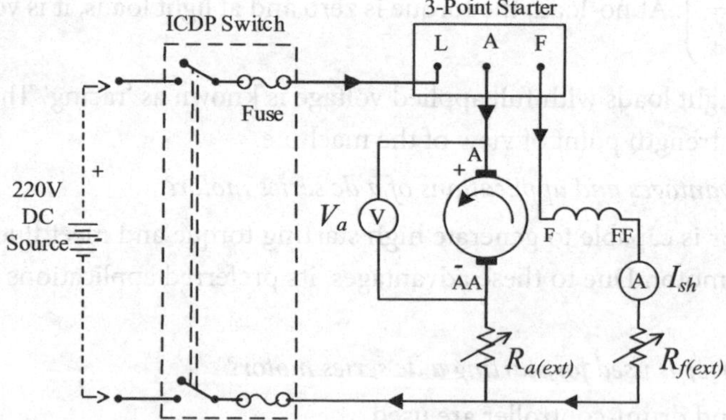

Fig. 2.39: Circuit connection for speed control of dc shunt motor

Theory:

This test is performed to control the speed of a *dc* shunt motor by various methods. Refer to Section 2.6.1 for details.

Precautions:

1. Make tight connections as shown and switch on the supply after they are verified by the instructor.
2. Ensure that you are standing on a rubber mat while performing the experiment.
3. Stand at a safe distance from the experimental setup when it is running.
4. Do not bend to note the readings. Stand straight and try to avoid parallax error while taking the readings.

Procedure:

1. Make connections as shown in circuit diagram.
2. Ensure the starter handle at 'OFF' position before switching on the supply.
3. Initially, set the armature variable rheostat $R_{a(ext)}$ at maximum resistance position and the field variable rheostat $R_{f(ext)}$ at minimum resistance position before switching on the supply.
4. Turn on the ICDP switch (Iron Clad Double Pole) and move the starter handle gradually towards 'RUN' position.

(a) *By Varying the Armature Circuit Resistance:*

(i) Keep the field rheostat undisturbed at its minimum resistance position so that the field current I_{sh} remains constant at its rated value.

(ii) Reduce the armature rheostat $R_{a(ext)}$ in steps and measure corresponding values of V_a and speed.

(b) *By Varying the Field:*

(i) Keep the armature rheostat undisturbed so that the armature voltage V_a remains constant at rated value.

(ii) Increase the field rheostat $R_{f(ext)}$ in steps and measure corresponding readings of I_{sh} and speed.

Observations and Calculated Results:

By Varying the Armature Circuit Resistance		
Shunt field current I_{sh} = A		
Sr. No	Armature voltage (V_a)	Speed (N)

By Varying the Field		
Armature voltage V_a = V		
Sr. No.	Field current (I_{sh})	Speed (N)

Model Graphs:

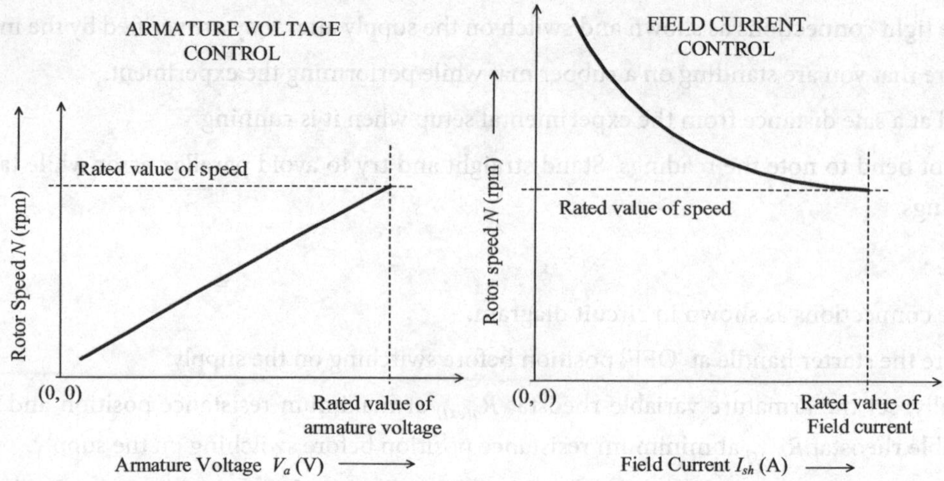

Fig. 2.40: Graph of Armature voltage versus speed and Field current versus speed

Result and Conclusion:

____

____

Discussion:

1. *What are the limitations of armature circuit resistance method and field control method of speed control?*

   By armature circuit resistance method, the rotor speed above rated value cannot be obtained whereas by field control method, the rotor speed below rated value cannot be obtained.

2. *Initially, why the field rheostat should be set at minimum resistance position?*
   If initially, the *field rheostat is set at maximum resistance position than, the field current and hence the main flux will be minimum. Since, the speed is inversely proportional to main flux, the motor will start to a high speed.*

3. *From the size of the winding terminals coming out of the machine, is it possible to identify the shunt field winding?*

   Yes, the leads of shunt field winding are thinner as compared to the leads of armature and series field windings.

4. *Why should the shunt field winding have a large number of turns whereas the series field winding have less number of turns?*

   The shunt field winding has to establish a major component of main field. At the same time, it is designed to carry less current so as to reduce the no-load current and the no-load losses. Therefore to produce high *mmf* (ampere.turns), the number of turns of shunt field winding are more than those of the series field winding. Secondly, the series winding is connected in series with armature winding and therefore, it carries the high armature current. Hence to produce lesser ampere.turns, its number of turns are less.

5. *Why the resistance of armature winding and series field winding are low?*

   The armature current flowing in these two windings is high in magnitude. To restrict the current density within prescribed limits, these two windings are made from thicker wires. Since resistance is inversely proportional to the area of cross section of the wire hence, resistance of armature winding and series field winding are low.

6. *How the losses affect the machine?*

   Loss means the loss of active power in watts. It causes heating of the machine. Due to temperature rise, the insulation is affected. If the temperature exceeds the specified value for insulation, the insulating material gets deteriorated and hence, the life of insulation as well as of machine decreases.

7. *What is a universal motor?*

   Universal motor is a special electrical machine that is designed to operate on *ac* as well as *dc* supply. It is basically a *dc* series motor with laminated magnetic circuits to minimize the iron loss when powered by *ac* supply.

8. *How it becomes possible for a dc series motor to work on ac supply?*

   In a *dc* series motor, both armature winding and field winding are connected in series and fed from a common supply. Hence, they carry the same current. If the supply voltage polarities are reversed than, the current in both windings will change simultaneously and so, the direction of main flux produced by field winding will also reverse. As the torque developed in the rotor is directly proportional to the product of armature current and main flux, it will remain unidirectional. Hence, the rotor continues to rotate in same direction even when alternating polarity supply is connected.

9. *Why a dc shunt motor cannot be used as universal motor?*

   In a *dc* shunt motor, the armature and field windings are connected in parallel and fed from the common source. If the polarity of voltage source is reversed than, the current in both windings will change and so, the direction of main flux produced by field winding will also reverse. Therefore, theoretically, it may be possible to run the shunt motor on *ac* supply.

   Practically, the field windings of a *dc* series motor and *dc* shunt motor are designed differently. In series motors, the field winding carries high armature current therefore, it is constructed from thick wire with a few number of turns so that, the voltage drop and power loss in this winding are low. In *dc* shunt motor, the field winding is designed to carry low current which is just sufficient to establish the required amount of main flux and therefore, it is constructed from thin wire with a large number of turns. But this also results in high resistance and high inductance of field winding in *dc* shunt motor as compared to those in *dc* series motor. If a shunt motor is connected across *ac* supply than, with every reversal of supply voltage, the currents in two windings will tend to reverse but they will not reverse simultaneously. In other words, the armature current and main flux will not be in phase with each other. The developed torque will not be uniform and even if the rotor rotates, it will run at low speed making noise.

10. *What is the difference between series rheostat and potential divider?*

    A series rheostat is connected in series with the circuit whose current is to be adjusted whereas a potential divider is connected across the circuit to vary the voltage. Normally, to use as a series rheostat two terminals of the variable resistor are sufficient but for use as potential divider, three terminals are necessary. Two are fixed terminals and the third one is variable.

KNOW MORE

Sir James Swinburne

The brake test studied in Section 2.8 is a direct method of determining efficiency of a *dc* motor. Though it is an accurate method, a considerable amount of power gets wasted in the form of heat in the brake drum due to the actual loading. Two alternate methods exist for the determination of efficiency. One is performed at no-load condition while the other one is performed on full-load. They are: (1) Swinburne's Test and (2) Hopkinson's Test respectively. These two tests are currently not in the syllabus of Diploma Programme. However, the fast learner curious students may explore more details on them and gain added knowledge.

Swinburne's Test was introduced by a British Electrical Engineer Sir James Swinburne. In this test, the *dc* shunt or *dc* compound machine is run as a motor at no-load and the no-load losses of the machine are determined simply by measuring the input current and input voltage. Then, the additional losses are estimated from the motor data. This is a very simple and economical test because it is performed at no-load and the machine efficiency can be determined in advance at any desired load.

Hopkinson's Test requires two identical *dc* shunt machines which are mechanically coupled and also electrically connected in parallel with each other. One is operated as a motor and the other one as generator. This method is useful for testing large capacity *dc* machines under loading conditions. The determined efficiency is more realistic in nature as it takes into account the stray load losses also. The Hopkinson's test is also known as regenerative test or back-to-back test or heat-run test. Though the machines are operated at full load condition, the power drawn from the supply is used only for supplying the losses. This test is used in those cases where the identical *dc* motor and *dc* generator are mechanically coupled (Motor-Generator set) and the combined iron losses are to be determined because their individual iron losses cannot be separated. Thus, in Hopkinson's test, a pair of both motor and a generator is required and the test is performed at full-load condition whereas in Swinburne's test, only one *dc* machine is required and the test is performed at no-load. Due to the operation at full-load, the temperature rise can be observed only in Hopkinson's test and not in Swinburne's test. Details of both tests can be studied from the references mentioned below.

In Unit-1 and Unit-2, we have studied the application of Fleming's right hand rule and Fleming's left hand rule. These simple methods to remember the relationships between magnetic field, current and force (or motion) in electrical machines were devised by a British Electrical Engineer Sir John Ambrose Fleming (1849 - 1945). He is the inventor of the oscillation diode valve or vacuum tube and also carried out a lot of research in electrical measurements and machinery. Some people also call him as Father of Electronics.

Professor Sir John Fleming

REFERENCES AND SUGGESTED READINGS

1. A.E. Fitzgerald, C. Kingsley Jr., A. Kusko, Electric Machinery 3e, McGraw Hill, 1971
2. T. Wildi, Electrical Machines, Drives, and Power Systems, Pearson Education Asia, 2001
3. P.K. Mukherjee, S. Chakravorti, Electrical Machines, Dhanpat Rai Publications, 2010
4. R.K. Srivastava, Electrical Machines 2e, Cengage Learning, 2013
5. D.P. Kothari, I.J. Nagrath, Basic Electrical Engineering 3e, McGraw Hill, 2010
6. G. K. Dubey, Fundamentals of Electrical Drives, Narosa Publishing House, 2001
7. P. S. Bhimbra, Electrical Machinery, Khanna Publishers, 2009.
8. A. Husain, Electric Machines, Dhanpat Rai & Co., 2022.
9. R. Krishnan, Permanent Magnet Synchronous and BLDC Motor Drives, CRC Press, 2010
10. S. G. Tarnekar, P. K. Kharbanda, S. B. Bodkhe, Laboratory Courses in Electrical Engineering, S. Chand & Co., 2013.
11. D. Mohanraj, R. Aruldavid, R. Verma, et. al., A Review of BLDC Motor: State of Art, Advanced Control Techniques, and Applications, IEEE Access, Vo. 10, 2022.
12. W. Gruber, W. Bäck and W. Amrhein, Design and Implementation of a Wheel Hub Motor for an Electric Scooter, IEEE Vehicle Power and Propulsion Conf., 2011, Chicago, USA
13. L. Shao, R. Navaratne , M. Popescu and G.Liu, Design and Construction of Axial Flux Permanent Magnet Motors for Electric Propulsion Applications- A Review, IEEE Access Vol. 9, 2021

QR Code for Further Reading

3 Single Phase Transformers

UNIT SPECIFICS

Through this unit the following aspects are discussed:

- Need and types of transformer;
- Construction and functions of different parts of transformer;
- Material used for different parts;
- Principle of operation;
- EMF equation and transformation ratio;
- Significance of transformer ratings;
- Winding resistances and leakage reactances;
- Phasor diagram at No-load and On-load conditions;
- Equivalent circuit of transformer, equivalent resistance and reactance;
- Voltage regulation, operating efficiency and all day efficiency;
- Direct loading test;
- Open circuit and short circuit (OC/SC) tests;
- Polarity test on single-phase transformer.

This unit presents a detailed explanation on different aspects of single phase transformer with sufficient number of diagrams and solved examples. With due consideration to the different cognitive levels of Bloom's taxonomy, multiple-choice questions, questions for short and long answers and numerical problems with final answers are provided for practice and self-assessment. References and a dynamic QR code are given at the end for suggestive self-learning.

Many laboratory experiments on single phase transformer are also included that explains the purpose, procedure, calculations, and interpretation of results. This will help students to correlate and validate the theoretical knowledge with experimental findings and write effective reports. At the end, a section "Know More" is included to arouse curiosity to explore more information on transformers.

RATIONALE

Imagine two electric circuits which are isolated electrically but still they work and perform in coordination with each other. Imagine an electric circuit in which you would like to maintain different voltage levels in different parts of the same circuit. Imagine a situation where you are required to measure current and voltage in the range of thousands of amperes and thousands of volts but the ammeter and voltmeter of those ratings are not available in the market. Imagine that the power generating station is located hundreds of kilometre away from the load centre and that the power station is required to provide power in megawatts at generated voltage level which will naturally require very thick transmission lines to carry huge currents.

Those who are new to the topic may get puzzled for the solution which can answer all the above questions. The answer common to all of them is 'Transformer'. Transformer is one of the vital components of Power System that is employed in generating stations, transmission and distribution networks. It also finds its role in different applications including traction system, measurements, welding, domestic appliances etc. Broadly, transformers can be classified on the basis of number of phases for which they are designed as single-phase type and multi-phase type. This unit will present and explain the various aspects of single phase transformer.

PRE-REQUISITES

Basics of Electrical Engineering

UNIT OUTCOMES

List of outcomes of this unit is as follows:

U3-O1: Discuss the uses of transformer and its classification based on different aspects;

U3-O2: Describe the construction, material used for different parts and function of every part;

U3-O3: Explain the principle of operation and derive voltage equations, EMF equation and transformation ratio;

U3-O4: Apply Right Hand Thumb Rule to determine the direction of flux and induced current;

U3-O5: Draw and explain the phasor diagrams at No-load and On-load conditions;

U3-O6: Draw the equivalent circuit and explain the significance of its parameters;

U3-O7: Calculate voltage regulation, operating efficiency, all-day efficiency and explain the significance of ratings;

U3-O8: Explain direct loading test and OC/SC tests and compare them for merits and demerits;

U3-O9: Use the OC/SC test observations to determine equivalent circuit parameters, efficiency, and regulation;

U3-O10: Perform laboratory experiments based on single phase transformer, analyse the results and interpret the findings to reach valid conclusions.

Unit-3 Outcome	EXPECTED MAPPING WITH COURSE OUTCOMES (1 - Weak Correlation; 2 - Medium Correlation; 3 - Strong Correlation)				
	CO-1	CO-2	CO-3	CO-4	CO-5
U3-O1	-	-	3	-	-
U3-O2	-	-	3	-	-
U3-O3	-	-	3	-	-

U3-O4	-	-	3	-	-
U3-O5	-	-	3	-	-
U3-O6	-	-	3	-	-
U3-O7	-	-	3	-	-
U3-O8	-	-	3	-	-
U3-O9	-	-	3	-	-
U3-O10	-	-	3	-	-

3.1 INTRODUCTION

Electrical energy is particularly useful because it can be transferred over long distances and distributed in a wider network and converted into other desired forms efficiently and economically. All this have become possible mostly due to the availability of a static electromagnetic device called as *Transformer*. It is a simple device which interlinks two or more electrical circuits by a common magnetic circuit. There are no moving parts. Transformers can increase or decrease the voltage or current in an *ac* circuit. They can electrically isolate different *ac* circuits from each other. In power system, it is said that in the entire world, after electricity is produced in power stations, it is transformed into different voltage levels three to four times before it finally reaches the consumer end. Fig. 3.1 shows a schematic of how transfer of electric power takes place from generating station to the user.

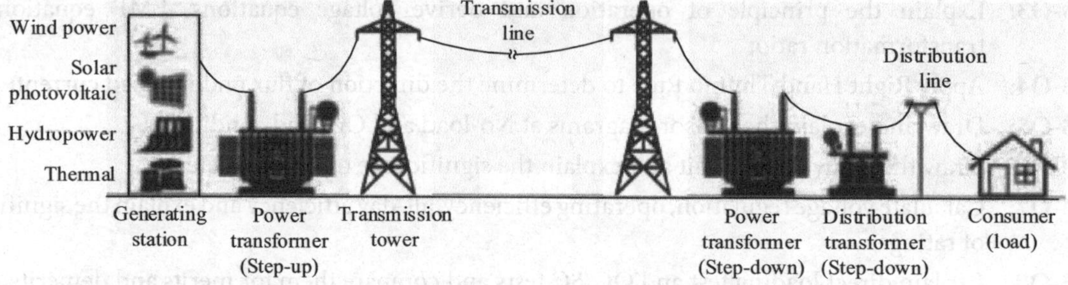

Fig. 3.1: Transfer of electric power from generating station to the user

Transformers work on the basic principle of *mutual induction*. It is known that, the flow of current in any circuit (or a conductor) is always associated with magnetic field surrounding that conductor. If the current is of alternating type then its magnetic field will also change in magnitude and direction in accordance with the changes in current with respect to time. This is the reason why we say that the magnetic flux produced by a current is in phase with that current. Now, if another circuit is placed in the vicinity of the first circuit which is carrying alternating current than, the second circuit will link with some of the alternating flux produced by first circuit and therefore an *emf* will be induced in the second circuit. This is called *mutual induction*. Because, it works on the principle of mutual induction, transformers are operated on *ac* supply and not on *dc*. As there are no moving parts, the power loss due to mechanical friction and windage is zero and little maintenance is required as compared to rotating machines. This results in high efficiency of transformers which lie between 94% - 96% approximately.

3.2 TRANSFORMER AND ITS TYPES

Transformer is a static electromagnetic device which transforms electrical potential from one level to another by mutual induction at same frequency on both sides i.e. input and output sides. It is mostly used in power system networks to step-up or step-down the voltage. Hence, the voltage levels on the two sides may or may not be equal depending upon for what application it is designed and fabricated. In electronic circuits, it is used to match the source and load impedances for maximum power transfer between them. It should be understood that, transformer is not an energy conversion device but a device that transforms electrical energy at same or different voltage and current levels. The size of transformer varies from very small ones that are used in audio, video and communication related electronic circuits to power transformers of thousands of MVA rating used in power system. Transformers can be classified on the basis of many aspects as shown in Fig. 3.2. In spite of these variations, all transformers work on the same principle and also, their basic philosophy of construction remains same.

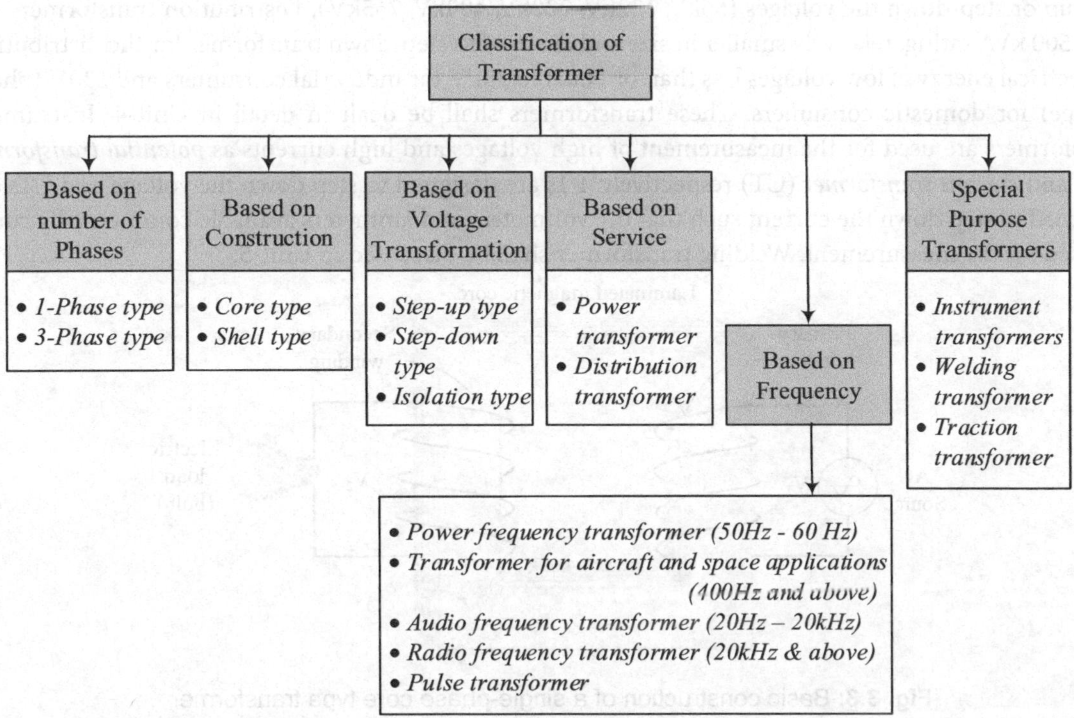

Fig. 3.2: Types of transformer

In its simplest form, transformer consists of a magnetic core on which windings are wound. A single-phase transformer has minimum two windings, one for the input side and one for the output side. A three-phase transformer has minimum six windings, three on each side which are either connected in star or delta configuration. The winding across which input *ac* source is connected is termed as *Primary winding* whereas the winding across which electrical load is connected is called as *Secondary winding.* The current in primary winding produces an alternating magnetic flux that travels through the core and links with the secondary winding where it induces *emf* by mutual induction. Similarly, the current in secondary winding establishes its own flux which links with primary winding. For a given transformer, though any side can be used as primary and secondary, care should be taken to limit the input voltage within the permitted voltage rating of the winding. Single phase transformers are normally used in electronic circuits, control circuits and domestic equipment. The three-phase transformers are employed in power system. A basic type of 1-phase transformer is shown in Fig. 3.3.

If the secondary voltage V_2 is greater than primary voltage V_1 than the transformer is called as *step-up transformer* whereas in case of $V_2 < V_1$, it is called as *step-down transformer*. In some low power applications, transformer is used just to isolate two circuits electrically and any change in voltage level is not required. In such cases $V_2 = V_1$. These transformers are called as *isolation transformer*. In radio transmitter and electronic communication devices that work on radio frequencies, the magnetic core is replaced by solid insulating substance. These transformers are called as *air-core transformer*.

The power and distribution transformers are used in power system. They differ in terms of power handling capacity, duty of operation and maximum efficiency condition. Power transformers come in ratings above 500 kVA. They are employed in sub-stations of power plant and transmission networks to step-up or step-down the voltages (66kV, 132kV, 220kV, 400kV, 765kV). Distribution transformers are upto 500 kVA rating, relatively smaller in size and are used as step-down transformer for the distribution of electrical energy at low voltages less than or equal to 33kV for industrial consumers and 230V (phase voltage) for domestic consumers. These transformers shall be dealt in detail in Unit-4. Instrument transformers are used for the measurement of high voltages and high currents as *potential transformer* (PT) and *current transformer* (CT) respectively. PTs are designed to step down the voltage and CTs are designed to step down the current such that the voltmeters and ammeters available commercially could be used for the measurement. Welding transformer shall be discussed in Unit-5.

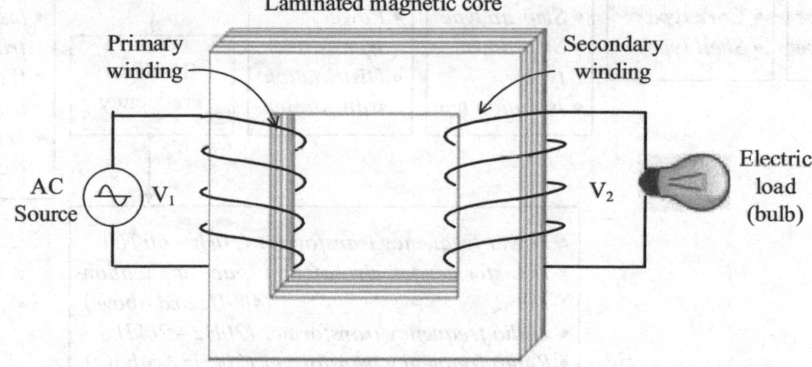

Fig. 3.3: Basic construction of a single-phase core type transformer

3.3 CONSTRUCTION OF 1-PHASE TRANSFORMER

The conventional transformer has following main elements:

(a) A magnetic circuit (core) comprising of limbs and yokes and

(b) Two electric circuits comprising of primary winding and secondary winding.

Based on the shape of magnetic core, transformer is classified as (a) Core type and (b) Shell type. The horizontal sections of magnetic core are called as *Yoke* and the vertical ones are called as *Limb*. In single-phase transformer, the core type construction has two limbs and the windings are wound on both the limbs. Shell type construction has three limbs and both windings, primary and secondary are wound on middle limb only. Fig. 3.4 shows the schematic diagrams of both types of transformer.

3.3.1 Construction of Magnetic Core

When the transformer is in operation, some amount of active power (watts) is lost in the magnetic core as well as in the windings. This power loss produces heat which gets dissipated to the surrounding area. Like any other electrical equipment, insulating items like electrical grade insulating paper, Bakelite sheet, pressboard, wood etc. are used in a transformer. Production of heat due to power loss imposes risk to these insulations. Therefore, all efforts are made during the manufacturing stage to ensure the power loss at minimum value. The power loss in magnetic core is called as *core loss, iron loss* or *No-load loss*. This core loss is further divided into two types as (a) *hysteresis loss* and (b) *eddy current loss*. Different aspects related to the construction of core are based on the minimization of this core loss.

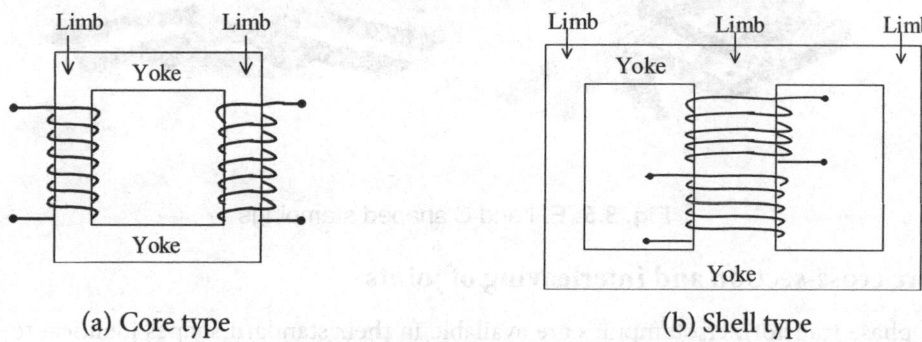

Fig. 3.4: Single-Phase Transformer (Basic construction)

(A) Use of silicon-steel alloy for reduction of hysteresis loss

Hysteresis loss (P_h) is produced due to continuous reversal in the direction of magnetization of the core because, the magnetic flux established in the core is of alternating type. To minimize hysteresis loss, the magnetic core should be made from a material that has high permeability. Commonly used material are CRGO silicon-steel alloy and amorphous alloy. Addition of silicon in iron increases its permeability. Addition of silicon content also increases the material resistivity which consequently helps in the reduction of eddy currents. A small amount of silicon (about 3%) is added in iron to produce the silicon-steel alloy. Higher content of silicon in iron tends to make the material brittle and difficult to cut in required shape for making the stampings. Different types of material used for magnetic core are dealt in detail in the section 3.4.

(B) Use of laminated core for reduction of eddy current loss

Eddy current loss results from the induction of eddy currents in the core when it carries alternating flux. These eddy currents whirl round and round in the form of concentric circles. This phenomenon can be visualized as identical to the eddies developed in water bodies. In order to minimize the power loss produced by eddy currents, the magnitude of eddy currents should be minimized by reducing the area of cross-section of the path where eddy currents are produced. For this purpose, the core is made from several pieces of thin sheets of silicon-steel instead of making it a solid one-piece core. The resulting core is called as *laminated core* and the thin sheets are called as *lamination* or *stamping*. Their thickness varies between 0.3 and 0.5 mm. To make a transformer core, first a thin coat of heat resistant enamel insulation like varnish, is applied on all sides of every stamping. Thereafter, they are stacked over each other and then tightly clamped together to form a solid core structure. As a result, eddy currents from one stamping cannot enter into the adjacent stamping and the sum total of eddy currents produced in all stampings is much less than the eddy current that would have been produced in a non-laminated core. Thus a laminated core results in reduced eddy current loss. Fig. 3.5 shows E, I and C shaped stampings.

Fig. 3.5: E, I and C shaped stampings

(C) Core cross-section and interleaving of joints

For single phase transformer, stampings are available in their standard shapes identical to the English letters 'L, I, C and E'. These individual pieces are joined together to form the core. To reduce the reluctance offered by magnetic core to the flux established in it, continuous joints in the core cross-section should be avoided so that there is no continuous air gap in the path of magnetic flux. This is ensured by interleaving the joints in adjacent layers of stampings. This is shown in Fig. 3.6(a). There are different methods of interleaving by adopting different types of joints. *Butt joints* as shown in Fig. 3.6(b) are suitable if Hot Rolled Steel (H.R.S.) is used for making the stampings. But for the Cold Rolled Grain Oriented steel (CRGO), such joints are not suitable. This is because at the place of joints, the flux path becomes at right angle to the direction of rolling which increases the magnetization current and power loss in the core. Therefore, for CRGO, *mitred joints* are used for interleaving as shown in Fig. 3.6(c). More information of CRGO and rolling process is provided in Section 3.4.

Most of the single-phase and 3-phase transformers are of core type. For small core type transformers, limbs with square or rectangular cross-section can be used. This is shown in Fig. 3.7(a). However in such arrangements, considerable amount of useful space remains unutilized. An improved arrangement is to employ cruciform cross-section as shown in Fig. 3.7(b). The cruciform limb requires stampings of different widths. Typical core sections having two and three steps are shown in Fig. 3.7(b) and (c) respectively. The core stepping is introduced such that the net cross-sectional area of the limb is maximum for the number of steps employed. The corners of the steps lie on the circle (of predetermined diameter) which is actually the inner periphery of the coils.

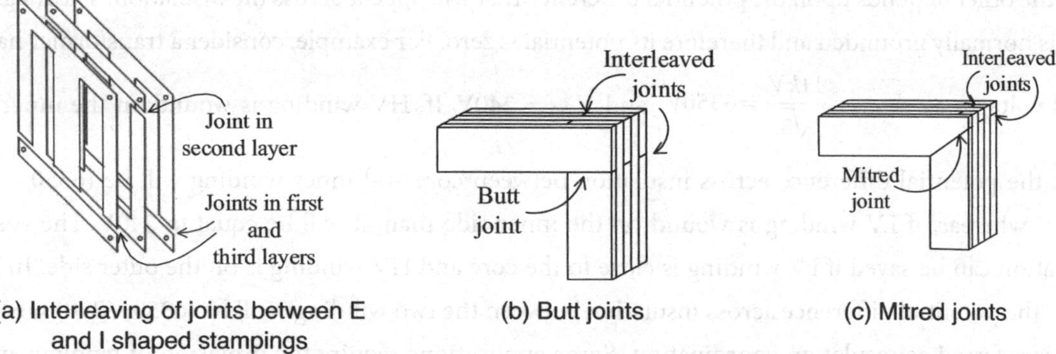

(a) Interleaving of joints between E and I shaped stampings (b) Butt joints (c) Mitred joints

Fig. 3.6: Interleaving of joints by different types of overlap

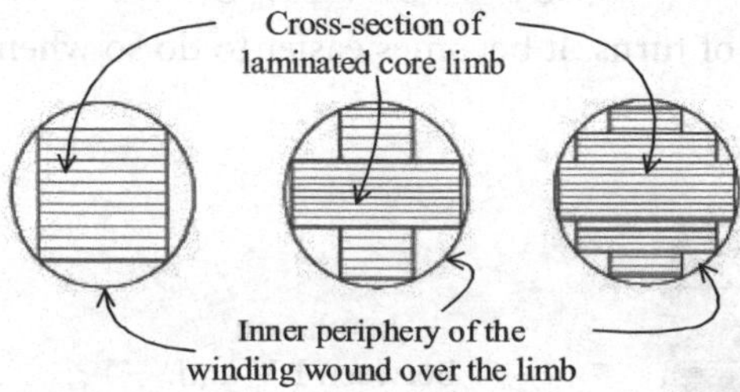

Fig. 3.7: Limbs of the core with different cross-sections: square, 2-step cruciform and 3-step cruciform

3.3.2 Single-Phase Core Type Transformer

Single-phase core type transformer has a magnetic core with two limbs. Two windings of concentric type also called as cylindrical type are wound spirally over the limbs. In the basic construction of core type transformer presented in Fig. 3.3, it was shown that, all the turns of primary winding are wound on one limb and all turns of secondary winding are wound on the second limb. This type of construction assumes that, all the flux produced by each winding will travel through the core and link with the second winding. Practically, some part of the total flux produced by each winding completes its path through air without travelling in the core and therefore, does not link with the other winding. This part of each flux which does not link with the other winding is called as *leakage flux* and it serves no useful work. Hence, if the two windings are wound away from each other on different limbs, they are not able to use all the flux established by each other. To minimize the effective leakage flux, both windings of a core type transformer are halved and wound on both limbs one over the other with an insulating paper maintained between them. Fig. 3.8 shows a practical core type transformer.

For step-up and step-down applications, the two windings are designed for different voltage ratings. One of them is a high voltage (HV) winding while the other one is the low voltage (LV) winding. The LV winding is wound on the inner side and the HV winding is wound on the outer side. This is primarily done to reduce the cost of insulation and to ease the provision of tappings. In case of On-load tap changing, tappings are provided on the HV winding. The thickness of insulating material that is provided between the core and inner winding as well as between the two windings which are wound one over the other depends upon the potential difference that will appear across the insulation. The magnetic core is normally grounded and therefore its potential is zero. For example, consider a transformer having rated voltages as, $V_{(hv)} = \frac{11kV}{\sqrt{3}} = 6350V$ and $V_{(lv)} = 240V$. If, HV winding is wound on the inner side than, the potential difference across insulation between core and inner winding will be (6350 – 0) = 6350V whereas, if LV winding is wound on the inner side than, it will be equal to 240V. The cost on insulation can be saved if LV winding is close to the core and HV winding is on the outer side. In both cases, the potential difference across insulation between the two windings will be (6350 – 240) = 6110V. This is referred as insulation coordination. Some applications require the provision of tappings in the transformer so as to obtain different voltage ratings. Tappings are provided on the HV winding because it has relatively higher number of turns. It becomes easier to do so when the HV winding is wound on the outer side.

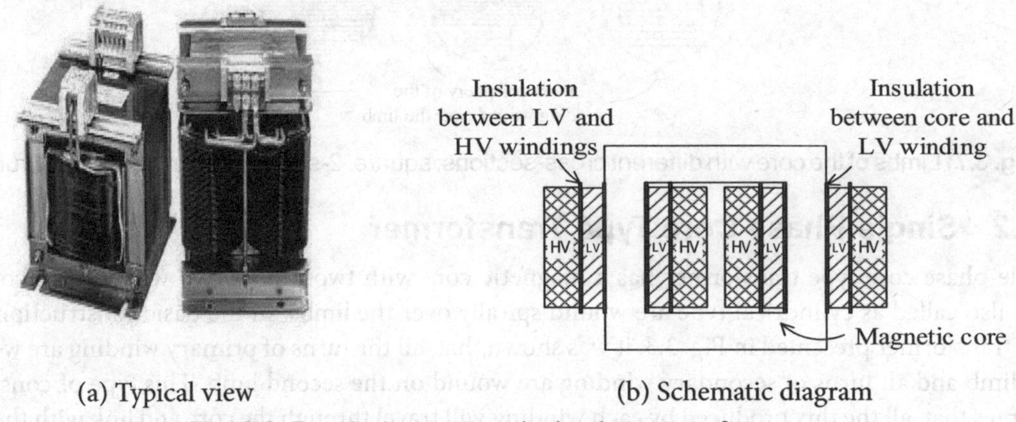

Fig. 3.8: Practical core type single-phase transformer

3.3.3 Single-Phase Shell Type Transformer

In shell type transformer, the LV and HV windings are wound over the central limb. These windings are of sandwich type as shown in Fig. 3.9(b). Each winding is subdivided into subsections. LV and HV subsections are alternately put in the form of sandwich. Since the core is grounded, LV winding is placed closer to the core so that the cost of insulation between the core and the winding closer to it is less. Note that, the bottom and top LV coils are of half the size of other LV coils.

Since both windings are on the central limb, it has total flux produced by both windings which then gets divided equally and returns back through the outer two limbs. Due to this reason, the area of cross-section of middle limb is twice that of the two outer limbs. As both windings are able to link with almost all the flux produced by each other, the leakage flux is less and hence the leakage reactance is also less. This results in better voltage regulation.

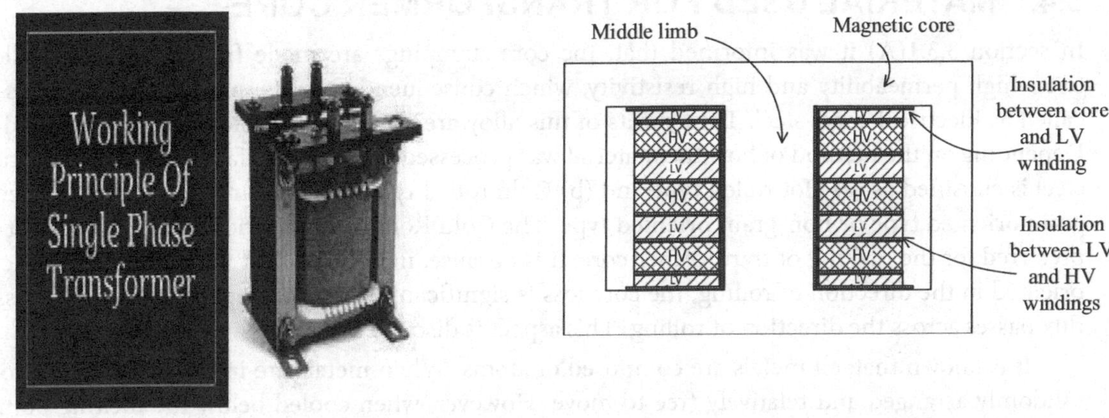

(a) Typical view (b) Schematic diagram

Fig. 3.9: Practical Shell type Transformer

3.3.4 Differences Between Core Type and Shell Type Transformers

Core type Transformer	Shell type Transformer
1. The core has two limbs and both LV and HV windings are wound on both limbs.	1. The core has three limbs but both windings are wound on middle limb only.
2. Windings are of concentric type also called as cylindrical type.	2. Windings are of sandwich type also called as disk type.
3. Windings surround the core therefore, it is easier to locate the winding faults.	3. Core surrounds the windings therefore, to locate the winding faults is difficult.
4. Easier to fabricate and repair.	4. Fabrication and repair is difficult.
5. For the same given output and voltage rating, core type transformer requires less iron.	5. For the same given output and voltage rating, core type transformer requires more iron.
6. Less costly.	6. Relatively, high cost.
7. Leakage flux is more compared to shell type construction therefore, the leakage reactance is more and hence voltage regulation is poor.	7. The magnetic and electric circuits are tightly coupled therefore leakage flux is less, the leakage reactance is less and hence voltage regulation is better.
8. The magnetic path is relatively longer and hence, the magnetizing current is more than that in a shell type transformer.	8. It provides shorter magnetic path and hence, the magnetizing current is less than that in a core type transformer.
9. Since windings surround the core, support against electromagnetic forces developed under fault conditions between current carrying conductors is less.	9. It gives better support against electromagnetic forces developed under fault conditions between the current carrying conductors.
10. Natural cooling is better as windings are on the outer side.	10. Natural cooling is poor due to the embedding of windings.
11. Core type transformer is preferred for high-voltage, high-power applications	11. Shell type transformer is preferred for low-voltage, low-power applications

3.4. MATERIAL USED FOR TRANSFORMER CORE

In section 3.3.1(A) it was informed that, the core stampings are made from silicon-steel alloy due to its high permeability and high resistivity, which consequently results in low core loss. This is also called as Electrical sheet steel. Thin sheets of this alloy are shear cut into stampings of required shape. Depending on the method of how the material was processed during manufacturing, the electrical sheet steel is classified as (a) Hot rolled type and (b) Cold rolled type. The cold-rolled sheet steel is either of grain oriented type or non-grain oriented type. The Cold Rolled Grain Oriented (CRGO) sheet steel is preferred for the making of transformer core. It is because, in CRGO sheet steel, if the magnetic flux is oriented in the direction of rolling, the core loss is significantly less as compared to the core loss when flux passes across the direction of rolling. This aspect is discussed below.

It is known that, all metals are composed of atoms. When metals are in liquid state, the atoms are randomly arranged and relatively free to move. However, when cooled below the melting point, they rearrange and pack themselves together to form regular arrays of atoms called crystals or grain or unit cell. These tiny crystals increase in size by the progressive addition of atoms. Thus metals and alloys have a strong tendency to crystallize so as to attain a solid state and therefore, the resulting solid metal is not one crystal but actually a huge bunch of many smaller crystals. For example, at room temperature the grain of iron is a body centred cubic (BCC) structure, which is an array with one atom at each corner of the cube, and one in the centre of this cube. Similarly, aluminium grain is a face centred cubic (FCC) structure. They are shown in Fig. 3.10. The structure, size and orientation of these grains result from the material composition (constituents of alloy) and the way the material is made (e.g. forging, casting or additive manufacturing etc.).

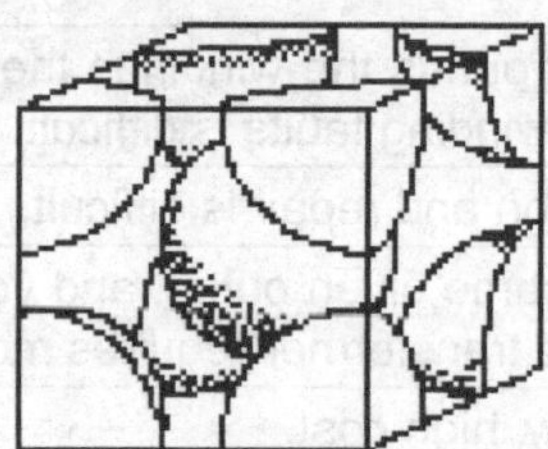

(a) Body Centered Cubic (BCC)

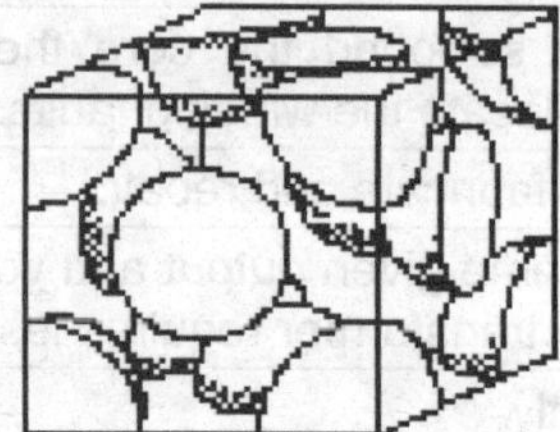

(b) Face Centered Cubic (FCC)

Fig. 3.10: Grain structures

Composition of the grains and their orientation determine the basic properties of metals and alloys. The metal forming process which is adopted in industries to manufacture thin sheets of metal or alloy is called as *'Rolling'*. It employs the use of many rollers to compress and alter the shape of raw material (billet), improve its uniformity and/or enhance the mechanical properties of the material. Such a rolled steel sheet can be categorized into two types as, (a) *Hot rolled steel* and (b) *Cold rolled* steel. Both demonstrate distinct characteristics that make them suitable for different applications. The main difference between hot rolled steel and cold rolled steel is the temperature at which they are processed. Hot rolled steel is rolled above recrystallization temperature of the material. At such high temperature, grains cannot be formed.

Cold rolled steel is essentially the hot rolled steel which is further processed below the material's recrystallization temperature. During the rolling process, it is possible to control and orient the grains in a desired direction. Accordingly, two types of sheet steels can be made as (a) *Non-Grain-Oriented* steel and (b) *Grain-Oriented* steel. The magnetic properties of a material depend on the orientation of the grains (i.e. crystals) of the material. The grain-oriented steel is costly and has high magnetic flux density to an extent of 30% in the rolling direction. It also has low area of the hysteresis loop, low core loss & high magnetic permeability.

3.4.1 Hot Rolled Steel

In the hot rolling process, a large rectangular slab of metal or alloy referred to as a *billet* (which is to be processed) is heated above its recrystallization temperature (usually 1700° F or greater) and then compressed into a large roll of thin sheet. While still hot, it is passed through a series of rotating rollers to achieve the required dimensions and then allowed to cool. During the process of hot rolling at temperature higher than the recrystallization temperature of the material, the crystals (grains) get damaged and the temperature does not allow 'recrystallization'. Recrystallization is a process where the damaged grains are replaced with a new set of grains during the rolling process. This can happen only when the metal is cool, so any metalworking process that involves forming the metal at temperatures above the recrystallization point is considered as "hot rolled".

It is known that, matter expands when it is hot and shrinks while cooling. In view of this fact, the dimensions of hot rolled steel cannot be controlled accurately and hence, it is used in those applications that do not require extremely tight dimensions, like agricultural equipment, construction material etc.

3.4.2 Cold Rolled Non-Grain-Oriented Steel (CRNGO)

Cold rolled steel is the hot rolled steel that has undergone additional processing to improve its dimensional and mechanical properties. During the cold rolling process, cooled hot rolled steel passes through another series of pairs of rollers at room temperature. Since the material is no longer hot and malleable, a significantly higher amount of pressure is required to compress it into the desired shape. While this process is more labour-intensive and costlier than the hot rolling process, it can achieve tighter dimensional tolerances and better surface qualities. Cold rolled has about 20% more physical strength than that of hot rolled steel. This makes it more suitable for use in high-stress applications.

A cold rolled non-grain-oriented-steel, usually has silicon content of 2 to 3.5% and has similar magnetic properties in all directions. In other words, it is isotropic in nature. CRNGO is less expensive than CRGO. It is used when cost is more important than efficiency and is preferred in applications where the direction of magnetic flux is not constant. Example, stator and rotor cores of rotating equipment like electric motors and generators. This is because, in rotating machines, the cross section is circular and hence the magnetic field at each point is tangential to circle. It means that the magnetic field changes its direction at every point.

3.4.3 Cold Rolled Grain-Oriented Steel (CRGO)

In cold rolled grain oriented steel, the silicon content is usually about 3%. By fully annealing during the cold rolling process, the grains are oriented in the direction of rolling. Consequently, the permeability is high and magnetic flux density is increased by 30% only in the direction of rolling compared to other directions. Thus CRGO is magnetically superior in the rolling direction. If the magnetic flux is oriented in the direction of rolling then it experiences less reluctance and hence the core loss is less. This alloy is commonly used for the cores of power and distribution transformers. CRGO is usually supplied by the steel mills in the form of roll or coil. It has to be shear cut into laminations of required size and shape to use them for transformer core.

In Fig. 3.6 (a) & (b), two types of joints between the stampings are shown. One is the butt joint and other is mitred joint. In butt joint, it can be seen that just before the place of joint when the flux coming through the yoke will bend by 90° to enter into the limb, it will be required to travel across the direction of rolling till it crosses the joint. This will increase the power loss in the joint area and increase the temperature of that part of the core. However in mitred joint, this problem will be minimum.

3.4.4 Amorphous Core

The core of the first practical transformer was made from carbon steel sheets by William Stanley in the year 1885. It was then replaced by Silicon-steel alloy in order to reduce the core losses. While CRGO silicon-steel is commonly used for the transformer core, introduction of amorphous alloy core is also receiving increasing popularity.

Amorphous core transformer is a kind of energy efficient transformer in which the core is made from very thin foils of amorphous alloy composed of about 80% iron and remaining about 20% is a mixture of boron and silicon. The amorphous alloy is a soft ferromagnetic material. It is formed by a process of rapid solidification of molten alloy. Due to high speed of solidification, the atoms get no opportunity to form crystalline structures (grains) and they solidify in a disordered manner (i.e. amorphous). This allows it to be magnetized and de-magnetized quickly in any direction with very less core loss. In other words, it is isotropic. Due to the non-crystalline structure, it has higher electrical resistivity which is about three times compared to the typical silicon-steel lamination sheets. Also, the thickness of amorphous foil is very less (about 0.025 mm). High resistivity and very less thickness of the foil results in significant reduction in eddy current loss [11]. Consequently, the core loss in amorphous core transformer is about one-third of that in conventional CRGO core transformer.

Instead of the hot rolling and cold rolling procedures for forming the steel sheets, the amorphous foil is produced by dripping hot molten alloy on an internally cooled, fast rotating wheel. When the molten alloy touches the cold wheel surface, the rapid cooling process of hot alloy fixes its iron molecules in their current position, leading to the amorphous structure without forming any crystalline array in the material. Hence, the amorphous alloy foil is also called as metal glass.

Although amorphous alloy core results in reduction of core loss, it presents some drawbacks like low saturation magnetization (i.e low saturation flux density) and increase in losses at high magnetic fields. Therefore the size of amorphous alloy core has to be bigger than that of silicon-steel core for the same value of magnetic flux. This increases the length of windings wound on the core and also the copper loss increases. Hence, the price of amorphous core transformer is higher compared to the traditional transformer [12]. A typical amorphous core is shown in Fig. 3.11.

Fig. 3.11: Amorphous alloy core

3.5 PRINCIPLE OF OPERATION

Consider a two-winding transformer with secondary winding *open* (not connected to load) as shown in Fig. 3.12. This is called as *no-load condition*. When an *ac* voltage V_1 is applied across the primary winding, some current starts flowing through it. This current is called as *No-load current* and is denoted by I_o. It sets up an alternating magnetic flux ϕ in the core which links with both the windings. By Faraday's law of electromagnetic induction, due to flux-linkage of this flux ϕ with the windings, it induces *emf* in both windings. Their RMS values are denoted by E_1 and E_2 for the primary and secondary windings respectively. E_1 is called, the *self-induced emf* and E_2 is called the *mutually-induced emf*. The self-induced *emf* E_1 opposes the applied voltage V_1 due to Lenz's law.

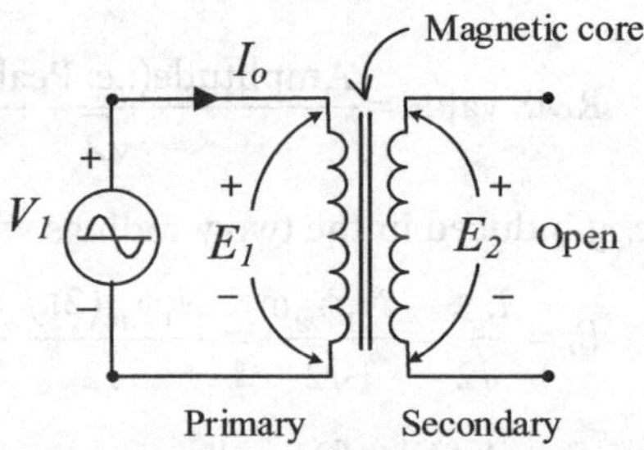

Fig. 3.12: Transformer at No-load

Now, if the secondary winding circuit is closed by connecting an electrical load across it than, the emf E_2 will drive a current I_2 in the secondary winding and in the load. Thus ac power from primary circuit is transformed to the secondary circuit by mutual induction. The polarities shown in Fig. 3.12 are the instantaneous polarities of *ac* voltages and emfs.

3.5.1 EMF Equations

The instantaneous value of alternating flux produced by the No-load current in primary winding is,

$$\phi = \phi_m \sin\omega t \quad \text{webers} \tag{3.1}$$

Where,

$$\omega = 2\pi f \text{ (in radians/second) and } f = \text{frequency (in Hz)}$$

By Faraday's law, the magnitude of *emf* induced in a coil (winding) due to the changing flux which links with the coil is directly proportional to the rate of change of flux-linkages. Mathematically, it can be expressed as,

$$e = -\frac{d\lambda}{dt} = -N\frac{d\phi}{dt} = -N\frac{d}{dt}(\phi_m \sin\sin\omega t)$$

Where,

$\lambda = N\phi$ = flux-linkage (volt.second) and Number of turns in the winding

Minus sign indicates the opposing nature of induced *emf* as per Lenz's law. The above equation can be used to write the equations for instantaneous value of *emf* induced in both windings. For primary side,

$$e_1 = -N_1\frac{d}{dt}(\phi_m \sin\sin\omega t) = -N_1\phi_m\omega\cos\cos\omega t$$

By trigonometric relation: $-\cos\omega t = \sin(\omega t - 90°)$

$$e_1 = N_1\, \phi_m\, \omega \sin(\omega t - 90°)$$

Let, $N_1\, \phi_m\, \omega = Em_1$ = amplitude (peak value) of e_1

$$\therefore \quad e_1 = Em_1 \sin(\omega t - 90°) \tag{3.2}$$

Similarly for the secondary winding, instantaneous value of induced *emf* will be,

$$e_2 = Em_2 \sin(\omega t - 90°) \tag{3.3}$$

Where, $E_{m2} = N_2\, \phi_m\, \omega$ = amplitude (peak value) of e_2

It is known that, the Root Mean Square (RMS) value of a sinusoidally alternating quantity is given by,

$$\text{RMS value} = \frac{\text{Amplitude(i.e. Peak value)}}{\sqrt{2}}$$

Therefore, the RMS values of *emf* induced in the two windings will be,

$$E_1 = \frac{E_{m1}}{\sqrt{2}} = \frac{N_1\phi_m\omega}{\sqrt{2}} = \frac{N_1\phi_m(2\pi f)}{\sqrt{2}}$$

Similarly, in secondary winding, $E_1 = 4.44\, \phi_m f N_1$ volts

$$E_2 = 4.44\, \phi_m f N_2 \text{ volts} \tag{3.4}$$

N_1 and N_2 are the number of turns in primary and secondary winding respectively.

3.5.2 Transformation Ratio (K)

Transformation ratio of a transformer is defined as the ratio of number of turns of secondary winding N_2 to the number of turns of primary winding N_1.

$$K = \frac{N_2}{N_1}$$

Using *emf* equations (3.4),

$$K = \frac{N_2}{N_1} = \frac{E_2}{E_1}$$

For a transformer, Input volt-amperes (V_1I_1) = Output volt-amperes (V_2I_2)

Also, for an ideal transformer, the winding impedances are zero therefore, $V_1 = E_1$ and $V_2 = E_2$

$$\therefore K = \frac{N_2}{N_1} = \frac{E_2}{E_1} = \frac{V_2}{V_1} = \frac{I_1}{I_2} \tag{3.5}$$

Where, V_1, E_1 and V_2, E_2 are the terminal voltages and induced *emfs* of primary and secondary windings respectively.

- For a step-up transformer: $V_2 > V_1$, $N_2 > N_1$, $\therefore K > 1$
- For a step-down transformer: $V_2 < V_1$, $N_2 < N_1$, $\therefore K < 1$
- For an isolation transformer: $V_2 = V_1$, $N_2 = N_1$, $\therefore K = 1$

Symbols of different types of transformer are shown in Fig. 3.13.

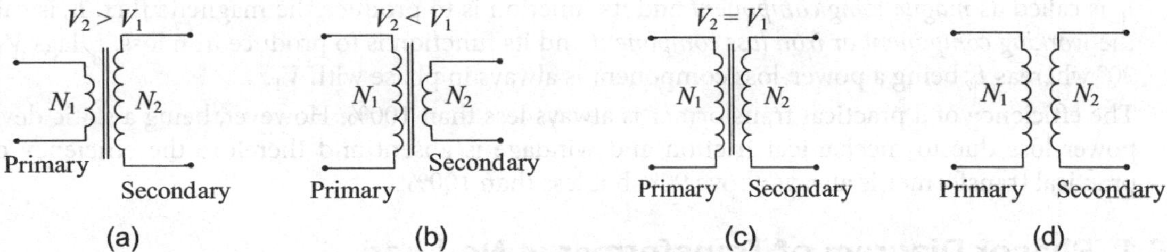

Fig. 3.13: Transformer symbols (a) step-up; (b) step-down; (c) isolation type; (d) air-core type

3.6 CONCEPT OF IDEAL AND PRACTICAL TRANSFORMERS

It is known that, ideal systems are those which have no losses. The output power delivered by them is equal to the input power drawn by them from the source and therefore, their efficiency is 100%. Practically, ideal systems do not exist but all practical systems should be designed to perform close to their ideal counterparts. An ideal transformer is considered to have the following properties:

- The resistances R_1 and R_2 of primary and secondary windings are zero. Therefore, the respective voltage drops (I_1R_1 and I_2R_2) and the copper losses (I_1^2 R_1 and I_2^2 R_2) are also zero.
- The complete magnetic flux produced by each winding travels through the magnetic core and links with the other winding. In other words, the leakage flux is zero.
- Permeability of the magnetic core is constant i.e., the core always remains in unsaturated state.
- At No-load condition when the secondary winding is open, the No-load current *Io* flowing in primary winding is utilized for the production of magnetic flux only. In other words, the No-load current *Io* is a purely magnetizing (i.e. purely inductive) current and lags the applied voltage V_1 by 90°. Hence, the input power taken by an ideal transformer at No-load condition from the source is $P_1 = V_1 I_0 \cos 90° = 0$.
- Both, iron loss and copper loss in an ideal transformer are zero at No-load as well as at On-load conditions. Therefore, the efficiency is always 100%.

A practical transformer deviates from ideal transformer in terms of all the above properties. Following are the comparative features of a practical transformer:

- The primary and secondary windings have non-zero resistances R_1 and R_2. Therefore, the respective voltage drops (I_1R_1 and I_2R_2) and the copper losses (I_1^2 R_1 and I_2^2 R_2) are not zero.
- Out of the total magnetic flux produced by each winding, some portion completes its path through air without travelling through the core and does not link with the other winding. This portion of total flux is called as leakage flux. It performs no useful work but results into a voltage drop (I_1X_1 and I_2X_2) in the winding which has produced it.
- Permeability of the magnetic core is not constant. It is directly proportional to the flux density. Normally, transformers are designed to operate at the knee point of their saturation curve.
- The No-load current Io flowing in primary winding is utilized for the production of magnetic flux as well as for producing the iron loss in the core. In other words, the No-load current Io performs two tasks and accordingly, it can be resolved into two components I_μ and I_w as,

$$I_0 = I_\mu + I_w \tag{3.6}$$

I_μ is called as *magnetizing component* and its function is to produce the magnetic flux. I_w is called the *working component or iron loss component* and its function is to produce iron loss. I_μ lags V_1 by 90° whereas I_w being a power-loss component is always in phase with V_1.

- The efficiency of a practical transformer is always less than 100%. However, being a static device, power loss due to mechanical friction and windage is absent and therefore the efficiency of a practical transformer is always above 90% but less than 100%.

3.6.1 Phasor Diagram of Transformer at No-Load

Phasor diagram of any *ac* circuit or an *ac* device is the graphical representation of its current and voltage equations where the RMS currents and RMS voltages are represented by phasors. Following general steps may be followed to draw the phasor diagram of any given *ac* device or *ac* machine.

- Understand the operation of the device.
- Write voltage and current equations of the device by using Kirchhoff's laws.
- Draw the phasors in the sequence identical to the sequence of their entry in the operating process. Use basic rules of phasor relationship between current and voltage for purely resistive, inductive and capacitive elements.
- Add or subtract the phasors as per the governing voltage and current equations.

Using above steps, let us draw the phasor diagram of an ideal and practical transformer at No-load condition as shown in Fig. 3.14 (a) and (b) respectively. Being a No-load condition, the secondary winding will remain open and therefore $I_2 = 0$. The operation will start with application of input voltage V_1 across primary and therefore V_1 shall be the first phasor to draw. It will produce a No-load current Io in primary winding. For ideal transformer, Io will lag V_1 by phase angle $\theta_0 = 90°$. In a practical transformer, $\theta_0 \approx 90°$. Therefore, Io should be resolved into two components, I_μ and I_w as explained in Section 3.6. In both cases, the magnetic flux ϕ will get established in the core. This flux should be shown in phase with the current that has produced it. From the flux and induced emfs of equations (3.1)-(3.3), it is clear that, E_1 and E_2 will lag flux by $\phi = 90°$.

$$\phi = \phi_m \sin \omega t$$
$$e_1 = E_{m1} \sin(\omega t - 90°)$$
$$e_2 = E_{m2} \sin(\omega t - 90°)$$

In a practical transformer, the No-load current I_0 is 1- 5% of Full-load primary current.

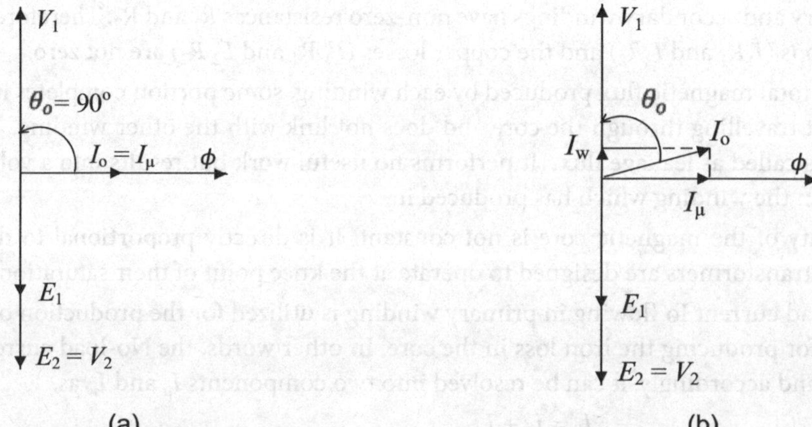

Fig. 3.14: No-load phasor diagram of (a) an ideal transformer and (b) a practical transformer

3.7 OPERATION OF TRANSFORMER IN ON-LOAD CONDITION

To begin with, consider a two winding transformer at No-load condition as discussed above. Let input voltage V_1 be connected across the primary. It will produce a No-load current I_0 in primary winding which then will establish a flux ϕ in the core as shown in Fig. 3.15(a). The direction of flux at any instant of time can be determined by applying the Right Hand Thumb rule. (See Appendix of Unit-1).

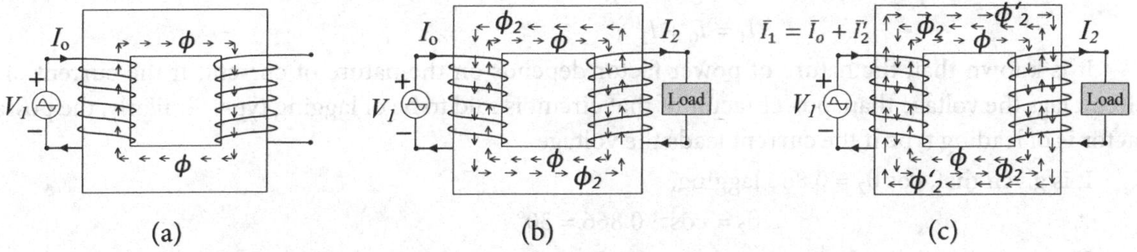

Fig. 3.15: Operation of a practical transformer On-load condition

Now, let the secondary side be closed by connecting a load across it than, the mutually induced *emf* will produce a current *I2* in the secondary winding. The direction of this current will be governed by Lenz's law according to which, *I2* should flow in a direction such that the flux ϕ_2 produced by I_2 will be in opposition to the original flux ϕ. This is shown in Fig. 3.15(b).

Now, there are two fluxes ϕ and ϕ_2 in the core. The net flux is $(\phi - \phi_2)$ which is less than the original net flux ϕ. Due to the reduction in net flux, the magnitude of self-induced *emf* will also reduce. If you observe the primary winding circuit where the input voltage V_1 has not changed its value but E_1 (which is in opposition to V_1) has reduced then consequently, you will find that, the input current delivered by V_1 has increased. Let the additional current flowing in primary winding be denoted by such that the resultant primary current is,

$$I_1 = I_o + I_2' \tag{3.7}$$

Since I_2' has come into existence due to the presence of load, it is called as *load component of primary current.* I_2' will flow in the same direction of I_o and will establish its own flux ϕ_2' in the same direction of original flux ϕ. Now, there are three fluxes in the core ϕ. ϕ_2 and ϕ_2'. ϕ_2 and ϕ_2' are in opposite directions. This is shown in Fig. 3.15(c). Since the *mmf* of primary side is equal to the *mmf* of secondary side, the magnitude of ϕ_2' is always equal to the magnitude of ϕ_2'.

$$N_1 I_2' = N_1 I_1 = N_2 I_2, \quad \therefore \; |\phi_2'| = |\phi_2|$$

Therefore, they will cancel each other and the flux left in the core is the original flux ϕ. Thus, the net flux in the core remains same (= ϕ). Practically, the variation in net flux is negligible, between 1% and 3% from No-load to Full-load conditions. Hence, transformer is also called as a *Constant Flux Machine.* The magnitude of I_2' is given by,

$$\left|I_2'\right| = \frac{N_2 I_2}{N_1} = \text{K}\left|I_2\right| \tag{3.8}$$

Where, K = transformation ratio.

Example 3.1: A 230/115 V, single phase transformer is supplying a load of 5A at a power factor of 0.866 lagging. The No-load current is 0.2A at a power factor of 0.208 lagging. Calculate the primary current and power factor.

Data:

$V_1 = 230V$, $V_2 = 115V$, $I_2 = 5$ A, $\cos\theta_2 = 0.866$ lagging, $I_0 = 0.2$ A,

$\cos\theta_o = 0.208$ lagging, $I_1 = ?$, $\cos\theta_1 = ?$

Solution:

To understand this example, refer to the phasor diagram as shown for lagging power factor load.

The current in primary winding is given by,

$$I_1 = I_0 + I_2'$$

It is known that, the nature of power factor depends on the nature of current. If the current of a circuit lags the voltage than, power factor of that circuit is said to be of lagging type. Similarly, the power factor is of leading type if the current leads the voltage.

It is given that, $\cos\theta_2 = 0.866$ lagging,

$$\therefore \quad \angle\theta_2 = \cos^{-1} 0.866 = 30°$$

Hence current I_2 can be expressed in polar frame as,

$I_2 = 5\angle - 30°$ A where the (–) sign indicates that this current is lagging behind the voltage V_2 by a phase angle of

Similarly, given that, $\cos\theta_o = 0.208$ lagging,

$$\therefore \quad \angle\theta_0 = \cos^{-1} 0.208 = 78°$$

Hence current I_0 can be expressed in polar frame as,

$$\bar{I}_0 = 0.2\angle - 78°\text{A}$$

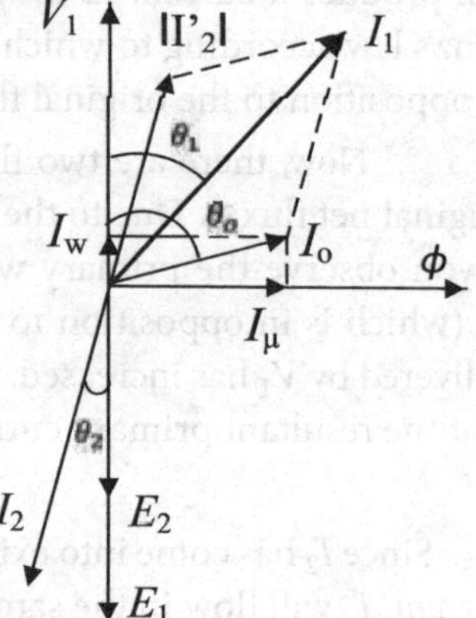

The magnitude of secondary current as referred to primary side is,

$$\left|I_2'\right| = KI_2 = \frac{V_2}{V_1} I_2 = \frac{115}{230} \times 5 = 2.5A$$

In polar frame it will become, $\bar{I}'_2 = |I'_2|\angle\theta_2 = 2.5\angle - 30°$ A

Substituting the values of $\bar{I}_0$ and $\bar{I}_2'$ in above equation.

$\bar{I}_2(0.2\angle - 78°) + (2.5\angle - 30°)$

Since it is convenient to perform phasor addition in rectangular reference frame, converting and from polar frame to rectangular frame.

$$\bar{I}_1 = (0.0415 - j\,0.1956) + (2.165 - j1.25) = (2.2065 - j1.4456)$$

Converting $\bar{I}_1$ from rectangular frame to the polar frame

$\therefore \; I_1 = 2.638\angle - 33.23°$ A

Thus, $|\bar{I}_1| = 2.638$A and $\cos\theta_1 = \cos 33.23° = 0.826$ lagging

3.8 WINDING PARAMETERS AND VOLTAGE EQUATIONS

Differences between an ideal transformer and a practical transformer were discussed in Section 3.6. The deviation between them is mainly due to the presence of winding resistances and leakage flux in a practical transformer. The resulting impedances comprising of winding resistances and leakage reactances are referred as winding parameters of the transformer.

3.8.1 Winding Resistances

In a practical transformer, the primary and secondary windings are made from copper which possesses finite magnitude of resistance uniformly distributed over the length of the windings. They can be denoted as R_1 and R_2 for the primary and secondary windings respectively. Although they are distributed throughout the winding length, for convenience, they are treated as lumped parameter.

3.8.2 Winding Leakage Reactances

In a practical transformer, the total flux produced by each winding has two components. One is the magnetizing component that travels through the magnetic core and links with the other winding to produce a *mutually-induced emf* in that winding. The other component is that which links with its own winding that has produced it but does not link with the other winding and hence is called as *leakage flux component.* Let the leakage flux components of the primary and secondary windings be denoted by ϕ_{L1} and ϕ_{L2}. Since they link with their own winding, ϕ_{L1} will produce a *self-induced emf* e_{L2} in primary and ϕ_{L2} will produce a *self-induced emf* e_{L2} in secondary. By Lenz's law, the polarity of e_{L2} will always be in opposition to the polarity of input voltage V_1. Similarly, the polarity of e_{L2} will be in opposition to the polarity of *mutually induced emf* E_2. In other words the *emfs* and induced by leakage fluxes will act like voltage-drops in primary and secondary winding respectively.

To account this effect of leakage flux in the transformer, let us assume two fictitious inductive reactances X_1 and X_2 in series with primary and secondary windings respectively. Their magnitudes should be such that, $e_{L1}= I_1X_1$ and $e_{L2}= I_2X_2$. These two fictitious reactances X_1 and X_2 which represent the effect of leakage flux are called as *leakage reactances* of the primary and secondary windings respectively.

In view of above, a practical transformer can be treated as a combination of an ideal transformer and winding parameters (R_1, X_1 and R_2, X_2) connected externally to the two windings. This is shown in Fig. 3.16 where all voltage polarities represent the instantaneous polarities.

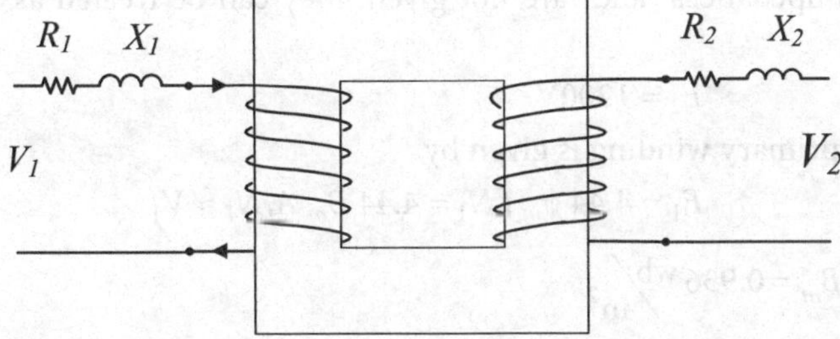

Fig. 3.16: Practical transformer = Ideal transformer + winding resistances + leakage reactances

3.8.3 Voltage Equations

From Fig. 3.16, the primary and secondary voltage equations can be developed by applying Kirchhoff's Voltage law (KVL).

- For Primary winding:

$$\bar{V}_1 = \bar{E}_1 + \bar{I}_1 (R_1 + jX_1)$$

$$= \bar{E}_1 + \bar{I}_1 \bar{Z}_1 \quad (3.9)$$

- For Secondary winding:

$$\bar{E}_2 = \bar{V}_2 + \bar{I}_2 (R_2 + jX_2)$$
$$= \bar{V}_2 + \bar{I}_2 \bar{Z}_2 \qquad (3.10)$$

Where,

V_1 is the input voltage connected across primary winding (in volts)

V_2 is the secondary terminal voltage that shall be available for the load (in volts)

E_1 and E_2 are the emfs produced in primary and secondary windings respectively (in volts)

I_1 and I_2 are the currents in primary and secondary winding respectively (in amperes)

R_1 and R_2 are the resistances of primary and secondary winding respectively (in Ω)

X_1 and X_2 are the leakage reactances of primary and secondary winding respectively (in Ω)

Z_1 and Z_2 are the impedances of primary and secondary winding respectively (in Ω)

Example 3.2: A single phase transformer has 350 primary and 1050 secondary turns. The net cross sectional area of the core is 55 cm². If the primary winding is connected to a 400V, 50 Hz single phase supply calculate, (i) voltage induced in secondary winding and (ii) maximum value of flux density in the core.

Data:

$N_1 = 350, N_2 = 1050, A = 55\ cm^2 = 55 \times 10^{-4}\ m^2, V_1 = 400\ V, f = 50\ Hz, E_2 = ?, B_m = ?$

Solution:

Transformation ratio $K = \dfrac{N_2}{N_1} = \dfrac{V_2}{V_1} = 3; \quad \therefore V_2 = KV_1 = 1200V$

The voltage equations of primary and secondary sides are,

$\bar{V}_1 = \bar{E}_1 + \bar{I}_1 (R_1 + jX_1)$ and $\bar{E}_2 = \bar{V}_2 + \bar{I}_2 (R_2 + jX_2)$

Since, the winding impedances and are not given, they can be treated as negligible and hence, $\bar{V}_1 \approx \bar{E}_1$ and $\bar{E}_2 \approx \bar{V}_2$

$$\therefore \qquad E_2 = 1200V$$

The emf induced in primary winding is given by,

$$E_1 = 4.44\ \phi_m f N_1 = 4.44\ B_m A_f N_1 \approx V_1$$

$$\therefore B_m = \frac{V_1}{4.44 AfN_1} \quad \therefore B_m = 0.936\ \mathrm{wb}/\mathrm{m}^2$$

Example 3.3: The required No-load voltage ratio in a single phase, 50 Hz core type transformer is 6600V/500V. Find the number of turns in each winding if the flux is to be 0.06 weber.

Data:

$f = 50\ Hz, V_1 = 6600\ V, V_2 = 500\ V, \phi_m = 0.06\ wb, N_1 = ?, N_2 = ?$

Solution:

The voltage equations of primary and secondary sides are,

$\bar{V}_1 = \bar{E}_1 + \bar{I}_1 Z_1$ and $\bar{E}_2 = \bar{V}_2 + \bar{I}_2 Z_2$

Since, the winding impedances and are not given, they can be treated as negligible and hence, $\bar{V}_1 \approx \bar{E}_1$ and $\bar{E}_2 \approx \bar{V}_2$

The emf induced in primary and secondary windings are,

$$\therefore \quad E_1 = 4.44\, \phi_m f N_1 \approx V_1$$

$$\therefore \quad E_2 = 4.44\, \phi_m f N_2 \approx V_2$$

$$N_2 = \frac{V_2}{4.44\phi_m f} = 37.53 \text{turns}$$

Similarly, $N_1 = 495.49$ turns

Since the number of turns cannot be a decimal, rounding to integers.

$$\therefore \quad N_1 = 496 \text{ turns and } N_2 = 38 \text{ turns}$$

Note: The value of N_1 can also be calculated using transformation ratio K.

3.9 PHASOR DIAGRAM ON LOAD CONDITION

After connection of load, new currents I_2 and I_2' come into existence and therefore, the phasor diagram of a practical transformer deviates from its No-load version shown in Fig. 3.14 (b). The current and voltage equations (3.6) – (3.10) are reproduced below.

$$\bar{I}_o = \bar{I}_\mu + \bar{I}_w$$

$$\bar{I}_1 = \bar{I}_0 + \bar{I}_2' \text{ where, } |I_2'| = K|I_2|$$

$$\bar{V}_1 = \bar{E}_1 + \bar{I}_1 (R_1 + jX_1)$$

$$\bar{E}_2 = \bar{V}_2 + \bar{I}_2 (R_2 + jX_2)$$

Recall that, in a purely resistive load, current I_2 will be in-phase with voltage V_2. In an inductive load, I_2 will lag behind V_2 by a load power-factor angle θ whereas, in a capacitive load, I_2 will lead V_2 by load-power factor angle θ. Following steps may be adopted to draw the phasor diagrams:

1. Draw flux ϕ as reference.
2. Considering equations (3.1)-(3.3), draw E_1 and E_2 lagging flux ϕ by 90°.
3. For convenience, reverse phasor E_1 on the upper side and consider its magnitude $|-E_1|$ only. This simplification allows to have all phasors belonging to primary winding on the upper side of flux ϕ and those belonging to secondary winding on the lower side of flux ϕ.
4. Considering equation (3.6), draw the magnetizing current (i.e flux producing current) $\bar{I}_\mu$ in phase with flux ϕ whereas $\bar{I}_w$ leading $\bar{I}_\mu$ by 90°. Add $\bar{I}_\mu$ and $\bar{I}_w$ by Parallelogram Method to get the resultant $\bar{I}_0$.
5. Draw voltage phasor V_2 as lagging E_2 by a small angle called as load-angle $\angle\delta$. Depending on the nature of load (resistive, inductive or capacitive), draw current I_2 as shown.
6. Considering equation (3.8), draw the load component of primary current $\bar{I}_2'$ in phase opposition (i.e. 180° apart) to I_2. Considering equation (3.7), add $\bar{I}_0$ and $\bar{I}_2'$ by Parallelogram Method to get the resultant primary current $\bar{I}_1$.
7. Now draw the primary voltage equation (3.9). Draw the resistive voltage drop $\bar{I}_1 R_1$ from the tip of $|-E_1|$ but in phase with I_1. Then draw the inductive voltage drop $\bar{I}_1 X_1$ from the tip of $\bar{I}_1 R_1$ but

leading I_1 by 90°. Finally by Polygon Method of phasor addition, add $|-E_1|$, $\bar{I}_1R_1$ and $\bar{I}_1R_1$ to get the input voltage $\bar{V}_1$.

8. Similarly draw the secondary voltage equation (3.9) to get the correct magnitude of its resultant phasor E_2.

Fig. 3.17 (a), (b) and (c) shows the phasor diagrams of a practical transformer supplying power to a purely resistive load, an inductive load and a capacitive load respectively.

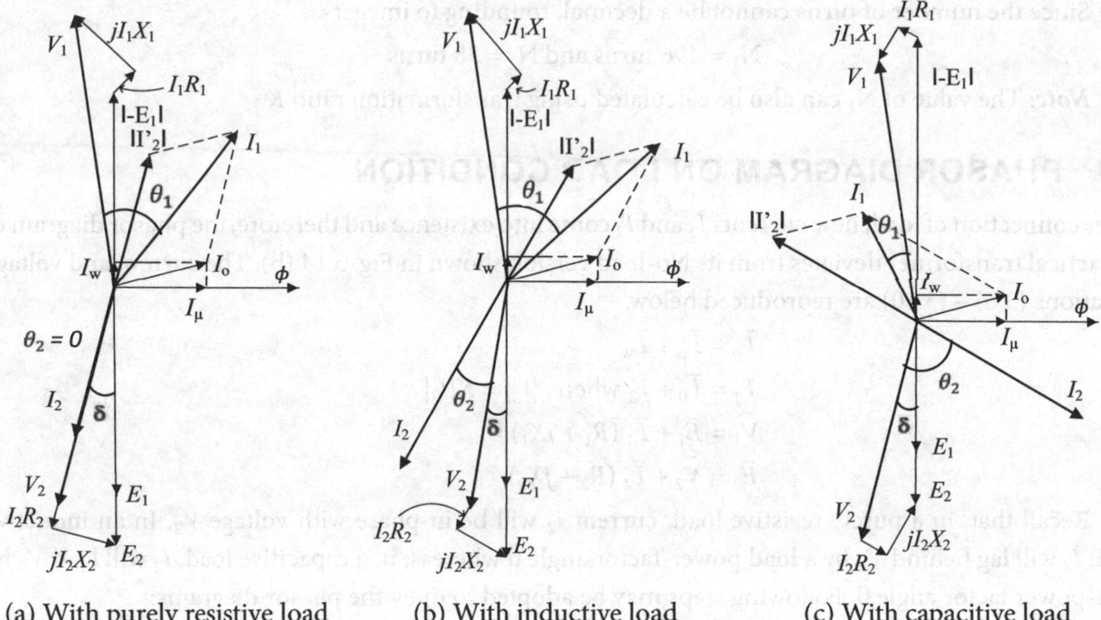

(a) With purely resistive load (b) With inductive load (c) With capacitive load

Fig. 3.17: Phasor diagram of a practical transformer on-load condition

3.10 EQUIVALENT CIRCUIT OF TRANSFORMER

Equivalent circuit of a transformer is an electrical model that can be used to predict the performance of the transformer at different loading conditions without actually loading it in real-time. The performance indicators may include No-load current, iron loss, copper loss, efficiency and voltage regulation. It is based on the same current and voltage equations (3.6)–(3.10) that we have used to draw the phasor diagrams in Section 3.9. Using equations (3.9) and (3.10), the practical transformer is represented by a combination of an ideal transformer, winding resistances and leakage reactances as shown in Fig. 3.16.

$$\bar{V}_1 = \bar{E}_1 + \bar{I}_1\,(R_1 + jX_1)$$

$$\bar{E}_2 = \bar{V}_2 + \bar{I}_2\,(R_2 + jX_2)$$

By equation (3.7), the input current $\bar{I}_1$ should be resolved into two components as,

$$\bar{I}_1 = \bar{I}_1 + \bar{I}'_2$$

Where, $\bar{I}_0$ is the No-load component of $\bar{I}_1$ while $\bar{I}'_2$ is the load component of $\bar{I}_1$. By equation (3.6), the No-load current $\bar{I}_0$ can be further resolved into two components as,

$$\bar{I}_0 = \bar{I}_\mu + \bar{I}'_w$$

Where, $\bar{I}_\mu$ is the magnetizing component (i.e. flux-producing component = inductive component) of $\bar{I}_0$ that produces the flux ϕ. This magnetizing effect of $\bar{I}_\mu$ can be represented by a *fictitious inductive reactance* X_o such that ($I_\mu^2 X_o$) will be the amount of reactive power (in volt-amperes-reactive) required to magnetize the core (i.e. to produce the flux ϕ).

$\bar{I}_w$ is the *working component or iron loss component of* $\bar{I}_0$ which decides the magnitude of iron loss in the core. Since iron loss is the loss of active power (in watts), $\bar{I}_w$ *is also called as active component of* $\bar{I}_0$ and its effect can be represented by a *fictitious resistor* R_o such that, ($I_w^2 X_o$) is the amount of active power that was consumed as iron loss in the core. With above additions, Fig. 3.16 can be modified to give a complete equivalent circuit of the transformer as shown in Fig. 3.18. ZL is the load impedance that can be connected across secondary.

$R_o = \frac{E_1}{I_w}$ = *No-load resistance* such that, iron loss = ($I_w^2 X_o$) watts

$X_o = \frac{E_1}{I_\mu}$ = *Magnetizing reactance* so that, magnetizing power = ($I_\mu^2 X_o$) VAR is required to produce ϕ.

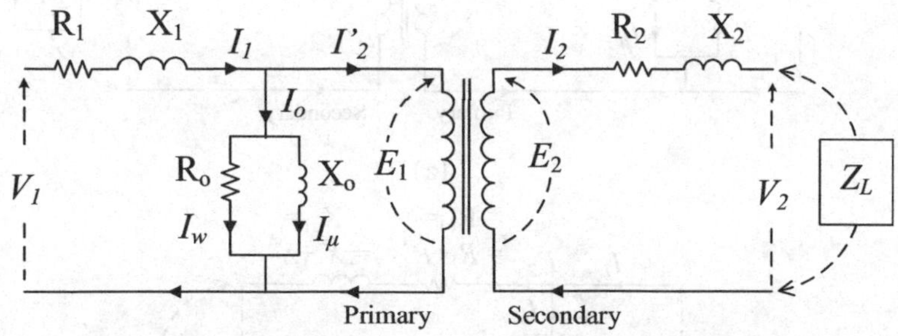

Fig. 3.18: Complete equivalent circuit of the transformer

3.10.1 Simplification of Equivalent Circuit

The complete equivalent circuit of Fig.3.18 has machine parameters on both sides, primary and secondary. It can be simplified by transferring all machine parameters on any one side.

A. *Simplified equivalent circuit as referred to primary side*

Step 1: The No-load branch (comprising of the parallel combination of R_0 and X_0 of Fig. 3.18 can be shifted towards source side with negligible loss of accuracy. This No-load branch is also called as *exciting branch.* The resulting form is shown in Fig. 3.19 (a).

Step 2: The secondary winding resistance R_2 and leakage reactance X_2 can be transferred on primary side as R_2' and X_2' respectively such that, $(I_2')^2 R_2' = I_2^2 R_2$ and $(I_2')^2 (X_2')^2 = I_2^2 X_2$.

- $\therefore R_2' = \left(\frac{I_2}{I_2'}\right)^2 R_2 = \frac{R_2}{K^2}$ = Secondary winding resistance as referred to primary side

- $X_2' = \left(\frac{I_2}{I_2'}\right)^2 X_2 = \frac{X_2}{K^2}$ = Secondary winding leakage reactance as referred to primary side

Since R_1 and R_2' are in series, they can be combined together. Similarly, the series combination of X_1 and X_2' can be combined together. The simplified equivalent circuit of transformer as referred to primary side is shown in Fig. 3.19 (b). Z_L is the load impedance that can be connected across secondary.

• Equivalent resistance of complete transformer as referred to primary side is,	$R_{o1} = (R_1 + R_2') = \left(R_1 + \frac{R_2}{K^2}\right)$	(3.11)
• Equivalent leakage reactance of complete transformer as referred to primary side is,	$X_{o1} = (X_1 + X_2') = \left(X_1 + \frac{X_2}{K^2}\right)$	(3.12)
• Equivalent impedance of complete transformer as referred to primary side is,	$Z_{o1} = (R_{o1} + jX_{o1})$	(3.13)

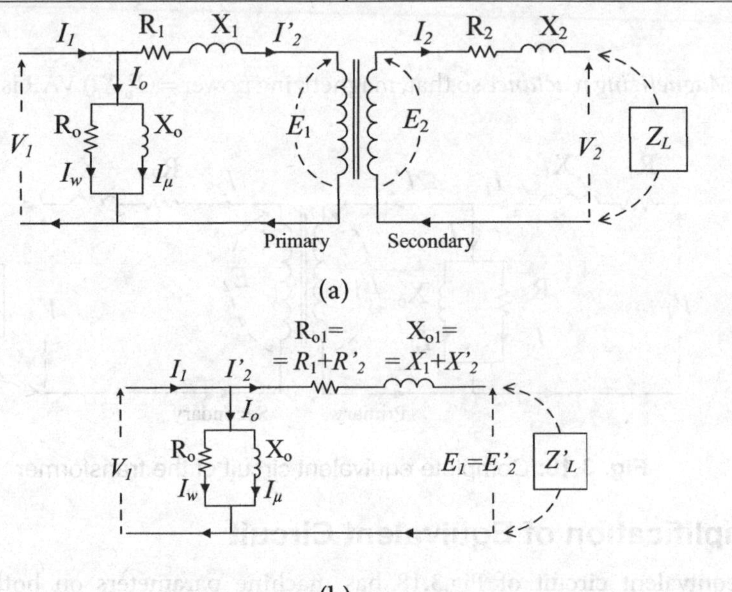

Fig. 3.19: Simplified equivalent circuit of transformer as referred to primary side

B. Simplified equivalent circuit as referred to secondary side

Conversely, the complete equivalent circuit of transformer (Fig. 3.18) can also be simplified by transferring all ohmic elements from primary side to the secondary side. The resulting elements as referred to secondary side will be:

• No-load resistance as referred to secondary side,	$R_o' = R_o K^2$	(3.14)
• Magnetizing reactance as referred to secondary side,	$X_o' = X_o K^2$	(3.15)
• Resistance of primary winding as referred to secondary side	$R_o' = R_1 K^2$	(3.16)
• Leakage reactance of primary winding as referred to secondary side	$X_o' = X_o K^2$	(3.17)

• Equivalent resistance of complete transformer as referred to secondary side is,	$R_{o2} = (R_2 + R_1') = (R_2 + R_1K^2)$	(3.18)
• Equivalent leakage reactance of complete transformer as referred to secondary side is,	$X_{o2} = (X_2 + X_1') = (X_2 + X_1K^2)$	(3.19)
• Equivalent impedance of complete transformer as referred to secondary side is,	$Z_{o2} = (R_{o2} + jX_{o2})$	(3.20)

Example 3.4: A 2400V/120V, 30kVA, 20Hz single phase transformer has a HV winding resistance of 0.1Ω and leakage reactance of 0.22Ω. The LV winding resistance is 0.035Ω and the leakage reactance is 0.012Ω. Find the equivalent winding resistance, reactance and impedance referred to the (i) HV side and (ii) LV side.

Data:

Rated voltages are: V_1 = 2400V and V_2 = 120V, Rated capacity (i.e. Full load capacity) = 30kVA, R_1 = 0.1Ω, X_1 = 0.22Ω, R_2 = 0.035Ω, X_2 = 0.012Ω,

(i) R_{o1} = ?, X_{o1} = ? (ii) R_{o2} = ?, X_{o2} = ?

Solution:

From the given voltage ratings of 2400V/120V, the given transformer is a step-down transformer because the higher voltage is written first. It means that, HV winding (2400V) will be the primary side and the LV winding (120V) will form the secondary side. Let the primary and secondary quantities be denoted by suffix '1' and '2' respectively.

$$\text{Transformation ratio } K = \frac{V_2}{V_1} = \frac{120}{2400} = \frac{1}{20}$$

The equivalent circuit parameters as referred to primary side (HV side) are:

$$R_{01} = \left(R_1 + R_2'\right) = \left(R_1 + \frac{R_2}{K^2}\right) = \left(0.1 + 0.035 \times 20^2\right) = 14.1\Omega$$

$$X_{01} = \left(X_1 + X_2'\right) = \left(X_1 + \frac{X_2}{K^2}\right) = \left(0.22 + 0.012 \times 20^2\right) = 5.02\Omega$$

$$Z_{01} = \sqrt{R_{01}^2 + X_{01}^2} = \sqrt{14.1^2 + 5.02^2} = 15\Omega$$

The equivalent circuit parameters as referred to secondary side (LV side) are:

$$R_{02} = \left(R_2 + R_1'\right) = \left(R_2 + R_1K^2\right) = \left(0.035 + 0.1 \times \frac{1}{20^2}\right) = 0.03525\Omega$$

$$X_{02} = \left(X_2 + X_1'\right) = \left(X_2 + X_1K^2\right) = \left(0.012 + 0.22 \times \frac{1}{20^2}\right) = 0.01255\Omega$$

$$Z_{02} = \sqrt{R_{02}^2 + X_{02}^2} = \sqrt{0.03525^2 + 0.01255^2} = 0.0374\Omega$$

Note: Once, R_{01}, X_{01} and Z_{01} are calculated, the values of R_{02}, X_{02} and Z_{02} can also be obtained directly as,

$$R_{02} = \left(R_{01}K^2\right) = \left(14.1 \times \frac{1}{20^2}\right) = 0.03525\Omega$$

$$X_{02} = \left(X_{01}K^2\right) = \left(5.02 \times \frac{1}{20^2}\right) = 0.01255\Omega$$

3.11 VOLTAGE REGULATION

With fixed applied voltage V_1 connected across primary winding, if the magnitude or power-factor of load connected on secondary side is changed than, a corresponding change in the secondary terminal voltage V_2 is noted. This variation in V_2 takes place due to the change in voltage drops across winding resistances and leakage reactances and the phenomenon is referred as '*Voltage Regulation.*'

Voltage regulation of a transformer is defined as the percentage change in secondary terminal voltage when the load at a given power-factor is changed from zero to Full-load value or vice-versa provided that, the applied voltage V_1 is held at constant RMS value.

Ideally, there should not be any change in the secondary terminal voltage even if changes in load has taken place. In other words, ideal value of voltage regulation is zero. Practically, a good transformer must keep its secondary terminal voltage as constant as possible under all conditions of variations of load. Hence, the voltage regulation should be as less as possible.

The amount of change in secondary terminal voltage due to change in load depends on the magnitude as well as on power-factor of the load. If a resistive load connected across secondary is increased, it results in rise in currents but corresponding drop in secondary terminal voltage. If the load is inductive and is increased then, the drop in secondary terminal voltage is more than that of resistive loads. However, with rising capacitive loads, there is a corresponding rise in secondary terminal voltage. Consequently, the voltage regulation becomes negative. Fig. 3.20 (a) shows the change in V_2 with load for different types of load. The variation of percentage regulation with power factor is shown in Fig. 3.20 (b). Mathematically, regulation can be calculated from secondary as well as from primary side.

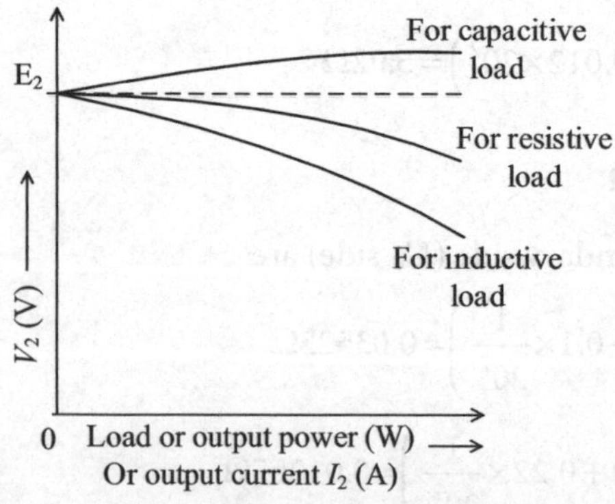

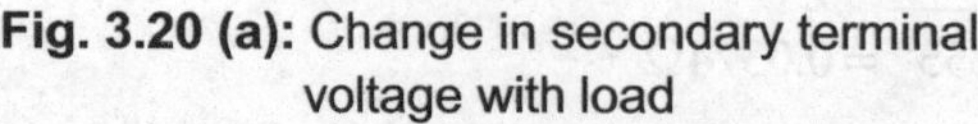

Fig. 3.20 (a): Change in secondary terminal voltage with load

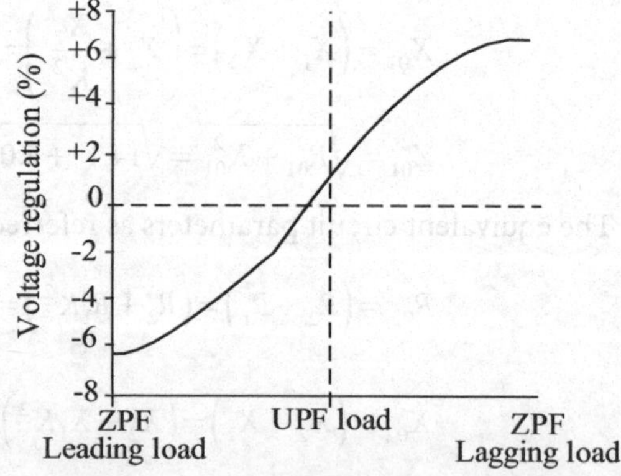

Fig. 3.20 (b): Change in voltage regulation with load power factor

3.11.1 Calculating Voltage Regulation from Secondary Side

If E_2 is the secondary terminal voltage at No-load (it also means the mutually induced *emf* in secondary winding) and V_2 is the secondary terminal voltage at Full-load (i.e. rated secondary voltage) than, the percentage regulation can be calculated as,

$$\%\text{ Voltage Regulation 'Up'} = \frac{E_2 - V_2}{V_2} \times 100$$

$$\%\text{ Voltage Regulation 'Down'} = \frac{E_2 - V_2}{E_2} \times 100 \tag{3.21}$$

Note that unless stated, the word '*Regulation*' should always be taken as *Regulation 'Down'*.

3.11.2 Calculating Voltage Regulation from Primary Side

Voltage regulation can also be calculated in terms of primary values. When the transformer is at no-load, V_1 is approximately equal to E_1 because at No-load the impedance drop I_oZ_1 is negligible. Using transformation ratio 'K', the corresponding values of secondary No-load voltage 'E_2' and secondary Full-load voltage 'V_2' can be obtained as referred to primary side as,

$$E_2' = \frac{E_2}{K} = E_1 \approx V_1$$

$$V_2' = \frac{V_2}{K}$$

$$\therefore \%\text{ Voltage Regulation} = \frac{V_1 - V_2'}{V_1} \times 100 \tag{3.22}$$

In equations (3.20) and (3.21), it was assumed that, the input voltage applied across primary winding is maintained at constant RMS value. As the transformer is loaded with resistive or inductive loads, the secondary terminal voltage falls and hence to restore it at rated value, the input voltage applied across primary winding should be increased. Hence,

Voltage regulation of a transformer is also defined as the percentage change in input voltage required to be applied across primary winding in order to maintain rated voltage at the terminals of secondary winding when the load at a given power-factor is changed from zero to Full-load value or vice-versa.

If the primary input voltage has to be changed from its rated value V_1 to V_1' than,

$$\therefore \%\text{ Voltage Regulation} = \frac{V_1' - V_1}{V_1} \times 100 \tag{3.23}$$

3.11.3 Calculating Approximate Voltage Drop

The numerator in equations (3.21)-(3.23) is the voltage drop from No-load to Full-load conditions. This voltage drop can be determined in terms of equivalent circuit parameters. Consider an approximate equivalent circuit as referred to secondary side. The No-load branch consisting of R_o and X_o is omitted with negligible loss of accuracy. This is shown in Fig. 3.21(a). The resulting voltage equation will be:

$$\bar{E}_2 = \bar{V}_2 + \bar{I}_2 (R_{02} + jX_{02})$$

Where, R_{o2} and X_{o2} are the equivalent resistance and leakage reactance of transformer as referred to secondary. The phasor diagram of this circuit supplying lagging power-factor load (inductive) is shown in Fig. 3.21(b).

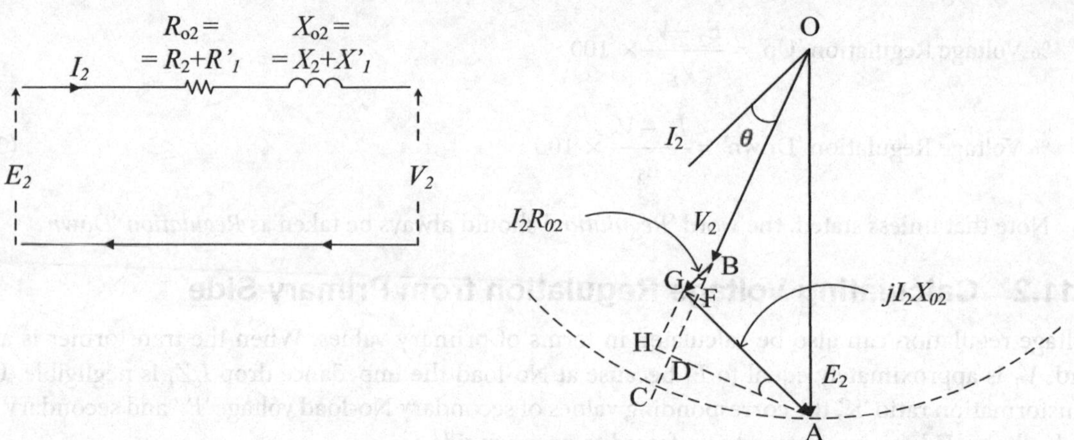

Fig. 3.21(a): Approximate equiv. circuit referred to secondary side

Fig. 3.21(b): Phasor diagram for equiv. circuit in Fig. 3.21(a)

On this phasor diagram, draw an arc of radius 'OA' ($= |E_2|$) considering 'O' as centre. Extrapolate the line 'OB' ($= |V_2|$) to cut this arc at some point 'C'. $\therefore l(OA) = l(OC)$. Now draw a line from 'A' perpendicular to 'OC'. Let them intersect at 'D'. From this diagram, the voltage drop ($E_2 - V_2$) will be approximately equal to:

$(E_2 - V_2) = [l(OA) - l(OB)] = [l(OC) - l(OB)] \approx [l(OD) - l(OB)]$

Construct ΔBGF and ΔAGH as shown. From these triangles, $[l(OD) - l(OB)] = [l(BF) + l(GH)]$

$\angle GBF = \angle GAH = \theta°$

$$l(BF) = I_2 R_{o2} \cos\theta \text{ and,}$$

$$l(GH) = I_2 X_{o2} \sin\theta$$

$$\therefore \quad (E_2 - V_2) = [l(BF) + l(GH)] = I_2 R_{o2} \cos\theta + I_2 X_{o2} \sin\theta$$

$$\therefore \quad (E_2 - V_2) = I_2 (R_{o2} \cos\theta + X_{o2} \sin\theta)$$

Similarly, for leading power-factor load (capacitive), the expression for voltage drop will be,

$$(E_2 - V_2) = I_2 (R_{o2} \cos\theta + X_{o2} \sin\theta)$$

Therefore in general, the voltage drop on secondary side is given as,

$$(E_2 - V_2) = I_2 (R_{o2} \cos\theta \pm X_{o2} \sin\theta) \tag{3.24}$$

And voltage regulation can be calculated as,

$$\%\text{ Voltage Regulation} = \frac{I_2\left(R_{o2} \cos\cos\theta \pm X_{o2} \sin\sin\theta\right)}{E_2} \times 100 \tag{3.25}$$

The approximate voltage drop of equation (3.23) can also be expressed as referred to primary.

$$\text{\% Voltage Regulation} = \frac{I_1 (R_{o1} \cos\cos\theta \pm X_{o1} \sin\sin\theta)}{V_1} \times 100 \tag{3.26}$$

Where, cos θ = load power factor. (+) sign is applicable for inductive loads (lagging power factor) and (-) sign is applicable for capacitive loads (leading power factor). For resistive loads, cos θ = 1 and sin θ = 0. ∴ $(E_2 - V_2) = I_2 R_{o2}$. From above equations, the resistive and reactive voltage drops are:

$$v_r = \frac{I_2 R_{o2}}{E_2} \times 100 = \frac{I_1 R_{o1}}{V_1} \times 100 = \text{\% R}$$

$$v_x = \frac{I_2 X_{o2}}{E_2} \times 100 = \frac{I_1 X_{o1}}{V_1} \times 100 = \text{\% X}$$

$$\text{\% Voltage Regulation} = v_r \cos\theta \pm v_x \sin\theta \tag{3.27}$$

Note that, from above values of % equivalent resistance (% *R*) and % equivalent leakage reactance (% *X*), their true values can be calculated as,

$$R_{o1} = \frac{\%R \times V_1}{I_1 \times 100} \quad ; \quad R_{o2} = \frac{\%R \times V_2}{I_2 \times 100}$$

$$X_{o1} = \frac{\%X \times V_1}{I_1 \times 100} \quad ; \quad X_{o2} = \frac{\%X \times V_2}{I_2 \times 100}$$

3.12 LOSSES AND EFFICIENCY

In a practical transformer, a certain fraction of input power is wasted in the form of heat and consequently, the output power of the transformer is less than its input power. This power loss is a loss of active power and can be classified in two categories (i) *Iron-loss* (*Pi*) and (ii) Copper-loss (*P*cu). As the names imply, iron-loss takes place in the magnetic core and therefore is also called as *Core Loss* or *No-load loss* or *Open-circuit loss.* The copper loss takes place in primary and secondary windings that are made from copper and is also called as *Ohmic loss.*

$$\text{Total loss in a transformer} = \text{Iron loss } (P_i) + \text{Copper loss } (P_{cu}) \tag{3.28}$$

3.12.1 Iron Loss

In Section 3.3.1, it was informed that the iron-loss is of two types, (i) Hysteresis loss (P_h) and (ii) Eddy current loss (P_e). Hysteresis loss is produced because the magnetic core is continuously subjected to reversal of magnetization by the alternating magnetic flux ϕ. And to minimize this loss, it is necessary that a highly permeable material like CRGO silicon-steel alloy or amorphous alloy should be used to fabricate the core laminations. More details about these alloys are given in Section 3.4.

Eddy current loss is produced due to the induction of eddy currents in the core by alternating magnetic flux. To minimize this loss, the core is made from thin insulated stampings or laminations and such a core is said to be a *laminated core.* For more details, revisit Section 3.3.1.

$$\text{Iron loss } (P_i) = \text{Hysteresis loss } (P_h) + \text{Eddy current loss } (P_e) \tag{3.29}$$

Mathematically, these two types of iron loss are expressed as,

$$P_h = K_h f B^n_{max} \text{ W/kg}$$
$$P_e = K_e (B_{max} f t)^2 \text{ W/kg} \quad (3.30)$$

where, K_h is the proportionality constant dependent on the material characteristics and volume of magnetic core, is the peak flux density (in Wb/m²), f represents the frequency of reversal of magnetization in Hz (which is the same as supply frequency) t is the thickness of lamination (in m) and the exponent n ranges between 1.5 and 2.5 depending upon the magnetic properties of the core material. It is called as Steinmetz constant (n). K_e is the proportionality constant whose value depends upon the volume and resistivity of core material, and the units employed.

3.12.2 Why Iron Loss is also Called as Constant Loss?

From equation (3.4), the *emf* equation is,

$$E_1 \approx V_1 = 4.44\, \phi_m f N_1 \text{ volts} = 4.44\, B_{max} A_i f N_1$$

$$\therefore B_{max} = \frac{V_1}{4.44 A_i f N_1}$$

Thus, maximum value of flux density B_{max} or flux ϕ in the core is directly proportional to the ratio of input voltage and input frequency f. B_{max} can be held constant by keeping the ratio $\left(\frac{V_1}{f}\right)$ constant.

Therefore from Equation 3.30,

$$P_h = K_h f B^n_{max} \propto f \quad (3.31)$$

In equations (3.31) the winding currents (I_1, I_2 and I'_2) are not present so it means that, both Hysteresis loss and Eddy current loss are independent of any changes in the load. However, they depend on input voltage V_1 and input frequency f.

Hence, for a constant input RMS voltage and frequency, the iron loss (P_i) is independent of load and its value remains same from No-load to Full-load. In other words, iron loss is a Constant loss. This can also be justified from Section 3.7 where we have concluded that, in a transformer, the net flux in the magnetic core remains constant from No-load to Full-load conditions. Hence, the iron loss that depends on flux also remains constant.

The value of iron loss can be determined experimentally by conducing Open-circuit test on the transformer (discussed ahead).

3.12.3 Copper Loss

The loss of active power that takes place in copper windings is called as copper loss. It is due to the ohmic resistances (R_1 and R_2) of the two windings. The total copper loss in a transformer is given by,

$$P_{cu} = I_1^2 R_1 + I_2^2 R_2 \text{ Watts}$$
$$= I_1^2 R_{o1} \text{ Watts}$$
$$= I_2^2 R_{o2} \text{ Watts} \quad (3.32)$$

From above equations, it is clear that the copper loss varies as square of current. Since the winding currents are directly proportional to load, it means that copper loss will vary with any changes in the magnitude of load. Hence *it is also called as Variable loss.* If $P_{cu(FL)}$ is the copper loss at Full load than at

Half Full-load it will be $\left(\frac{1}{2}\right)^2 P_{cu(FL)}$; at one-third Full-load it will be $\left(\frac{1}{3}\right)^2 P_{cu(FL)}$ and so on.

At No-load, it will be very small because at No-load, the primary current I_o is very small and the secondary current I_2 is absent. From equation (3.27), the winding resistance of complete transformer is given by,

$$\%R = v_r = \frac{I_1 R_{o1}}{V_1} \times 100\ \%$$

Similarly, if we compute it in terms of secondary referred quantities than,

$$\%R = v_r = \frac{I_2 R_{o2}}{V_2} \times 100\ \%$$

Multiply and divide the right hand side of above equations by their respective currents,

$$\text{Copper loss} = \frac{I_1^2 R_{o1}}{V_1 I_1} \times 100\% = \frac{I_2^2 R_{o2}}{V_2 I_2} \times 100\ \% \tag{3.33}$$

Note that, using the above equations, the value of equivalent resistance can be obtained as,

$$R_{o1} = \frac{\%\text{Copper loss} \times V_1}{I_1 \times 100};\ R_{o2} = \frac{\%\text{Copper loss} \times V_2}{I_2 \times 100}$$

Experimentally, the value of copper loss can be measured by conducing *Short-circuit test* on the transformer (discussed ahead).

3.12.4 Efficiency (η)

Efficiency of a transformer is defined as the ratio of active output power to active input power. Hence, it is also referred as *Power efficiency* or *Commercial efficiency.* Since efficiency is based on active power (in watts) and not on the apparent power (in volt-amperes), it depends on power-factor (cos ϕ°). For a given volt-ampere load, efficiency is highest at unity power-factor.

$$\%\ \eta = \frac{\text{Output power(W)}}{\text{Input power(W)}} \times 100 = \frac{S \cos\cos\theta}{S \cos\cos\theta + \text{Losses}} \times 100$$

$$\%\ \eta = \frac{\text{Output power}}{\text{Output power} + P_i + P_{cu}} \times 100 = \frac{\text{Input power} - P_i - P_{cu}}{\text{Input power}} \times 100 \tag{3.34}$$

Where,

P_i and P_{cu} = iron loss and copper loss respectively,

cos θ = load power-factor and

$S = V_2 I_2$ = apparent power in voltamperes.

Experimentally, efficiency of a transformer can be determined by performing any of the following tests: (a) Direct loading test and (b) Open-circuit test and short-circuit test, (c) Back to back or Sumpner's test.

3.12.5 Condition for Maximum Efficiency

The variation of efficiency with output power and load power-factor cos θ is shown in Fig. 3.22. At No-load, efficiency is zero. With rise in the magnitude of load, output current and power increases and so there is a rise in efficiency also. At some value of load, the peak value of efficiency is reached. For any further rise in load, the rise in P_{cu} is more significant than the rise in output power and therefore, the efficiency falls. Let us now derive the condition which when fulfilled should yield maximum efficiency. From equation (3.34),

$$\eta = \frac{\text{Input power} - P_i - P_{cu}}{\text{Input power}}$$

i.e.

$$\eta = \frac{V_1 I_1 \cos\cos\theta_1 - P_i - I_1^2 R_{01}}{V_1 I_1 \cos\cos\theta_1}$$

$$= 1 - \frac{P_i}{V_1 I_1 \cos\cos\theta_1} - \frac{I_1^2 R_{01}}{V_1 I_1 \cos\cos\theta_1}$$

Where, is the input power factor.

For maximum value of η; $\frac{d\eta}{dI_1} = 0$

$$\frac{d\eta}{dI_1} = 0 + \frac{P_i}{V_1 I_1^2 \cos\cos\theta_1} - \frac{R_{01}}{V_1 \cos\cos\theta_1} = 0$$

∴

$$P_i = I_1^2 R_{01} \tag{3.35}$$

Fig. 3.22: Change in efficiency with output power

The above equation is the condition for maximum efficiency. If means that, to get maximum efficiency, the magnitude of load applied (or output power) should be such that, at that value of load, the iron loss and copper loss in the transformer are equal. The load or output power in volt-amperes ($S_{max\,\eta}$) which gives maximum efficiency is given by,

$$S_{max\ \eta} = S_{FL} \times \sqrt{\frac{P_i}{P_{cu(FL)}}} \quad (3.36)$$

S_{FL} = *Rated capacity in voltamperes,* Pi = rated iron loss and $P_{cu(FL)}$ = Full-load copper loss.

3.13 ALL-DAY EFFICIENCY OR ENERGY EFFICIENCY

Transformers used in distribution networks supply power to commercial and residential consumers. Their primaries are energized throughout twenty-four hours. The load connected on the secondary side however keeps on changing between very low values or No-load and Full-load. Consequently, the output power and the copper loss also changes with load but the iron loss occurs throughout the day. In Section 3.12.2, we have studied that, iron loss does not depend on load and it remains constant between No-load and Full-load conditions. However as seen in Section 3.12.2, actual iron loss depends on input voltage V_1. Any change in V_1 will change the iron-loss. Distribution transformers are therefore designed to have low value of iron loss.

Equation (3.34) cannot give correct information about the efficiency of distribution transformers or any other transformer which has its primary energized all the time but the load (i.e. output power) on its secondary side is not constant. Therefore in place of power efficiency, the concept of *all-day efficiency* also called as *energy efficiency* is introduced which is measured for a period of 24 hours.

$$\eta_{all-day} = \frac{\text{Output energy in Watthours}}{\text{Input energy in Watthours}} \text{ (For 24 hours)} \quad (3.37)$$

Example 3.5: A 50 kVA, 11000/440V, 50 Hz single phase transformer has primary winding resistance of 7.8Ω and secondary winding resistance of 0.0085Ω. Calculate the equivalent resistances as referred to primary and secondary sides and from them calculate the Full-load copper loss.

Data:

Rated voltages are: V_1 =11000V and V_2 = 440V, Rated capacity (i.e. Full load capacity) = 50kVA, f = 50 Hz, R_1 = 7.8Ω, R_2 = 0.0085Ω, (i) R_{o1} = ?, (ii) R_{o2} = ?, $P_{cu(FL)}$ = ?

Solution:

As explained in Example 3.4 above, the HV winding of 11000V will be the primary side and the LV winding of 440V will be the secondary side. Let the primary and secondary quantities be denoted by suffix '1' and '2' respectively.

$$\text{Transformation ratio } K = \frac{V_2}{V_1} = \frac{440}{11000} = 0.04$$

The equivalent resistance as referred to primary side (HV side) is:

$$R_{01} = \left(R_1 + R_2'\right) = \left(R_1 + \frac{R_2}{K^2}\right) = \left(7.8 + \frac{0.0085}{0.04^2}\right) = 13.11\Omega$$

The equivalent resistance as referred to secondary side (LV side) is:

$R_{02} = (R_2 + R_1') = (R_2 + R_1 K^2) = (0.0085 + 7.8 \times 0.04^2) = 0.02098\Omega$

The copper loss that takes place in both windings can be determined by using any of the following methods:

Method 1: $P_{cu} = I_1^2 R_1 + I_2^2 R_2$ watts

Method 2: $P_{cu} = I_1^2 R_{01}$ watts

Method 3: $P_{cu} = I_2^2 R_{02}$ watts

Note: The values of primary and secondary currents I_1 and I_2 should correspond to the same load condition at which the copper loss is to be calculated.

Example: To find copper loss at Full-load, the currents should be Full-load currents. If copper loss at Half-Full load is to be determined than the currents should be Half-Full-load currents.

Full load currents (also called as rated currents) are calculated as,

$$I_{1(FL)} = \frac{\text{Full load capacity of transformer in voltamperes}}{\text{rated primary voltage in volts}} = \frac{50000}{11000} = 4.545A$$

$$I_{2(FL)} = \frac{\text{Full load capacity of transformer in voltamperes}}{\text{rated secondary voltage in volts}} = \frac{50000}{440} = 113.64A$$

$$\therefore \quad P_{cu(FL)} = I_{1(FL)}^2 R_{01} = 4.545^2 \times 13.11 = 270.91 \text{ W}$$

Note: Students are advised to calculate $P_{cu(FL)}$ by the other two methods also and verify that the result for practice.

Example 3.6: A 25 kVA, 2200V to 220V single phase transformer has primary resistance of 1 Ω and secondary resistance of 0.01Ω. Find the equivalent resistance as referred to secondary and the Full-load efficiency at 0.8 lagging power factor if the iron loss of the transformer is 80% of the Full-load copper loss.

Data:

Rated capacity (i.e. Full load capacity) = 25kVA, Rated voltages are: V_1 =2200V and V_2 = 220V, R_1 = 1 Ω, R_2 = 0.01Ω, (i) R_{o2} = ?, (ii) $\eta_{(FL,0.8)}$ = ?

Solution:

$$\text{Transformation ratio } K = \frac{V_2}{V_1} = \frac{220}{2200} = 0.1$$

The equivalent resistance as referred to secondary side (LV side) is:

$$R_{02} = (R_2 + R_1') = (R_2 + R_1 K^2) = (0.01 + 1 \times 0.1^2) = 0.02\Omega$$

Full load secondary current (also called as rated secondary current) is,

$$I_{2(FL)} = \frac{\text{Full load capacity of transformer in voltamperes}}{\text{rated secondary voltage in volts}} = \frac{25000}{220} = 113.64A$$

$$\therefore \quad P_{cu(FL)} = I_{2(FL)}^2 R_{02} = 113.64^2 \times 0.02 = 258.264 \text{ W}$$

It is given that, $P_i = 0.8\, P_{cu(FL)}$

$$\therefore \quad P_i = 0.8 \times 258.264 = 206.611\text{W}$$

Efficiency is given as,

$$\eta = \frac{\text{Output power(W)}}{\text{Input power(W)}} = \frac{\text{VA capacity} \times \text{power factor}}{\text{VA capacity} \times \text{power factor} + \text{iron loss} + \text{copper loss}}$$

At Full load and 0.8 power factor, it will be,

$$\eta_{(FL,0.8)} = \frac{\text{Full load VA capacity} \times 0.8}{\text{Full load VA capacity} \times 0.8 + P_i + P_{cu(FL)}}$$

$$\therefore \eta_{(FL,0.8)} = \frac{25000 \times 0.8}{25000 \times 0.8 + 206.611 + 258.264} \times 100 = 97.72\%$$

Example 3.7: A 100 kVA single phase transformer has a maximum efficiency of 98% at Full load and a power factor of 0.8 lagging. The daily load cycle of the transformer is as follows:

- 20kW load at 0.5 lagging power factor ….. 12 hours
- 80kW load at 0.8 lagging power factor ….. 06 hours
- 100kW load at 0.9 lagging power factor ….. 06 hours

Calculate the all-day efficiency of transformer.

Data:

Full load capacity = 100 kVA, η_{max} is at Full-load condition, $\eta_{(max\,(FL,0.8)}$ = 98%

Solution:

Transformer output power at Full load is, $=100 \times 10^3 \times 0.8 = 80000$ W = 80kW

$$\text{Since, efficiency} = \frac{\text{Output power(W)}}{\text{Output power (W)} + \text{losses(W)}}$$

$\therefore$ Total Loss at FL and 0.8 power factor = 1.632 kW = $P_i + P_{(cu(FL)}$

It is given that the η_{max} happens at full-load condition. Also, it is known that at η_{max}, iron loss = copper loss. Therefore, in this example,

$$\therefore P_i = P_{cu(FL)} = \frac{1.632 \times 10^3}{2} = 816.33\text{W}$$

Above value corresponds to Full-load of 100 kVA. $\therefore P_{cu(100kVA)}$ = 816.33W

To obtain the copper loss during the given three load conditions, convert the corresponding value of output power from kW to kVA.

- The 20 kW load atcos θ = 0.5 will correspond to 20/(0.5 = 40 kVA)
- The 80 kW load atcos θ = 0.8 will correspond to 80/(0.8 = 100 kVA)
- The 100 kW load atcos θ = 0.9 will correspond to 100/(0.9 = 111.11 kVA)

The energy lost in copper loss for given durations in different intervals will be,

- During 20 kW, 0.5 pf (i.e. 40 kVA) load for 12 hours,

$$E_{cu(20kW)} = 816.33\text{W} \times \left(\frac{40\text{kVA}}{100\text{kVA}}\right)^2 \times 12 = 1567.4\text{Wh}$$

- During 80 kW, 0.8 pf (i.e. 100 kVA) load for 6 hours,

$$E_{cu(20kW)} = 816.33\text{W} \times \left(\frac{100\text{kVA}}{100\text{kVA}}\right)^2 \times 6 = 4899.96\text{Wh}$$

- During 100 kW, 0.9 pf (i.e. 111.11 kVA) load for 6 hours,

$$E_{cu(20kW)} = 816.33\text{W} \times \left(\frac{111.11\text{kVA}}{100\text{kVA}}\right)^2 \times 6 = 6046.76\text{Wh}$$

∴ Energy lost in Copper loss during 24 hours is,

$$E_{cu} = 1567.4 + 4899.96 + 6046.76 = 12514.12 \text{ Wh} = 12.514 \text{ kWh}$$

Energy lost in Iron loss during 24 hours is,

$$E_i = 816.33 \times 24 = 19591.92 \text{ Wh} = 19.591 \text{ kWh}$$

∴ Total energy lost in 24 hours is $E_{loss} = E_{cu} + E_i = 32.105$ kWh

Output energy in 24 hours = (20 × 12 + 80 × 6 + 100 × 6) = 1320 kWh

$$\therefore \quad \text{All-day efficiency} = \frac{\text{Output energy during 24hours}}{\text{Output energy during 24hours} + \text{losses during 24hours}}$$

$$= \frac{1320}{1320 + 32.105} \times 100 = 97.62\%$$

3.14 SIGNIFICANCE OF TRANSFORMER RATINGS

The name plate of a transformer is fixed on its outer body (transformer tank). It bears the apparent power rating (in kVA or MVA), primary and secondary voltage ratings, frequency, polarity, maximum ambient temperature, cooling method, percentage impedance, name of the manufacturer, year of manufacturing Indian Standard (IS) or International Standard (IEC) Number and other details. In addition, a three-phase transformer name-plate also includes winding connection diagram, vector group, type of insulating oil, and weight.

The kVA or MVA rating represents maximum permissible capacity of that transformer up to which it can be loaded to operate safely. It is interesting to note that this rating is expressed in volt-amperes (VA) and not in watts (W) although the load or output power is measured in watts which is given by,

$$P = VI \cos\theta \text{ watts}$$

Let it be known that, the maximum loading capacity of any electrical device is decided by the maximum amount of heat that can be produced and safely dissipated by that device without causing any harm to itself. It depends upon the class of insulating material employed in the device and the cooling arrangement made available to dissipate the heat to the surrounding atmosphere.

In Section 3.12, we have seen that in a transformer, heat is produced due to two types of power loss: (i) iron loss and (ii) copper loss. In the same section we have also studied that, iron loss is dependent on voltage and the copper loss is dependent on winding currents. None of these two losses are directly dependent on power factor cos θ. This is one of the prime reasons why the power rating of a transformer is provided in kVA or MVA and not in kW or MW.

Secondly, the power rating in kVA or MVA simplifies the calculation of rated currents in the two windings without considering the load power-factor. However, for the calculation of actual currents, this information about load is necessary. The given kVA or MVA rating mentioned on the name plate is the Full-load power rating and is applicable to both primary and secondary sides. If 'kVA' is the given power rating and $V1$, $V2$ are the given voltage ratings of a single-phase transformer than the Full-load currents (also called as rated currents) are calculated as,

$$I_{1(FL)} = \frac{kVA \times 10^3}{V_1} \text{ Amperes}$$

$$I_{2(FL)} = \frac{kVA \times 10^3}{V_2} \text{ Amperes}$$

The maximum ambient temperature represents the allowable surrounding temperature limit. The percentage impedance indicates the change in secondary terminal voltage at full-load. If the secondary No-load voltage is 433V and percentage impedance is 6.5% than at Full-load (100% loading), the secondary terminal voltage will be volts. [433 – (0.065 × 433)] = 404.85

3.15 DIRECT LOADING TEST

Direct loading test on a transformer is one of the basic laboratory experiment performed to determine the following:

(a) Total loss and efficiency

(b) Change in terminal voltage and voltage regulation

As the name implies, the transformer is made to supply power to an actual variable load that is connected across its secondary winding. On both sides of the transformer, voltage, current and power are measured with the help of measuring instruments i.e. voltmeters, ammeters and wattmeters. A single-phase *ac* source is connected across the primary winding. Fig. 3.23 shows the schematic to conduct this test. The variable load may be of resistive, R-L or R-C type.

After connecting the circuit, the input source is switched ON with load maintained in switched-OFF condition. The current, voltage and power readings V_1, I_1, W_1 and V_2, I_2, W_2 on primary and secondary sides respectively are noted. Obviously, at No-load, I_2 and W_2 will be zero. Let the secondary terminal voltage during this first set of reading at No-load be designated as $V_{2(NL)} = E_2$. The load is then increased in steps upto its Full-load value and all the above quantities are measured every time. Let the input power, output power and secondary terminal voltage at Full-load be denoted by $W_{1(FL)}$, $W_{2(FL)}$ and $V_{2(FL)}$ respectively.

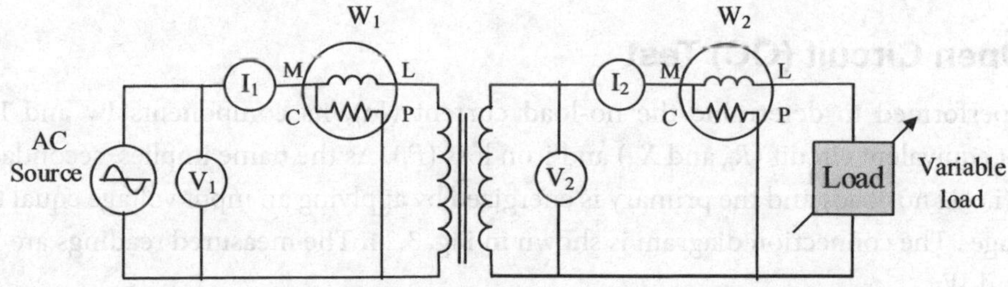

Fig. 3.23: Direct loading test on single-phase transformer

The Full-load efficiency is then calculated as,

$$\% \, \eta_{(FL)} = \frac{W_{2(FL)}}{W_{1(FL)}} \times 100 \quad (3.38)$$

The percentage voltage regulation at full load is,

$$\% \text{ Regulation} = \frac{V_{2(NL)} - V_{2(FL)}}{V_{2(NL)}} \times 100 \tag{3.39}$$

Total loss at Full-load is, $= W_{1(FL)} - W_{2(FL)}$

Voltage drop at Full-load is, $= V_{2(NL)} - V_{2(FL)}$

Due to the direct loading of transformer by a real load, the accuracy of results of this test is high. However, significant amount of electrical energy is consumed which is directly proportional to the transformer rating and the time required to complete the test. Secondly, upon loading, V_1 drops. It has to be maintained at rated value using an autotransformer. Hence, this test is preferred only on small size transformers.

3.16 DETERMINATION OF EQUIVALENT CIRCUIT PARAMETERS

The equivalent circuit parameters of a transformer are determined by performing *Open-circuit (OC) and Short-circuit (SC) tests.* The parameters so obtained can be used to predetermine the efficiency and regulation at any load condition. In Section 3.15, we have studied the direct loading method to determine efficiency and regulation. Let us start with the basic differences between these two methods. The former method is direct and makes use of an actual load whereas, the *OC-SC test is an indirect loading method* and is analytical. It does not require any actual load to be connected across secondary. Though the accuracy of direct loading method is slightly higher, the OC-SC tests offer following relative advantages.

- The OC-SC test gives no-load current (I_o), rated iron-loss (P_i), rated copper loss ($P_{cu(FL)}$) and all parameters of equivalent circuit in addition to efficiency and regulation.
- The OC-SC test is used to predict the performance of transformer at any desired magnitude and power-factor of load without actually using a real load.
- The direct loading method is lossy since it involves power consumption equal to the rating of transformer. While, the OC-SC test that is based on equivalent circuit consumes only a small fraction of its power rating. Hence it is cost-effective.

3.16.1 Open Circuit (OC) Test

This test is performed to determine the no-load current (Io), its components Iw and Iμ, no-load parameters of equivalent circuit (R_o and X_o) and iron loss (P_i). As the name implies, secondary winding is kept open (i.e. at no-load) and the primary is energized by applying an input voltage equal to the rated primary voltage . The connection diagram is shown in Fig. 3.24. The measured readings are V_o (= rated voltage), I_o and W_o.

Since, the secondary side is open-circuited, the secondary current $I_2 = 0$, $I'_2 = 0$ and therefore the output power is zero. Still the wattmeter on primary side indicates some amount of input power (W_o) drawn from the supply. Obviously, this input power is being utilized to produce power loss only. From equation (3.7), the input current will be equal to the no-load current I_o. From equation 3.6, has two components, the loss producing active component I_w and the flux producing magnetizing component I_μ. They can be determined from measured readings as follows.

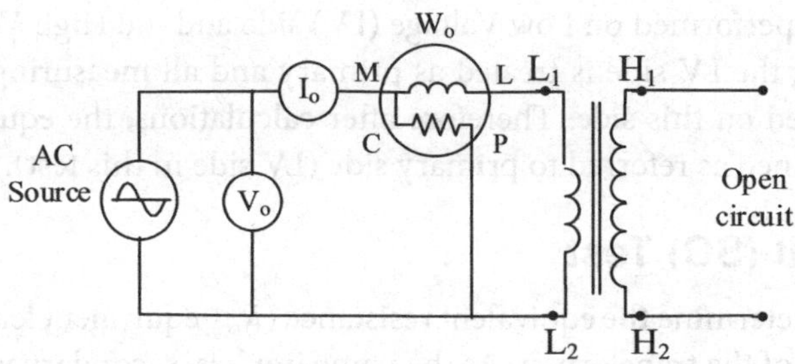

Fig. 3.24: Open-circuit test on a two-winding transformer

From the phasor diagram of practical transformer in Fig. 3.14(b) we have,

$$I_w = I_o \cos\theta_o$$
$$I_\mu = I_o \sin\theta_o \quad (3.40)$$

Where, the no-load power-factor is, $\cos\theta_o = \dfrac{W_o}{V_o I_o}$

From the simplified equivalent circuit of Fig. 3.19(b), the no-load resistance R_o and magnetizing reactance X_o will be,

$$R_o = \frac{V_o}{I_w} = \frac{V_o}{I_o \cos\cos\theta_o}$$

$$X_o = \frac{V_o}{I_\mu} = \frac{V_o}{I_o \sin\sin\theta_o} \quad (3.41)$$

During this test the load is zero therefore, the variable loss (i.e. copper loss) which is dependent on load is also zero. It means that, the input power (W_o) taken from supply is solely consumed to produce iron loss. Since, the applied input voltage (V_o) is equal to rated primary voltage, the iron loss will also be of rated value. Hence, the wattmeter reading (W_o) during open-circuit test is treated as rated iron-loss (P_i).

$$W_o = P_i \quad (3.42)$$

Actually during the open-circuit test, in addition to iron loss a small amount of copper-loss also takes place in primary winding. Since, the magnitude of I_o is very small (I_o is 1 - 5% of full-load current) the resulting copper loss is negligible as compared to rated iron loss. Hence, wattmeter reading (W_o) is considered as rated iron-loss (*Pi*). The equivalent circuit for OC test is shown in Fig. 3.25.

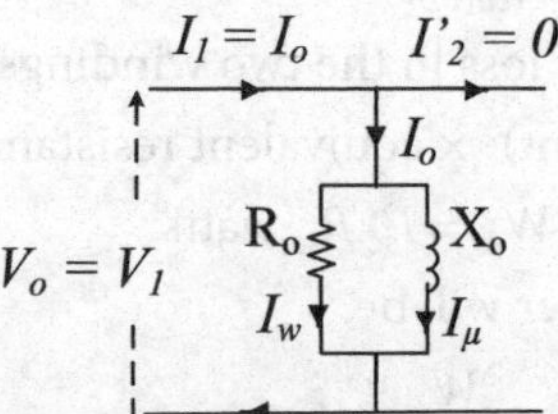

Fig. 3.25: Transformer equivalent circuit during open-circuit test

This test is normally performed on Low Voltage (LV) side and the High Voltage (HV) side is open-circuited. In other words, the LV side is treated as primary and all measuring instruments along with input source are connected on this side. Therefore after calculations, the equivalent circuit parameter values R_o and X_o are obtained as referred to primary side (LV side in this test).

3.16.2 Short Circuit (SC) Test

This test is performed to determine the equivalent resistance (R_{sc}), equivalent leakage reactance (X_{sc}), and rated copper loss ($P_{cu(FL)}$) of the transformer. As the name implies, secondary winding is short-circuited and the primary is energized by applying a *reduced input voltage* (V_{sc}) which will be just sufficient to pass *rated currents* (i.e. full-load currents) in the two windings. The connection diagram is shown in Fig. 3.26. The measured readings are V_{sc}, I_{sc} (= rated current), and W_{sc}.

The SC test is normally performed on High Voltage (HV) side and the Low Voltage (LV) side is short-circuited. In other words, all measuring instruments as well as the input source are connected on HV side. Therefore after calculations, the equivalent circuit parameter values R_{sc} and X_{sc} are obtained as referred to HV side (which will be the secondary side as per the convention adopted in Section 3.16.1: OC test).

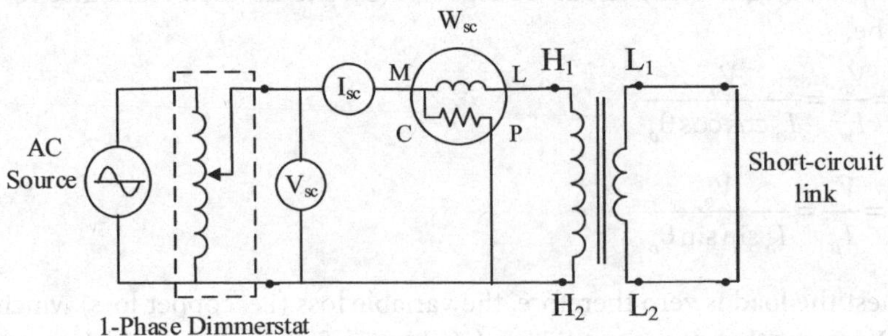

Fig. 3.26: Short-circuit test on a two-winding transformer

Since, the secondary side is short-circuited, the secondary terminal voltage and therefore the output power is zero. Still the wattmeter on primary side indicates some amount of input power (W_{sc}) drawn from the supply. Obviously, this input power is being utilized to produce power loss only. During SC test, both iron-loss and copper-loss in the two windings will take place. But the iron-loss is negligibly small because, the applied input voltage (V_{sc}) is very less. Whereas, due to the presence of rated magnitude currents in the two windings, the copper-loss has rated magnitude. Hence, the wattmeter reading (W_{sc}) during short-circuit test is treated as full-load copper-loss ($P_{cu(FL)}$).

$$W_{sc} = P_{(cu(FL)} \tag{3.43}$$

From equation (3.32), total copper loss in the two windings is defined as,

total copper loss = (winding current)^2 × equivalent resistance of transformer

$$P_{cu(FL)} = W_{sc} = I_{sc}^2 R_{sc} \text{ watts}$$

Equivalent resistance of transformer will be,

$$R_{sc} = \frac{W_{sc}}{I_{sc}^2} \ \Omega \tag{3.44}$$

The equivalent impedance of transformer will be,

$$Z_{sc} = \frac{V_{sc}}{I_{sc}} \ \Omega \tag{3.45}$$

∴ Equivalent leakage reactance of transformer will be,

$$X_{sc} = \sqrt{Z_{sc}^2 - R_{sc}^2} \ \Omega \tag{3.46}$$

Since, the HV side on which short-circuit test was performed is the secondary side (as per the convention adopted in Section 3.16.1: OC test),

∴ $R_{sc} = R_{o2}$ and $X_{sc} = X_{o2}$

Once, the equivalent circuit parameters and losses are obtained, efficiency and voltage regulation can be calculated using equations (3.25) and (3.34) for any load current I2 and power-factor . The equivalent circuit for SC test is shown in Fig. 3.27.

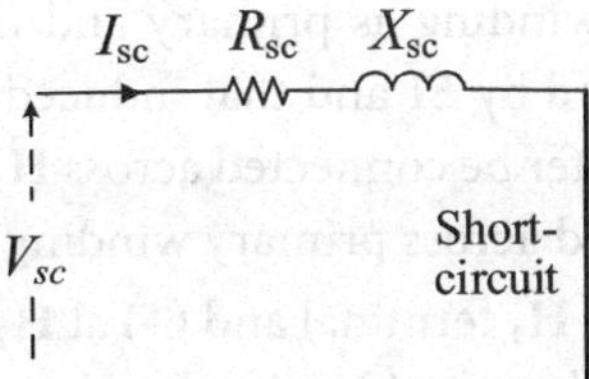

Fig. 3.27: Transformer equivalent circuit during short-circuit test

3.17 POLARITY MARKING

At any instant of time in each winding of a transformer, one terminal is always positive with respect to the other. There is a rise in voltage from negative terminal (lower potential terminal) to the positive terminal (higher potential terminal). When two or more transformers are to be connected in parallel for their joint operation, the terminals having similar polarity should be connected to each other otherwise, it may result into a short-circuit.

In Section 3.7 it was explained that by Lenz's law, the secondary flux is always in opposite direction to the flux produced by primary winding. At any instant of time, the polarity of secondary induced *emf* E_2 and the direction of induced current I_2 are then determined from the direction of by applying right hand thumb rule. It is worth noting that the direction in which the two windings are wound on the limb play an important role in deciding the instantaneous direction of E_2 and I_2. This is demonstrated by two examples shown in Fig. 3.28 (a) and (b). In both of them, the direction of primary winding and the polarities of input source V_1 are maintained same. However, the directions in which secondary winding is wound on the limb are different. After the application of right hand thumb rule, it will be observed that, the direction of I_2 and E_2 are different in the two examples. In other words, they have different polarities. Hence, before interconnecting the windings of different transformers for their parallel operation, it is essential to identify the similar polarity terminals of all windings.

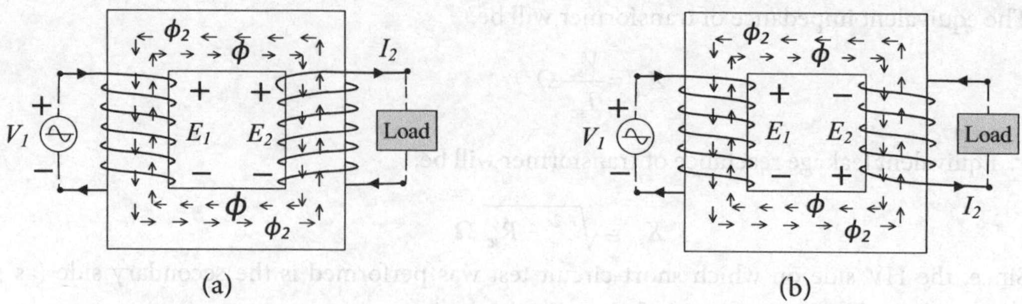

Fig. 3.28: Effect of change in winding direction on the direction of E_2 and I_2

Polarity test is performed to determine the terminals having the same instantaneous polarity. In a transformer, polarity means relative direction of induced *emfs* in the primary and secondary windings. There are two types of polarities: (i) *Subtractive* and (ii) *Additive.*

Consider a two-winding transformer having HV winding terminals $\mathbf{H_1}$, $\mathbf{H_2}$ and LV winding terminals $\mathbf{L_1}$, $\mathbf{L_2}$. If we consider HV winding as primary and LV winding as secondary than, the *emf* induced in HV winding will be denoted by $E1$ and that induced in LV winding by E_2. Let the terminals $\mathbf{H_2}$, $\mathbf{L_2}$ be joined together and a voltmeter be connected across $\mathbf{H_1}$, $\mathbf{L_1}$. Let the secondary winding remain open and an *ac* input voltage be applied across primary winding as shown in Fig. 3.29 (a) and (b).

Imagine arbitrary polarities (+) at $\mathbf{H_1}$ terminal and (–) at $\mathbf{H_2}$ terminal on the primary side. Observe the voltmeter reading from which we have to identify the corresponding polarities at terminals on the secondary side. For ease of understanding, assume that the transformer voltage rating is 2200V / 220V.

Subtractive polarity:

If the voltmeter reading = $(E_1 - E_2)$ = (2200 – 220) = 1980V than, the transformer is said to have subtractive polarities. From the unfolded view of transformer with subtractive polarity in Fig. 3.29 (a), it can be easily found out that on the secondary side, the polarity (+) must be at L_1 terminal and (–) at L_2 terminal. So, the terminals H_1 and L_1 have similar polarities. Similarly, the polarities at H_2 and L_2 are similar. For polarity marking, instead of assigning the (+) and (–) signs, usually the symbols like ●, ■, ▲ etc. are used to represent terminals having similar polarities.

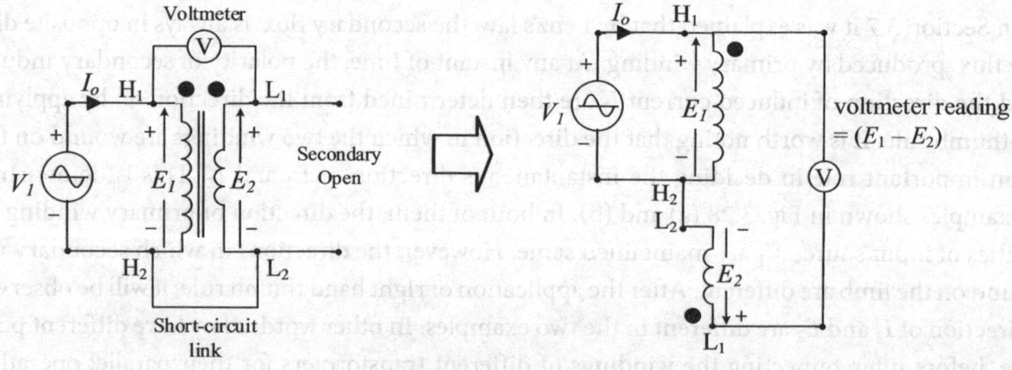

(a) Subtractive polarity

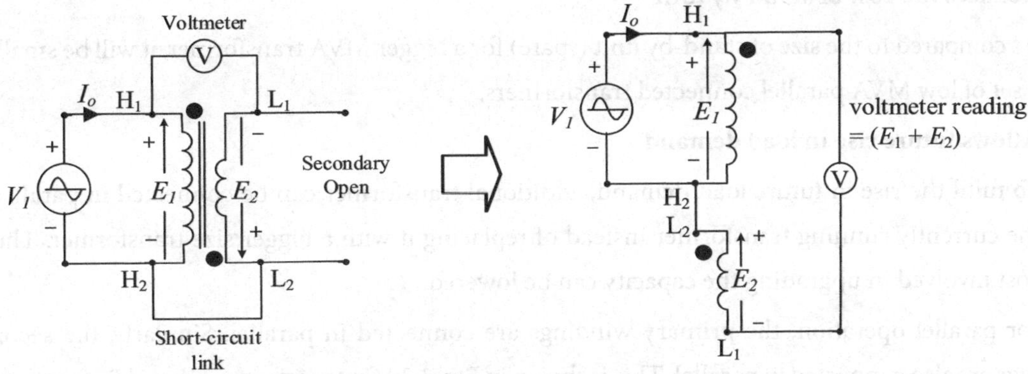

(b) Additive polarity

Fig. 3.29: Polarity test on single-phase transformer

Additive polarity:

If the voltmeter reading = $(E_1 + E_2) = (2200 + 220) = 2420V$ than, the transformer is said to have additive polarities. From the unfolded view of transformer with additive polarity shown in Fig. 3.29 (b), on the secondary side, the polarity (+) must be at L_2 terminal and (−) at L_1 terminal. So, the terminals H_1 and L_2 have similar polarities. Similarly, the polarities at H_2 and L_1 are similar. The AC (+) polarities are represented by symbol dot (•) on the diagrams.

3.18 PARALLEL OPERATION OF SINGLE PHASE TRANSFORMERS

In Section 3.1 we have seen that, for the transfer of electricity from generating station to load centres, power transformers and distribution transformers are necessary. The MVA capacity of these transformers should be sufficient enough to handle the power flow and peak load demand satisfactorily. In many cases, it is preferred to install and operate two or more transformers of lesser MVA rating in parallel instead of running a single transformer unit of higher MVA rating. Though multiple number of low MVA transformers connected in parallel will require more space and a bigger protection system against faults as compared to a single unit of higher MVA transformer, the former choice yields following overriding advantages.

- **Improves system reliability**

  In a set of transformers operating in parallel, if fault is developed on any one of them than it can be disconnected for repair work while others can continue and maintain the process of power transfer (at reduced level). On the contrary, if there is a single unit of bigger MVA transformer and if it has a fault than, when removed from the circuit, total power supply to the load will get disconnected.

- **Improves system economy and efficiency**

  When operating in parallel, power transformers can be turned ON and OFF depending upon the availability of load. During light load periods, some transformers can be disconnected and others can supply the total load. This will nullify the iron-loss in disconnected transformer thus improving the overall efficiency of the system. Also, the cost incurred for power wastage in iron-loss will be saved thus enhancing the system economy.

- **Reduces the cost of stand-by unit**

  As compared to the size of stand-by unit (spare) for a bigger MVA transformer, it will be smaller for a set of low MVA parallel connected transformers.

- **Allows future rise in load demand**

  To fulfil the rise in future load demand, additional transformer can be connected in parallel with the currently running transformer instead of replacing it with a bigger size transformer. Thus the cost involved in upgrading the capacity can be lowered.

For parallel operation, the primary windings are connected in parallel. Similarly, the secondary windings are also connected in parallel. This is shown in Fig. 3.30 for transformers A and B. Together, they transfer power from common source to the common load. It is necessary that the sum of MVA capacity of all transformers that are running together in parallel should not be less than the peak power demand of the load. Fig. 3.31 presents the simplified equivalent circuit of two parallel connected transformers. The *emfs* induced in their secondary windings E_{2A} and E_{2B} are represented as voltage sources for clarity. Z_{o2A} and Z_{o2B} are the equivalent impedances of two transformers as referred to their secondary sides. Note that for the sake of simplicity, the no-load exciting branches comprising of R_o and X_o are omitted from the equivalent circuit. (Refer to Section 3.10.1(B) and Fig. 3.21 (a) in Section 3.11.3 for details).

Thus the total load current and total apparent power (in kVA or MVA) delivered to the load are,

$$I_2 = I_{2A} + I_{2B} \text{ amperes} \tag{3.47}$$

$$S = S_A + S_B = V_2 I_{2A} + V_2 I_{2B} \text{ voltamperes}$$

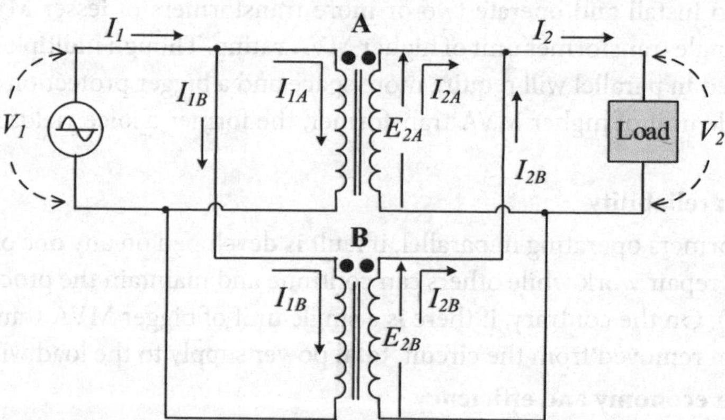

Fig. 3.30: Conceptual diagram of two single-phase transformers A and B in parallel

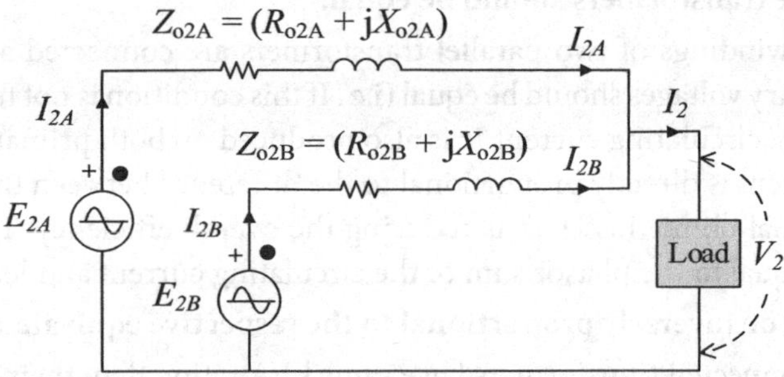

Fig. 3.31: Simplified equivalent circuit of two parallel connected transformers referred to secondary side

3.18.1 Conditions for Parallel Operation of Single-Phase Transformers

Before connecting two or more transformers in parallel, it should be verified that after connecting their primaries in parallel across the same source, their no-load secondary voltages will match each other in magnitude as well as in phase. This can be ensured by fulfilling the following conditions.

- **Transformers should have similar polarities:**

  This is the most important condition which must be satisfied before interconnecting two transformers in parallel. Let us examine the situation when transformers A and B with different polarities are connected in parallel across a common load. This is shown in Fig. 3.32.

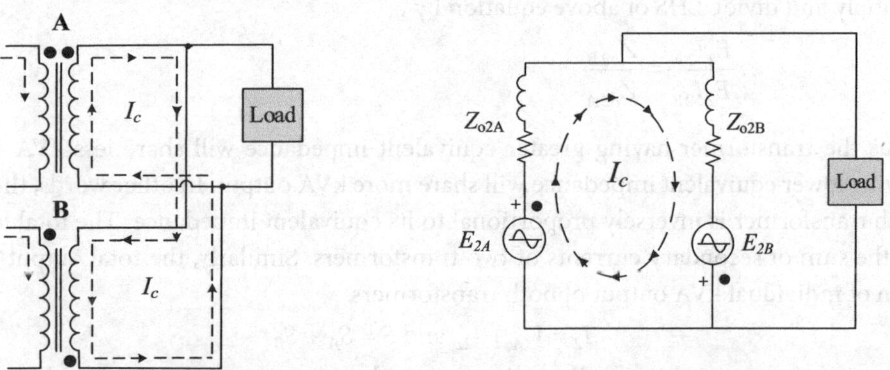

(a) Circulating current in secondary windings (b) Equivalent circuit representation

Fig. 3.32: Transformers with different polarities connected in parallel

It is clear that due to different instantaneous polarities, the secondary emfs E_{2A} and E_{2B} will add together and produce a heavy circulating current *Ic* within the secondary loop and the transformer may get damaged. This results in dead short circuit.

$$I_c = \frac{E_{2A} + E_{2B}}{Z_{o2A} + Z_{o2B}}$$

- **Voltage ratios of the transformers should be equal:**

  When the primary windings of two parallel transformers are connected across the same voltage source, their secondary voltages should be equal (i.e . If this condition is not fulfilled than in addition to the load current I_2, circulating current Ic is also produced on both primary and secondary sides. This circulating current is directly proportional to the difference between the *emfs* E_{2A} and E_{2B} and gives rise to additional ohmic losses thus reducing the overall efficiency. The resultant current in any transformer is equal to the phasor sum of the circulating current and load current shared by it.

- **kVA ratings should be inversely proportional to the respective equivalent impedances:**

  If the two parallel connected transformers have same kVA rating than their equivalent impedances Z_{o2A} and Z_{o2B} should also be equal. But if the transformers having different kVA ratings are to be operated in parallel than their kVA ratings should be inversely proportional to their equivalent impedances Z_{o2A} and Z_{o2B}. This can be proved as follows. Refer to the simplified equivalent circuit in Fig. 3.31. By Kirchoff's Voltage law (KVL), the voltage equations will be,

  $$E_{2A} - I_{2A}\, Z_{o2A} = V_2$$

  $$E_{2B} - I_{2B}\, Z_{o2B} = V_2$$

  $$\therefore \quad E_{2A} - I_{2A}\, Z_{o2A} = E_{2B} - I_{2B}\, Z_{o2B}$$

  Assuming that the voltage ratios of the two transformers are equal so that, $E_{2A} = E_{2B} = E_2$

  $$\therefore \quad I_{2A}\, Z_{o2A} = I_{2B}\, Z_{o2B}$$

  $$\frac{I_{2A}}{I_{2B}} = \frac{Z_{o2B}}{Z_{o2A}}$$

  Multiply and divide LHS of above equation by ,

  $$\therefore \frac{E_2 I_{2A}}{E_2 I_{2B}} = \frac{Z_{o2B}}{Z_{o2A}}$$

  Thus the transformer having greater equivalent impedance will share less kVA output and that having lower equivalent impedance will share more kVA output. In other words, the kVA shared by each transformer is inversely proportional to its equivalent impedance. The total load current will be the sum of secondary currents of two transformers. Similarly, the total output kVA will be the sum of individual kVA output of both transformers.

  $$I_2 = I_{2A} + I_{2B} \text{ and } S = S_A + S_B$$

  In per unit system, this condition is expressed as '*for parallel operation, the per unit leakage impedances of the transformers based on their own kVA ratings should be equal*'.

- **Ratio of equivalent leakage reactance to equivalent resistance should be equal:**

  For parallel operation, the ratio of equivalent leakage reactance to equivalent resistance of the two transformers should be equal.

  $$\frac{X_{o2A}}{R_{o2A}} = \frac{X_{o2B}}{R_{o2B}} \text{ or } \frac{X_{o1A}}{R_{o1A}} = \frac{X_{o1B}}{R_{o1B}}$$

Since, $\tan^{-1}\left(\frac{X_{o2}}{R_{o2}}\right)=\theta$ and $\cos\theta$ = power factor,

It means that, if the ratio of $\frac{X_{o2}}{R_{o2}}$ of two transformers are not equal than, the one having higher $\frac{X_{o2}}{R_{o2}}$ ratio will operate at poor power factor as compared to the other transformer having lower $\frac{X_{o2}}{R_{o2}}$ ratio. Obviously, the total kVA output delivered to the load will be less than the sum of individual kVA outputs of the two transformers.

Important: Amongst the above four conditions of parallel operation, the first one related with polarities is mandatory. The second one related to equal voltage ratios should be satisfied as accurately as possible since different secondary voltages will give rise to undesired circulating currents even at No-load. For the last two conditions, some deviation is tolerable.

3.19 LOAD SHARING BETWEEN PARALLEL CONNECTED TRANSFORMERS

The sharing of load current and apparent power between two parallel connected transformers can be discussed for the following two situations:

(i) When the voltage ratios are equal ($\therefore E_{2A} = E_{2B}$)

(ii) When the voltage ratios are not equal (i.e. $E_{2A} \neq E_{2B}$)

3.19.1 Load Sharing when Voltage Ratios are Equal

Consider two transformers A and B connected in parallel as shown in Fig. 3.30. The simplified equivalent circuit as referred to secondary side is given in Fig. 3.31. The voltage equations will be,

$$E_{2A} - I_{2A}\, Z_{o2A} = V_2$$

$$E_{2B} - I_{2B}\, Z_{o2B} = V_2$$

$$\therefore \quad E_{2A} - I_{2A}\, Z_{o2A} = E_{2B} - I_{2B}\, Z_{o2B}$$

Since voltage ratios are equal, the no-load secondary voltages (i.e. induced emfs) will also be equal.

$$E_{2A} = E_{2B}$$

$$\therefore \quad I_{2A}\, Z_{o2A} = I_{2B}\, Z_{o2B}$$

The total load current, $I_2 = I_{2A} + I_{2B}$

$\therefore$ $I_{2B} = (I_2 - I_{2A})$ Put this value in above equation and simplify.

$$I_{2A} = (I_2 - I_{2A})\frac{Z_{o2B}}{Z_{o2A}}$$

$$\therefore \quad I_{2A} = I_2\frac{Z_{o2B}}{(Z_{o2A} + Z_{o2B})}$$

Similarly, $I_{2B} = I_2 \dfrac{Z_{o2A}}{(Z_{o2A} + Z_{o2B})}$ (3.48)

Equations (3.48) represent the load currents shared by two parallel connected transformers when their voltage ratios are equal. To determine the sharing of apparent power (voltamperes), multiply both sides of equations (3.48) by V_2 and simplify. Since,

$V_2 I_{2A} = S_A$ = Apparent power output supplied by transformer A to the load

$V_2 I_{2B} = S_B$ = Apparent power output supplied by transformer B to the load

$V_2 I_2 = S = (S_A + S_B)$ = Total apparent power consumed by the load

$$S_A = S\frac{Z_{o2B}}{(Z_{o2A} + Z_{o2B})}$$

$$S_B = S\frac{Z_{o2A}}{(Z_{o2A} + Z_{o2B})} \quad (3.49)$$

3.19.2 Load Sharing when Voltage Ratios are not Equal

Again consider the same two transformers A and B connected in parallel as shown in Fig. 3.30 and their simplified equivalent circuit referred to secondary side as given in Fig. 3.31.

Let, Z_L Load impedance, $\therefore \quad V_2 = I_2 Z_L = (I_{2A} + I_{2B}) Z_L$

The voltage equations will be as given below where $E_{2A} \neq E_{2B}$ because the voltage ratios are not equal.

$$E_{2A} - I_{2A} Z_{o2A} = V_2 = I_2 Z_L = (I_{2A} + I_{2B}) Z_L \quad (3.50)$$

$$E_{2B} - I_{2B} Z_{o2B} = V_2 = I_2 Z_L = (I_{2A} + I_{2B}) Z_L \quad (3.51)$$

$$\therefore \quad E_{2A} - I_{2A} Z_{o2A} = E_{2B} - I_{2B} Z_{o2B}$$

$$\therefore \quad E_{2A} - E_{2B} = I_{2A} Z_{o2A} - I_{2B} Z_{o2B}$$

$$\therefore I_{2B} = -\frac{\left[(E_{2A} - E_{2B}) - I_{2A} Z_{o2A}\right]}{Z_{o2B}}$$

To determine I_{2A}, eliminate I_{2B} from equation (3.50) by substituting above value of I_{2B} and simplify.

$$E_{2A} - I_{2A} Z_{o2A} = \left\{ I_{2A} - \frac{\left[(E_{2A} - E_{2B}) - I_{2A} Z_{o2A}\right]}{Z_{o2B}} \right\} Z_L$$

$$I_{2A} = \frac{E_{2A} Z_{o2B} + Z_L (E_{2A} - E_{2B})}{Z_{o2A} Z_{o2B} + (Z_{o2A} + Z_{o2B}) Z_L}$$

Similarly, $I_{2B} = \dfrac{E_{2B} Z_{o2A} + Z_L (E_{2B} - E_{2A})}{Z_{o2A} Z_{o2B} + (Z_{o2A} + Z_{o2B}) Z_L}$ (3.52)

The total load current is, $I_2 = (I_{2A} + I_{2B}) = \dfrac{E_{2A}Z_{o2B} + E_{2B}Z_{o2A}}{Z_{o2A}Z_{o2B} + (Z_{o2A} + Z_{o2B})Z_L}$ (3.53)

To derive expression for terminal voltage V2, divide the numerator and denominator of equation (3.54) by $Z_{o2A}\, Z_{o2B}$ and simplify,

$$I_2 = \frac{\left(\dfrac{E_{2A}}{Z_{o2A}} + \dfrac{E_{2B}}{Z_{o2B}}\right)}{\left(1 + \dfrac{Z_L}{Z_{o2B}} + \dfrac{Z_L}{Z_{o2A}}\right)}$$

Multiply both sides by Z_L,

$$V_2 = I_2 Z_L = \frac{\left(\dfrac{E_{2A}}{Z_{o2A}} + \dfrac{E_{2B}}{Z_{o2B}}\right)}{\left(\dfrac{1}{Z_{o2A}} + \dfrac{1}{Z_{o2B}} + \dfrac{1}{Z_L}\right)} \quad (3.54)$$

Apparent power output supplied by transformer A to the load $S_A = V_2\, I_{2A}$
Apparent power output supplied by transformer B to the load $S_B = V_2\, I_{2B}$
Total apparent power consumed by the load $S = (S_A + S_B)$

Example 3.8: A single phase transformer takes 10A on no-load at a power factor of 0.2 lagging. The turns ratio is 4:1 (step down). If the load on the secondary is 200A at a power factor of 0.85 lagging, find the primary current and power factor. Neglect the voltage drop in windings.

Data:

$N_1 : N_2 = 4 : 1$, $I_2 = 200A$, $\cos\theta_2 = 0.85$ lagging, $I_0 = 10A$, $\cos\theta_o = 0.2$ lagging, $I_1 = ?$, $\cos\theta_1 = ?$

Solution:

Refer to the phasor diagram of Example 3.3 again.

The current in primary winding is given by,

$$\bar{I}_1 = \bar{I}_0 + \bar{I}'_2$$

It is given that, $\cos\theta_2 = 0.85$ lagging, $\therefore \theta_2 = \cos^{-1} 0.85 = 31.78°$

Hence current I2 can be expressed in polar frame as,

$$\bar{I}_2 = 200\angle -31.78°$$

A where the (-) sign indicates that this current is lagging behind the voltage V2 by a phase angle of

Similarly, given that, $\cos\theta_o = 0.2$ lagging, $\therefore\ \theta_o = \cos^{-1} 0.2 = 78.46°$

Hence current I0 can be expressed in polar frame as, $\bar{I}_0 = 10\angle -78.46°$ A

The magnitude of secondary current as referred to primary side is,

$$\left|I_2'\right| = KI_2 = \frac{N_2}{N_1} I_2 = \frac{1}{4} \times 200 = 50A$$

In polar frame it will become, $\bar{I}_2' = |\bar{I}_2'|\angle\theta_2 = 50\angle -31.78°$ A

Substituting the values of $\bar{I}_0$ and $\bar{I}_2'$ in above equation.

$\therefore \qquad \bar{I}_1 = (10\angle -78.46°) + (50\angle -31.78°)$

Converting $\bar{I}_0$ and $\bar{I}_2'$ from polar frame to rectangular frame (Refer Appendix),

$\therefore \qquad \bar{I}_1 = (2 - j9.797) + (42.5 - j26.33) = (44.5 - j36.129)$

Converting from rectangular frame to the polar frame

$\therefore \qquad \bar{I}_1 = 57.32\angle -39.07°$

Thus, $|\bar{I}_1|| = 57.32A$ and $\cos\theta_1 = \cos 39.07 = 0.77$ lagging

Example 3.9: The following readings were obtained from OC and SC tests performed on a 10 kVA, 450V/120V, single phase transformer.

OC test: 120V, 4.2A, 80W ..... on LV side with HV side open circuited

SC test: 9.65V, 22.2A, 120W ..... on HV side with LV short circuited

Calculate the approximate equivalent circuit constants referred to primary side.

Data:

Rated capacity (i.e. full load capacity) = 10kVA, Rated voltages are: $V_1 = 450V$ and $V_2 = 120V$,

$R_o = ?, X_o = ?, R_{o1} = ?, X_{o1} = ?$

Solution:

As per given voltage ratings (450V/120V), the transformer is of step down type. Therefore HV side will be the primary and LV will be the secondary side.

From OC Test: As mentioned in the question, OC test readings were observed on LV side i.e. secondary side. Hence, the OC test calculation results will be as referred to secondary. Since the constants are required to be determined as referred to primary, necessary change in calculation will be required.

$$\cos\theta_o = \frac{W_o}{V_o I_o} = 0.1587$$

$\therefore \angle\theta_o = 80.86°$ and $\sin\theta_o = 0.9873$

The no-load parameters of equivalent circuit will be,

$$R_o = \frac{V_o}{I_w} = \frac{V_o}{I_o \cos\cos\theta_o} = 180.034\Omega$$

$$X_o = \frac{V_o}{I_\mu} = \frac{V_o}{I_o \sin\sin\theta_o} = 28.938\Omega$$

As referred to primary, they will be,

$$R_o' = \left(\frac{R_o}{K^2}\right) = \left[\frac{180.034}{\left(120/450\right)^2}\right] = 2532\ \Omega$$

$$X_o' = \left(\frac{X_o}{K^2}\right) = \left[\frac{28.938}{\left(120/450\right)^2}\right] = 407.14\ \Omega$$

From SC Test:

$$Z_{sc} = \frac{V_{sc}}{I_{sc}} = 0.4346\ \Omega$$

$$R_{sc} = \frac{W_{sc}}{I_{sc}^{\ 2}} = 0.2434\ \Omega$$

$$X_{sc} = \sqrt{Z_{sc}^2 - R_{sc}^2} = 0.3599\ \Omega$$

Since, the SC test was performed on HV side which is the primary side in this problem hence,

$$R_{o1} = R_{sc} = 0.2434\ \Omega$$

$$X_{o1} = X_{sc} = 0.3599\ \Omega$$

Example 3.10: For the transformer and OC/SC test results of Example 3.9, calculate:

(i) Efficiency at full load and at half full load, 0.8 power factor lagging

(ii) Voltage regulation at full load and at half full load, 0.8 power factor lagging

Data:

Rated capacity (i.e. full load capacity) = 10kVA, Rated voltages are: V_1 = 450V and V_2 = 120V, V_o = 120V, I_o = 4.2A, W_o = 80W, V_{sc} = 9.65V, I_{sc} = 22.2A, W_{sc} = 120W,

(i) $\eta_{(FL,0.8)} = ?,\ \eta_{(\frac{1}{2}FL,0.8)} = ?$ (ii) $V.R._{(FL,0.8)} = ?,\ V.R._{(\frac{1}{2}FL,0.8)} = ?$

Solution: The results of Example 3.9 are: R_{o1} = 0.2434 Ω; X_{o1} = 0.3599 Ω

Calculating primary currents:

(a) At Full load, $I_{1(FL)} = \dfrac{\text{Full load voltamperes}}{\text{rated primary voltage}} = \dfrac{10000}{450} = 22.2A$

(b) At Half full load, $I_{1(\frac{1}{2}FL)} = \dfrac{\text{Half Full load voltamperes}}{\text{rated primary voltage}} = \dfrac{5000}{450} = 11.1A$

Note: The half full load current can also be determined as $I_{1(\frac{1}{2}FL)} = \frac{1}{2} I_{1(FL)}$

Calculating Efficiencies:

During OC test, voltage applied on LV side was equal to the rated voltage of LV side.

i.e. $V_o = V_2$ = 120V. Hence, the wattmeter reading W_o will represent rated iron loss.

∴ Iron loss $P_i = W_o$ = 80W

Similarly, during SC test, the current flowing in HV winding (i.e. primary winding in this example) was equal to the full load primary current. i.e. $I_{sc} = I_{1(FL)}$ = 22.2A. Therefore, the wattmeter reading W_{sc} will represent full-load copper loss.

∴ Full load copper loss $P_{cu(FL)} = W_{sc}$ = 120W

(i) Efficiency at full load and 0.8 power factor is,

$$\eta_{(FL,0.8)} = \frac{\text{Full load VA}\times 0.8}{\text{Full load VA}\times 0.8 + P_i + P_{cu(FL)}} = \frac{10000\times 0.8}{10000\times 0.8 + 80 + 120}\times 100 = 97.56\%$$

(ii) Efficiency at half full load and 0.8 power factor is,

$$\eta_{\left(\frac{1}{2}FL,0.8\right)} = \frac{\text{Half Full load VA}\times 0.8}{\text{Half Full load VA}\times 0.8 + P_i + P_{cu\left(\frac{1}{2}FL\right)}}$$

$$= \frac{5000\times 0.8}{5000\times 0.8 + 80 + \left(\frac{1}{2}\right)^2\times 120}\times 100 = 97.32\%$$

Calculating Voltage Regulation (V.R.):

(i) At full load, 0.8 lagging, the voltage drop as referred to primary side is,

Voltage drop = $I_{1(FL)}\ (R_{o1}\cos\theta + X_{o1}\sin\theta) = 22.2\ (0.2434\times 0.8 + 0.3599\times 0.6) = 9.11\text{V}$

$$\therefore V.R_{(FL,0.8lag)} = \frac{\text{voltage drop as referred to primary}}{\text{rated primary voltage}}\times 100 = \frac{9.11}{450}\times 100 = 2.02\%$$

(ii) At half full load, 0.8 lagging, the voltage drop as referred to primary side is,

$$\text{Voltage drop} = I_{1(\frac{1}{2}FL)}\left(R_{o1}cos\theta + X_{o1}\sin\theta\right) = 11.1\left(0.2434\times 0.8 + 0.3599\times 0.6\right) = 4.55V$$

$$\therefore V.R_{(FL,0.8lag)} = \frac{\text{voltage drop as referred to primary}}{\text{rated primary voltage}}\times 100 = \frac{4.55}{450}\times 100 = 1.01\%$$

Example 3.11: A 10 kVA, 500V/250V, single phase transformer gave following test results.

OC test (LV side): 250V, 3A, 200W

SC test (LV side): 15V, 30A, 300W

Calculate the efficiency and voltage regulation at full load, 0.8 power factor lagging.

Data:

Rated capacity (i.e. full load capacity) = 10kVA, Rated voltages are: V_1 = 500V and V_2 = 250V, V_o = 250V, I_o = 3A, W_o = 200W, V_{sc} = 15V, I_{sc} = 30A, W_{sc} = 300W,

(i) $\eta_{(FL,0.8)}$ = ?, (ii) $V.R._{(FL,0.8)}$ = ?

Solution:

If this example is compared with Examples 3.9 and 3.10, you will find that in this example, both OC and SC tests are performed on LV side. In other words, the measuring instruments were connected on LV side during both tests and the HV side was open-circuited or short-circuited.

From given voltage rating (500V/250V), the LV side is secondary side. Therefore, all calculation results of OC and SC tests will be the values as referred to secondary side.

Calculating equivalent resistance and equivalent leakage reactance:

From SC test observations:

$$Z_{sc} = Z_{02} = \frac{V_{sc}}{I_{sc}} = 0.5\ \Omega$$

$$R_{sc} = R_{02} = \frac{W_{sc}}{I_{sc}^2} = 0.333\ \Omega$$

$$X_{sc} = X_{02} = \sqrt{Z_{sc}^2 - R_{sc}^2} = 0.3727\ \Omega$$

Calculating Efficiency:

During OC test, voltage applied on LV side was equal to the rated voltage of LV side i.e. $V_o = V_2 =$ 250V. Hence, the wattmeter reading Wo will represent rated iron loss.

$\therefore$ Iron loss $P_i = W_o = 200\text{W}$

However, during SC test, the current ($I_{sc} = 30\text{A}$) flowing in LV winding is not equal to the full load secondary current because, the full load secondary current is,

$$I_{2(FL)} = \frac{\text{Full load voltamperes}}{\text{rated secondary voltage}} = \frac{10000}{250} = 40\text{A}$$

Therefore, the wattmeter reading Wsc will not represent full-load copper loss and it should be calculated separately.

Full load copper loss = $P_{cu(\text{FL})} = I_{2\,(\text{FL})}^2 R_{02} = 40^2 \times 0.3333 = 533.28\text{W}$

(i) Efficiency at full load and 0.8 power factor is,

$$\eta_{(FL,0.8)} = \frac{\text{Full load VA}\times 0.8}{\text{Full load VA}\times 0.8 + P_i + P_{cu(FL)}} = \frac{10000\times 0.8}{10000\times 0.8 + 200 + 533.28}\times 100 = 91.6\%$$

Calculating Voltage Regulation (V.R.):

(i) At full load, 0.8 pf lagging, the voltage drop as referred to secondary side is,

Voltage drop = $I_{2(\text{FL})}\,(R_{o2}\cos\theta + X_{o2}\sin\theta) = 40(0.333 \times 0.8 + 0.3727 \times 0.6) = 19.6\text{V}$

$$\therefore V.R_{(FL,0.8lag)} = \frac{\text{voltage drop as referred to secondary}}{\text{rated secondary voltage}}\times 100 = \frac{19.6}{250}\times 100 = 7.84\%$$

Example 3.12: A 50 kVA single phase transformer has on full-load, the copper loss of 600W and iron loss of 500 W. Calculate the maximum efficiency and the load at which it appears. Assume unity power factor load.

Data:

Rated capacity (i.e. full load capacity) = 50kVA, $P_{cu(\text{FL})} = 600$ W, $Pi = 500$ W, $\cos\theta = 1$, (i) $S_{max\,\eta} = ?$ (ii) $\eta_{max} = ?$

Solution: From eq. (3.34),

(i) To get maximum efficiency, the load or output power ($S_{max\,\eta}$) in kVA should be:

$$S_{max\,\eta} = S_{FL} \times \sqrt{\frac{P_i}{P_{cu(FL)}}}$$

Where, S_{FL} = Rated capacity (i.e.full load capacity) = 50 kVA

$$\therefore S_{max\,\eta} = S_{FL} \times \sqrt{\frac{P_i}{P_{cu(FL)}}} = 50\times\sqrt{\frac{500}{600}} = 45.64\text{kVA}$$

(ii) Maximum efficiency is expressed as,

$$\eta_{max} = \frac{P_{max\ \eta}(\text{in W})}{P_{max\ \eta}(\text{in W}) + 2P_i} = \frac{S_{max\ \eta}(\text{in VA}) \times \text{load powerfactor}}{S_{max\ \eta}(\text{in VA}) \times \text{load powerfactor} + 2P_i}$$

$$\therefore \eta_{max} = \frac{45640 \times 1}{45640 \times 1 + 2 \times 500} \times 100 = 97.85\%$$

Example 3.13: The efficiencies of a 400 kVA, single phase transformer are 98.77% when delivering full-load at 0.8 power factor and 99.13% at half-load and unity power factor. Calculate the iron loss and full-load copper loss.

Data: Rated capacity (i.e. full load capacity) = 400 kVA,

$$\eta_{(FL,0.8)} = 98.77\%, \eta_{\left(\frac{1}{2}FL,UPF\right)} = 99.13\%,$$

(i) P_i = ? (ii) $P_{cu(FL)}$ = ?

Solution:

Efficiency of a transformer at any fraction 'x' of the load is given by,

$$\eta_{(x,cos\varnothing)} = \frac{x\ \text{Full load VA} \times \cos}{x\ \text{Full load VA} \times \cos\ + P_i + x^2 P_{cu(FL)}}$$

In given case 1: $x = 1$, $\cos\theta = 0.8$, $\eta_x = 98.77\%$

Substitute all given values in above equation,

$$Pi + P_{cu(FL)} = 3.985 \text{ W} \quad \text{(i)}$$

In given case 2: $x = 1/2$, $\cos\theta = 1$, $\eta_x = 99.13\%$

Substitute all given values in above equation,

$$\therefore \quad Pi + 0.25\, P_{cu(FL)} = 1.755 \text{ W} \quad \text{(ii)}$$

Subtracting eq.(ii) from eq. (i) we get,

$$P_{cu(FL)} = 2.973 \text{ kW and } P_i = 1.012 \text{ kW}$$

Example 3.14: A single phase transformer with a ratio 5:1 has primary resistance of 0.4 Ω and reactance of 1.2 Ω and the secondary winding resistance of 0.01 Ω and reactance of 0.04 Ω. Determine the percent regulation when delivering 125A at 600V and power factors of (i) 0.8 lagging, (ii) unity and (iii) 0.8 leading.

Data:

$N_1 : N_2 = 5:1$, $R_1 = 0.4\ \Omega$, $X_1 = 1.2\ \Omega$, $R_2 = 0.01\Omega$, $X_2 = 0.04\Omega$, $I_2 = 125\text{A}$, V_2(at 125A load) = 600V,
Find V. R. at (i) $\cos\theta = 0.8$ lagging, (ii) $\cos\theta = 1$, (iii) $\cos\theta = 0.8$ leading

Solution:

$$\text{Transformation ratio } K = \frac{N_2}{N_1} = \frac{1}{5}$$

The equivalent circuit parameters as referred to secondary side are:

$$R_{02} = \left(R_2 + R_1'\right) = \left(R_2 + R_1 K^2\right) = \left(0.01 + 0.4 \times \frac{1}{5^2}\right) = 0.026\Omega$$

$$X_{02} = \left(X_2 + X_1'\right) = \left(X_2 + X_1 K^2\right) = \left(0.04 + 1.2 \times \frac{1}{5^2}\right) = 0.088\Omega$$

Now, calculate the voltage drops referred to secondary side,

$$\text{Voltage drop} = I_2 \left(R_{o2} \cos\theta \pm X_{o2} \sin\theta\right)$$

Where (+) sign is for inductive loads (lagging power factor) and (–) sign is for the capacitive loads (leading power factor). Given that at a load current of I_2 = 125A, the secondary terminal voltage is 600V.

In given case (i): cos θ = 0.8 lagging,

$$\text{Voltage drop} = 125 \times (0.026 \times 0.8 + 0.088 \times 0.6) = 9.2 \text{ V}$$

It is known that, when inductive load is connected across the secondary winding, the secondary terminal voltage drops from its no-load value (i.e. rated value) to some value on-load which is 600V in this example. Since a voltage drop of 9.2V has taken place, it means that the no-load voltage for this case must be = 600 + 9.2 = 609.2 V

$$V.R_{(125A,0.8lag)} = \frac{\text{voltage drop as referred to secondary}}{\text{secondary No-load voltage}} \times 100 = \frac{9.2}{609.2} \times 100 = 1.51\%$$

In given case (ii): cos ϕ = 1

$$\text{Voltage drop} = 125 \times (0.026 \times 1 + 0.088 \times 0) = 3.25 \text{ V}$$

It is known that, with resistive load, the secondary terminal voltage drops from its no-load value to some value on-load which is 600V in this example. Since a voltage drop = 3.25V, it means that no-load voltage for this case must be = 600 + 3.25 = 603.25 V

$$V.R_{(125A,UPF)} = \frac{\text{voltage drop as referred to secondary}}{\text{secondary No-load voltage}} \times 100 = \frac{3.25}{603.25} \times 100 = 0.538\%$$

In given case (iii): cos ϕ = 0.8 leading,

$$\text{Voltage drop} = 125 \times (0.026 \times 0.8 - 0.088 \times 0.6) = -4 \text{ V}$$

It is known that, with capacitive load, the secondary terminal voltage rises from its no-load value to some value on-load which is 600V in this example. Since a voltage drop = – 4V, it means that the no-load voltage for this case must be = 600 – 4 = 596 V

$$V.R_{(125A,0.8lead)} = \frac{\text{voltage drop as referred to secondary}}{\text{secondary No-load voltage}} \times 100 = \frac{-4}{596} \times 100 = -0.671\%$$

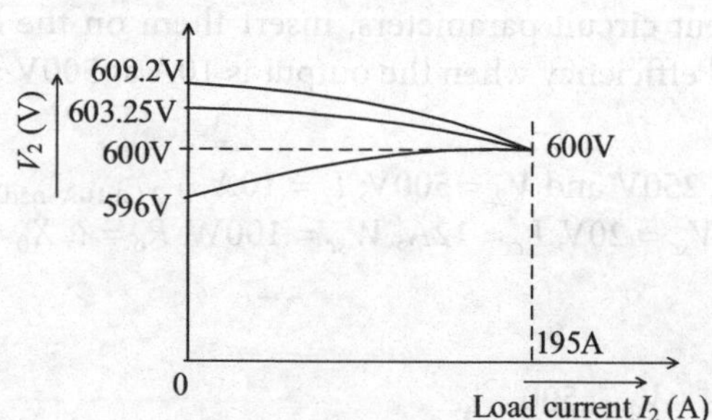

Example 3.15: A 2300V/230V single-phase transformer has the primary and secondary winding resistances of 2 Ω and 0.02 Ω respectively. Its iron loss at normal supply voltage is 600W. Calculate the secondary current at which the efficiency will be maximum and also calculate the maximum efficiency when the load power factor is 0.9.

Data:

Rated voltages are: $V_1 = 2300V$ and $V_2 = 230V$, $R_1 = 2\ \Omega$, $R_2 = 0.02\Omega$, $P_i = 600$ W, $\cos\theta = 0.9$, (i) $I_{2(max\,\eta)} = ?$ (ii) $\eta_{max} = ?$

Solution:

The equivalent resistance as referred to secondary side (LV side) is:

$$R_{02} = \left(R_2 + R_1'\right) = \left(R_2 + R_1 K^2\right) = \left[R_2 + R_1\left(\frac{V_2}{V_1}\right)^2\right]$$

$$= \left[0.02 + 2 \times \left(\frac{230}{2300}\right)^2\right] = 0.04\ \Omega$$

From eq. (3.33),

At maximum efficiency, the iron loss and copper loss are equal i.e.,

$$P_i = P_{cu(max\,\eta)} = 600\text{ W}$$

The copper loss of transformer at maximum efficiency condition can be expressed as,

$$P_{cu(max\,\eta)} = I^2_{2(max\,\eta)}\, R_{02}$$

$$\therefore I_{2(max\,\eta)} = \sqrt{\frac{P_{cu(max\,\eta)}}{R_{02}}} = \sqrt{\frac{600}{0.04}} = 122.47\text{A}$$

The output power in watts at maximum efficiency condition and 0.9 power factor is,

$$P_{(max\,\eta)} = V_2\, I_{2(max\,\eta)} \cos\theta = 230 \times 122.47 \times 0.9 = 25352.21\text{ W}$$

$$\therefore \eta_{max} = \frac{P_{max\,\eta}}{P_{max\,\eta} + 2P_i} = \frac{25352.21}{25352.21 + 2\times 600} \times 100 = 95.48\%$$

Example 3.16: A 250V/500V single phase transformer gave following test results:

SC Test: 20V, 12A, 100W with LV winding short-circuited.

OC Test: 250V, 1A, 80W on LV side

Determine the equivalent circuit parameters, insert them on the equivalent circuit diagram and calculate applied voltage and efficiency when the output is 10A at 500V and 0.8 power factor lagging.

Data:

Rated voltages are: $V_1 = 250V$ and $V_2 = 500V$; $I_2 = 10A$, $V_{2\,(\text{at 10A load})} = 500V$, $\cos\theta = 0.8$ lagging, $V_o = 250V$, $I_0 = 1A$, $W_0 = 80W$, $V_{sc} = 20V$, $I_{sc} = 12A$, $W_{sc} = 100W$, $R_o = ?$, $X_o = ?$, $R_{01} = ?$, $X_{01} = ?$, $V_{1(\text{at 10A load})} = ?$, $\eta_{(\text{at 10A load})} = ?$

Solution:

$$\text{Transformation ratio } K = \frac{V_2}{V_1} = \frac{500}{250} = 2$$

As per given voltage ratings (250V/500V), the transformer is of step-up type. Therefore LV side will be the primary and HV will be the secondary side.

From OC Test: As per question, the OC test readings were observed on LV side i.e. primary side. Hence, the OC test calculation results will be as referred to primary.

$$\cos\theta_o = \frac{W_o}{V_o I_o} = \frac{80}{250\times 1} = 0.32, \therefore \angle\phi_o^o = 71.33° \text{ and } \sin\theta_o = 0.9474$$

The no-load parameters of equivalent circuit will be,

$$R_o = \frac{V_o}{I_w} = \frac{V_o}{I_o \cos\phi_o^o} = 781.25\Omega \;\; ; X_o = \frac{V_o}{I_\mu} = \frac{V_o}{I_o \sin\phi_o^o} = 263.88\Omega$$

From SC Test: As per question, during the SC test, LV winding was short circuited. It means that the input supply was fed to HV side and readings were observed on HV side i.e. secondary side. Hence, the SC test calculation results will be as referred to secondary. They should be converted as referred to primary to represent them on the simplified equivalent circuit which will be as referred to primary.

$$Z_{sc} = Z_{02} = \frac{V_{sc}}{I_{sc}} = 1.666\ \Omega;$$

$$R_{sc} = R_{02} = \frac{W_{sc}}{I_{sc}^2} = 0.694\Omega;\ X_{sc} = X_{02} = \sqrt{Z_{sc}^2 - R_{sc}^2} = 1.514\ \Omega$$

Now, refer these values to primary side.

$$R_{01} = \left(\frac{R_{02}}{K^2}\right) = \left(\frac{0.694}{2^2}\right) = 0.1735\Omega;\ X_{01} = \left(\frac{X_{02}}{K^2}\right) = \left(\frac{1.514}{2^2}\right) = 0.3785\Omega$$

$R_{01}=$ 0.1735 Ω $X_{01}=$ 0.3785 Ω

I_1 I'_2 I_o

V_1 Ro = 781.25 Ω Xo = 263.88 Ω $E_1=E'_2$

Calculating efficiency:

During OC test, the applied value of input voltage was equal to the rated voltage = 250V hence, the wattmeter reading during OC test will represent rated iron loss.

Iron loss $P_i = W_o = 80\text{W}$

The copper loss at $I_2 = 10\text{A}$ will be $= P_{cu} = I_2^2\, R_{02} = 10^2 \times 0.694 = 69.4\ \text{W}$

$$\eta = \frac{\text{Output power(W)}}{\text{Input power(W)}} = \frac{V_2 I_2 \cos\cos\theta}{V_2 I_2 \cos\cos\theta + P_i + P_{cu}}$$

$$= \frac{500\times 10\times 0.8}{500\times 10\times 0.8 + 80 + 69.4}\times 100 = 96.39\%$$

Calculating $V_{1(\text{at 10A load})}$:

During the load current of I_2 = 10A, voltage drop (VD) at 0.8 lagging power factor and as referred to secondary side is,

$VD_2 = I_2 (R_{o2} \cos \theta + X_{o2} \sin \theta) = 10 \times (0.694 \times 0.8 + 1.514 \times 0.6) = 14.636$ V

Referring this secondary voltage drop to primary side using transformation ratio,

$$\therefore VD_1 = \frac{VD_2}{K} = \frac{14.636}{2} = 7.31\text{V}$$

It is known that, with inductive load, the secondary terminal voltage drops from its no-load value to some value on-load which is 500V in this example. It means that to obtain rated secondary voltage of 500V during the inductive load of I_2 = 10A, input voltage higher than its rated value of 250V should be applied on primary side.

$$\therefore \quad V_{1(\text{at 10A load})} = V_1 + VD_1 = 250 + 7.31 = 257.31\text{V}$$

UNIT SUMMARY

- Transformers work on the principle of mutual induction. They have primary and secondary windings that are *electrically isolated* from each other. As these windings are wound on the same magnetic core, they are *magnetically coupled* to each other.
- The magnetic core is made from the laminations of either CRGO silicon steel or the amorphous alloy.
- The winding to which input *ac* source is connected is called as 'Primary winding' and that across which the electrical load is connected is called as 'Secondary winding'. Any winding of a given transformer can be used as primary and secondary.
- Transformers transform electrical potential from one level to another *at same frequency*. In other words, frequency remains same on both sides.
- In an ideal transformer, voltages are transformed in the direct ratio of turns, $\frac{V_2}{V_1} = \frac{N_2}{N_1}$

  Currents are transformed in inverse ratio of turns, $\frac{I_1}{I_2} = \frac{N_2}{N_1}$

  Impedances are transformed in direct ratio squared, $\frac{Z_2}{Z_1} = \left(\frac{N_2}{N_1}\right)^2$
- Power (watts) and volt-amperes in an ideal transformer remains same on both sides.
- In an ideal transformer, No-load current is used only for magnetizing the core. Therefore in an ideal transformer, $I_o = I_\mu$ and $I_w = 0$.
- Two types of power loss takes place in a transformer. (i) Iron loss takes place in the magnetic core and (ii) Copper loss takes place in the copper windings.
- Iron loss remains constant for all values of load (from No-load to Full-load) provided that the input voltage and input frequency supplied to the transformer are not changed. Hence, iron loss is also called as 'No-load loss' and 'Constant loss'.

- Copper loss varies as square of currents which in turn varies with changes in load. Hence, it is also called as 'Variable loss'.
- Iron loss comprises of (i) Hysteresis loss and (ii) Eddy current loss.
- Being a static device, mechanical loss (i.e. friction and windage) is absent in transformers hence, its efficiency is always greater (> 90%) than the efficiency of rotating machines.
- To reduce the hysteresis loss, core laminations are made from highly permeable material like CRGO silicon steel or amorphous alloy.
- To reduce the eddy current loss, the magnetic core is made from a stack of thin laminations which are insulated from each other before forming the stack.
- Transformer works with maximum efficiency (i.e minimum losses) at that value of load when "Iron loss = Copper loss".
- Since, the transformers employed in power system operate continuously for 24 hours a day, All-day efficiency (also called as energy efficiency) is calculated for them.
- In core type transformer, the windings surround the core and therefore winding repair is relatively easy.
- In shell type transformer, the magnetic core surrounds the windings hence, winding repair is relatively not easy.
- The *emfs* induced an primary and secondary windings are given by: $E_1 = 4.44\ \phi_m f N_1$ volts and $E_2 = 4.44\phi_m f N_2$ volts
- Transformation ratio is: $K = \dfrac{N_2}{N_1} = \dfrac{E_2}{E_1} = \dfrac{V_2}{V_1} = \dfrac{I_1}{I_2}$
- The current equations are: $\bar{I}_o = \bar{I}_\mu + \bar{I}_w$ and $\bar{I}_1 = \bar{I}_o + \bar{I}_o'$ where, $|I_2'| = K|I_2|$
- The net flux in the magnetic core remains same at all values of load (from No-load to Full-load). Hence, transformer is also called as 'Constant flux machine'.
- The voltage equations on the primary and secondary sides are:
  $\bar{V}_1 = \bar{E}_1 + \bar{I}_1 (R_1 + j X_1)$ and $\bar{V}_2 = \bar{E}_2 + \bar{I}_2 (R_2 + j X_2)$
- Equivalent circuit of a transformer is useful for pre-determining analytically, the performance of that transformer at any magnitude and any power factor of the load without connecting an actual load across its secondary. For doing so, the values of simplified equivalent circuit parameters i.e. Core loss resistance R_o, Magnetizing reactance X_o, Equivalent resistance (R_{o1} or R_{o2}) and Equivalent leakage reactance (X_{o1} or X_{o2}) must be known.
- The values of R_o and X_o can be determined from Open-circuit test and values of (R_{o1} or R_{o2}) and (X_{o1} or X_{o2}) can be determined from Short-circuit test.
- In Open-circuit test, the input power indicated by wattmeter reading W_o is treated as equal to rated iron-loss P_i provided that, rated value of input voltage V_1 should be applied.
- In Short-circuit test, the input power indicated by wattmeter reading W_{sc} is treated as equal to rated copper-loss P_{cu} provided that, the windings are made to carry rated value of their currents.
- OC test is normally performed on LV side (with HV side open-circuited) and SC test is normally performed on HV side (with LV side short-circuited). This is not a mandatory requirement. It is practised so that the measuring instruments of commercially available range could be used.

- Voltage regulation of a transformer is defined as the percentage change in secondary terminal voltage when the load at a given power-factor is changed from zero to full-load value or vice-versa provided that, the applied voltage V1 is held at constant RMS value.
- Voltage regulation of a transformer is also defined as the percentage change in input voltage required to be applied across primary winding in order to maintain rated voltage at the terminals of secondary winding when the load at a given power-factor is changed from zero to full-load value or vice-versa.
- With rise in resistive load, there is fall in secondary terminal voltage V_2. If the load is of inductive type or R-L type, V_2 drops by larger magnitude. However, if the load is of capacitive then with rise in magnitude of load, there is a rise in V_2 also. Hence for capacitive load, the transformer's regulation is negative.
- Transformer full-load capacity is expressed in terms of apparent power (kVA or MVA) and not in terms of active power (kW or MW). This is because, the iron loss depends on voltage and copper loss depends upon current. None of them depends on power-factor.
- Experimentally, efficiency and voltage regulation can be determined by performing any of the two tests: (i) Direct loading test or (ii) OC/SC test. In direct loading test, an actual variable load is connected across secondary. In OC/SC test, first, the parameters of equivalent circuit are calculated from the observed readings. Then from these parameters, efficiency and voltage regulation are determined.
- Polarity marking is very important when two or more transformers are to be connected in parallel. Polarity marking is important in multiphase systems also.
- Transformers are required to be connected in parallel when the power capacity of an existing transformer is insufficient to cater the increased load demand.
- Mandatory condition that should be fulfilled before connecting two or more transformers in parallel is that, they should have similar polarities. Otherwise, heavy circulating current is produced within the transformers that may damage them even at No-load.
- Other desirable conditions to be fulfilled for parallel operation are (i) voltage ratios should be equal, (ii) the ratio of their kVA (or MVA) ratings should be inversely proportional to the ratio of their equivalent leakage impedances and (ii) their ratios of equivalent leakage reactance to equivalent resistance should be equal.

EXERCISES

Multiple Choice Questions

1. In an ideal transformer,

   (a) iron losses are absent (b) winding resistances are zero
   (c) magnetic core has infinite permeability (d) all the above options

2. Magnetic core in transformers in primarily used to,

   (a) reduce iron loss (b) provide low reluctance path to the flux
   (c) reduce eddy current loss (d) all the above options

3. Transformer magnetic core is made from laminations to,
   (a) reduce eddy current loss (b) reduce cost of transformer
   (c) simplify the construction (d) all the above options
4. The self and mutually induced emfs E_1 and E_2 in a transformer are always,
   (a) in-phase with each other (b) out of phase with each other
   (c) equal in magnitude (d) determined by magnitude of load
5. A step-down transformer decreases the,
   (a) voltage (b) current
   (c) frequency (d) all the above options
6. In a transformer, leakage flux is always proportional to the winding currents because,
   (a) the two windings are electrically isolated (b) leakage path does not saturate due to air in it
   (c) mutual flux is confined to the core (d) all the above options
7. In a 2-winding transformer, emf per turn in primary is always ________ emf per turn in secondary
   (a) equal to (b) K times
   (c) 1/K times (d) K^2 times
8. While performing SC test on a transformer, ________ winding is usually short-circuited.
   (a) High voltage (b) Low voltage
   (c) any
9. The regulation of transformer is negative when the load is of ________ type.
   (a) resistive (b) inductive
   (c) capacitive (d) R-L
10. No-load test on a transformer is performed to determine the,
    (a) magnetizing current and iron loss (b) No-load current and rated copper loss
    (c) magnetizing current and hysteresis loss (d) No-load current and eddy current loss
11. During SC test, iron loss is neglected because,
    (a) iron loss is actually absent
    (b) iron loss is constant at all values of load
    (c) due to reduced voltage, mutual flux is just a small fraction of normal flux
12. Transformers are rated in kVA and not in kW because,
    (a) load power factor is normally unknown
    (b) kW is the unit of active power
    (c) all losses depend only on voltage and current
    (d) It is convenient to express in kVA
13. At lighter loads, transformer efficiency is poor because,
    (a) transformer losses are high (b) constant loss is high in proportion to output
    (c) copper loss is low (d) all the above options
14. The efficiency of a transformer is maximum when,
    (a) Output power is greater than input power
    (b) Output power is equal to input power
    (c) both iron loss and copper loss are zero
    (d) its iron loss is equal to the copper loss

15. Energy efficiency of a transformer primarily depends on,
    (a) copper loss (b) magnitude of load
    (c) duration of load (d) options (b) & (c)
16. The essential condition for parallel operation of two or more transformers is that their,
    (a) polarities should be equal (b) voltage ratio should be equal
    (c) kVA ratings should be equal (d) all the above options
17. A 200V/400V transformer has 1000 primary turns. The number of turns in secondary are,
    (a) 500 (b) 4000
    (c) 2000 (d) 400
18. A 220/1100 V transformer has primary resistance = 0.1Ω. As referred to secondary, its value is,
    (a) 2.5 Ω (b) 0.5 Ω
    (c) 0.02 Ω (d) 0.004 Ω
19. During SC test, the applied voltage is 2% of rated voltage. The core loss during SC test will be,
    (a) 0.04% of rated core loss (b) 4% of rated core loss
    (c) 0.4% of rated core loss (d) 40% of rated core loss
20. A 200kVA transformer has iron loss = 1kW and Full-load copper loss = 2kW. It will have maximum efficiency at ______ kVA load.
    (a) 50 (b) 100
    (c) 70.6 (d) 141.4
21. If copper loss at 3/4th Full-load is 3600W then its value at Full-load will be,
    (a) 2025 W (b) 6400 W
    (b) 2700 W (d) 4800 W
22. A transformer has core loss = 100W and equivalent resistance referred to secondary side = 0.25 Ω. Its secondary current at maximum efficiency will be,
    (a) 5A (b) 20A
    (c) 25A (d) 10A
23. The primary and secondary rated currents of a 5 kVA, 220/1100V single phase transformer are:
    (a) 4.54A, 22.72A (b) 5A, 10A
    (c) 22.72A, 4.54A (d) 10A, 5A

ANSWERS															
1.	(d)	2.	(b)	3.	(a)	4.	(a)	5.	(a)	6.	(b)	7.	(a)	8.	(b)
9.	(c)	10.	(a)	11.	(c)	12.	(c)	13.	(b)	14.	(d)	15.	(d)	16.	(a)
17.	(c)	18.	(a)	19.	(a)	20.	(d)	21.	(b)	22.	(b)	23.	(c)		

Short Answer Questions

1. Stating the assumptions show that, $K = \frac{N_2}{N_1} = \frac{E_2}{E_1} = \frac{V_2}{V_1} = \frac{I_1}{I_2}$.
2. What is the order of magnitude of No-load current in a transformer?

3. What is the significance of magnetizing current in a transformer?
4. Name the law and rule that are used to determine the direction of induced emf in a transformer.
5. What is the difference between magnetizing reactance and leakage reactance?
6. State whether true or false: Transformer is used to connect two circuits at different frequencies.
7. Define equivalent resistance and equivalent leakage reactance of a transformer.
8. What are the different losses incurred in a transformer? How they can be reduced?
9. State the condition for maximum efficiency.
10. What is the difference between direct loading test and indirect loading tests of a transformer?
11. Justify: 'Regulation of transformer is negative when load is capacitive'.
12. How is the magnetic leakage reduced in practical transformers?
13. Justify: Parallel operation of transformers improves the system reliability'.
14. What is the difference between hot rolled and CRGO steels?
15. Compare ideal transformer with practical transformer.
16. Why is the core made from several thin laminations?

Long Answer Questions

1. Explain the construction of a transformer. Discuss the material used for each of them
2. What are the benefits of cruciform cross section and mitred joints in the core?
3. Explain the working of a transformer at No-load and when load is connected.
4. Discuss: 'Transformer operates at poor input power-factor when working at No-load'. Hint: Refer to the phasor diagrams at No-load and On-load.
5. Draw and explain phasor diagram of a transformer supplying power to an R-L load.
6. Discuss OC/SC tests. Compare with direct loading test.
7. Explain: 'Transformer is a constant flux machine'.
8. Develop complete equivalent circuit of a transformer. Simplify it as referred to (i) primary side and (ii) secondary side.
9. Explain how the primary current I1 increases if the secondary current I2 is increased?
10. In a transformer, explain how impedances are transferred from primary to secondary and vice-versa.
11. Why iron is used for making the transformer core? What will happen if the core is made from non-magnetic material?
12. What will happen if a transformer designed for 50 Hz is operated on a 100 Hz supply?
13. What will happen if the primary winding is connected across dc supply?
14. What are the different conditions for parallel operation of transformers? Explain wht will happen if they are not fulfilled?
15. What do you mean by additive polarity and subtractive polarity of a transformer?
16. What are the different types of insulations used in a transformer? What is responsible for their ageing and damage?

Numerical Problems

1. A single-phase transformer is rated as 600/200V, 25kVA, 50Hz. Calculate (i) magnitudes of primary and secondary currents when it is fully loaded, (ii) maximum core flux and output voltage when the transformer is excited from HV side. Number of turns of HV winding = 60. Assume ideal transformer.
2. A 230/460V transformer has a primary resistance of 0.2 Ω and leakage reactance of 0.5 Ω. The corresponding values of secondary side are 0.75 Ω and 1.8 Ω respectively. Find the secondary terminal voltage when supplying 10A at 0.8 lagging power-factor.
3. A 5kVA, 200/1000V, 50Hz, 1-phase transformer gave following test results.

   OC test (on LV side):200V, 1.2A, 90W; SC test (on HV side): 50V, 5A, 110W

   Compute the parameters of approximate equivalent circuit referred to LV side and find percentage voltage regulation when the transformer is delivering 3kW at 0.8 lagging power factor. The input primary voltage is 200V.
4. The SC test results (on HV side) of a 200V/400V, 50 Hz transformer are: 15V, 10A, 85W. Calculate secondary voltage when delivering 5kW at 0.8 lagging power factor, the primary voltage being 200V.
5. In a 25kVA, 2000/200V transformer, the iron and copper loss are 350W and 400W respectively at rated load. Calculate the efficiency on unity power factor at (i) Full-load and (ii) Half full load. Also determine the load to get maximum efficiency.
6. A 10kVA, 500/250V single phase transformer has maximum efficiency of 94% when delivering 90% of its rated output at unity power factor. Estimate the efficiency when delivering rated output at 0.8 lagging power factor.
7. A 50kVA single-phase transformer of 2300/230V rating has the primary and secondary winding resistances of 2 Ω and 0.02 Ω respectively. The iron loss is 412W. Calculate the efficiency at (i) Half full-load and (ii) Full-load when the load power factor is 0.8.
8. A 220/400V, 10kVA, 50Hz single-phase transformer has at Full-load, a copper loss of 120W. If it has an efficiency of 98% at Full-load unity power factor, determine the iron loss.
9. The efficiency of a 1000kVA, 110/220V, 50Hz 1-phase transformer is 98.5% at Half full-load and 0.8 leading power factor. The efficiency of same transformer at Full-load, UPF is 98.8%. Determine (i) iron loss, (ii) Full-load copper loss and (iii) maximum efficiency at maximum power factor.
10. Find the all-day efficiency of a 500kVA distribution transformer whose copper loss and iron loss at Full-load are 4.5kW and 3.5kW respectively. During the 24 hours duration, it is loaded as given below.

Number of hours	Loading in kW	Power-factor
06	400	0.8
10	300	0.75
04	100	0.8
04	0	—

Answers of Numerical Problems

1. *41.66A, 125A, 0.045Wb, 200V*
2. *424.8V*
3. $R_o = 444.44\Omega$, $X_o = 179.85\ \Omega$, $R_{o1} = 0.176\ \Omega$, $X_{o1} = 0.359\ \Omega$, *3.34%*
4. *377.8V*
5. *97.08%, 96.52%, 23.38kVA*
6. *92.57%*
7. *94.56%, 95.76%*
8. *84.08W*
9. *4.073kW, 8.073kW, 98.9%*
10. *97.6%*

PRACTICALS

Experiment No: 1

Aim: To check the functioning of single-phase transformer

Transformer specifications:

Refer to the name plate of transformer and note down its specifications whichever are available.

kVA rating: 1 kVA	HV side rated voltage: 230V	LV side rated voltage: 115V
Frequency: 50Hz	Ambient temperature:	Cooling method:
% Impedance:	IS Standard:	Year of manufacturing:
Make:		

Apparatus:

Device	Make	Range	Quantity
1-phase dimmerstat		Input: 0-240 V, 5A	01
AC Voltmeter		0-150-300 V	02
Multimeter			01
Insulation tester			01
Thermometer			01
ICDP Switch			01

Circuit connections:

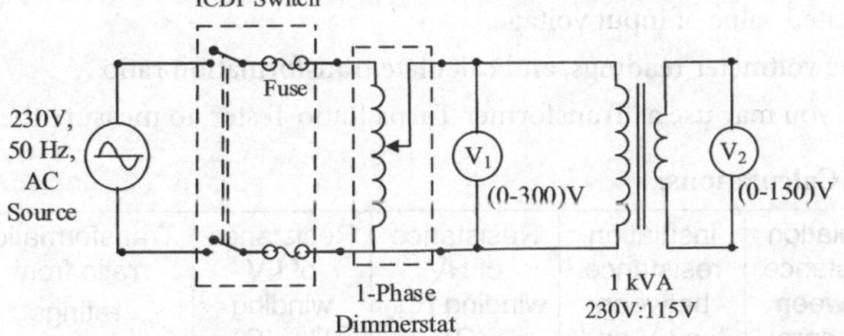

Fig. 3.33: Connection diagram for transformation ratio of single-phase transformer

Theory:

(i) The objective of this experiment is to,

(ii) Inspect the transformer for external damages if any.

(iii) Measure the insulation resistance at available ambient temperature

(iv) Check the continuity of two windings.

(v) Measure the winding resistances at available ambient temperature

(vi) Measure transformation ratio.

Precautions:

1. Make tight connections as shown and switch on the supply after they are verified by the instructor.
2. Ensure that you are standing on a rubber mat while performing the experiment.
3. Stand at a safe distance from the experimental setup when it is running.
4. Do not bend to take readings. Stand straight and avoid parallax error while taking the readings.
5. Ensure that the dimmerstat output voltage setting is at zero setting before switching ON the supply.
6. Do not apply input voltage higher than the rated primary voltage of transformer.
7. Select voltmeters of appropriate range if the transformer ratings are different from those given here.

Procedure:

1. Inspect the exterior parts of given single phase transformer for visible damages if any.
2. Check the insulation resistance:

   For this purpose, first short-circuit each winding by joining its both terminals to each other. Then connect the leads of insulation tester (i) between windings and core, and (ii) between windings to measure the insulation resistance. Also note the ambient temperature using thermometer.
3. Check winding continuity:

   Use multimeter or an ohmmeter to check separately, the continuity of each winding.
4. Measure winding resistances:

   Using the same multimeter or ohmmeter, measure *dc* resistances (R_{HV} and RLV) of both windings.
5. Measure transformation ratio (K):
   - (i) Connect the circuit as shown above.
   - (ii) Ensure that the dimmerstat is at zero voltage position and turn ON the ICDP switch.
   - (iii) Apply rated value of input voltage.
   - (iv) Note the voltmeter readings, and calculate transformation ratio.

   Alternatively, you may use a 'Transformer Turns Ratio Tester' to measure the value of K.

Observations and Calculations:

Insulation resistance between the core and HV winding (MΩ)	Insulation resistance between the core and LV winding (MΩ)	Insulation resistance between the LV and HV winding (MΩ)	Resistance of HV winding R_{HV} (Ω)	Resistance of LV winding R_{LV} (Ω)	Transformation ratio from ratings $K = \frac{V_{2(rated)}}{V_{1(rated)}}$	Transformation ratio from measurements $K = \frac{V_{2(measured)}}{V_{1(measured)}}$

Ambient temperature (°C): ____________________

External damages observed if any: ____________________

Conclusions:

____

Discussion:

1. *Why is it important to check the insulation?*

   Insulation resistance is primarily checked to test the integrity of the winding insulation. Better insulation implies lower leakage current and safe performance.

2. *Which factors are primarily responsible for the deterioration of insulation?*

   Temperature and the potential difference acting across the insulator are primarily responsible for the deterioration of insulation.

3. *What are the different classes of insulation?*

   As per Indian Standard IS: 1271 (1958), insulation is classified on the basis of thermal stability (maximum temperature that it can sustain safely) as given below.

Class of insulation	Temperature (°C)	Examples
Y	90	Cotton, silk, paper cellulose, wood without impregnation. These material are not suitable for electrical machines due to fast deterioration.
A	105	Material of Class Y with impregnation or coating of a dielectric oil
E	120	Synthetic resin, enamels, cotton and paper laminates
B	130	Inorganic material like mica, glass fibre, asbestos etc
F	155	Class B material with suitable bonding material of high thermal stability
H	180	Glass fibre, asbestos and mica with silicon resins
C	> 180	Porcelain, glass, quartz etc

4. *In a practical transformer, where is insulation employed?*

   (i) Insulation between core and windings

   (ii) Enamel varnish coated on the surface of copper wires

   (iii) Interlayer insulation

   (iv) Inter-winding insulation between primary and secondary

   (v) Outer layer insulation on HV winding

   (vi) Laminations of the core are insulated from each other to avoid the flow of eddy currents between them.

   (vii) Insulation between transformer assembly and the container (i.e transformer tank)

   (viii) Transformer oil in an oil-immersed transformer acts as a insulation between the transformer assembly and tank. It also acts as medium to transfer heat from core and windings to the outer tank body.

5. *Why in a core type transformer, both windings are wound one over the other on both limbs?*

   As explained in Section 3.3.2, both windings are halved and wound over each other on both limbs so that almost whole of the magnetic flux produced by each of them links with the other winding. In other words, it reduces the leakage flux.

6. *Which winding is wound first over to the core and why?*

   The LV winding is wound first over the core and then the HV winding is wound on the LV winding. This practice minimizes the cost of insulation between the core and its nearest winding. Read Section 3.3.2 for more details.

7. *In the magnetic core made from CRGO silicon steel, in which part of the core, iron-losses are more?*
   In a magnetic core made from CRGO silicon steel, iron-loss is more wherever the flux is not oriented in the direction of grains. For example, (i) around the holes punched in the laminations where bolts are inserted, (ii) at corners where the direction of flux is required to change by 90° etc.
8. *In distribution and power transformers, what is the potential of the core?*
   The magnetic core is normally grounded hence its potential is zero volts.

Experiment No: 2

Aim: To determine regulation and efficiency of a single-phase transformer by direct loading

Transformer specifications:

Refer to the name plate of transformer and note down its specifications whichever are available.

kVA rating: 1 kVA	HV side rated voltage: 230V	LV side rated voltage: 115V
Frequency: 50Hz	Ambient temperature:	Cooling method:
% Impedance:	IS Standard:	Year of manufacturing:
Make:		

Apparatus:

Device	Make	Range	Quantity
1-phase dimmerstat		Input: 0-240 V, 5A	01
AC Voltmeter		0-300 V	01
AC Voltmeter		0-150 V	01
AC Ammeter		0-10 A	01
AC Ammeter		0-5 A	01
1-Phase Wattmeter		0-5 A, 0-300 V	01
ICDP Switch (including fuse)		10 A	01
Variable resistive load with DPST switch		0 - 1 kW	01

Circuit connections:

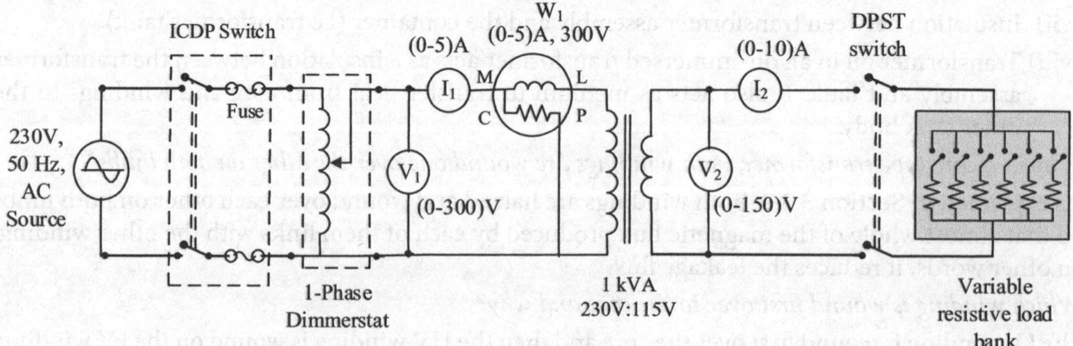

Fig. 3.34: Connection diagram for direct loading test on single-phase transformer

Theory:

Refer to Section 3.15 for the theoretical explanation on polarity marking.

Precautions:

1. Make tight connections as shown and switch on the supply after they are verified by the instructor.
2. Ensure that you are standing on a rubber mat while performing the experiment.
3. Stand at a safe distance from the experimental setup when it is running.
4. Do not bend to take readings. Stand straight and avoid parallax error while taking the readings.
5. Ensure that the dimmerstat output voltage setting is at zero setting before switching ON the supply.
6. Do not apply input voltage higher than the rated primary voltage of transformer.
7. Do not increase load above rated value by limiting the winding currents upto rated magnitude.
8. Select measuring instruments of appropriate range if the transformer ratings are different from those given here.

Procedure:

1. Make connections as shown in Fig. 3.34.
2. Set the dimmerstat at zero voltage position and the load bank switches at OFF position.
3. Calculate the rated values of currents from the transformer ratings and keep them ready with you.
4. Turn ON the ICDP switch.
5. Gradually increase the dimmerstat output voltage upto the rated primary voltage of transformer and keep it undisturbed. Note the readings of all measuring instruments. These are the no-load readings.
6. Now turn ON the DPST switch and increase the load in steps upto rated currents.
7. For every setting of the load, note down the readings of all measuring instruments.
8. Turn OFF the load and reduce the dimmerstat output voltage to zero.
9. Turn OFF the ICDP switch. Calculate efficiency and regulation for every set of reading.
10. Draw graph of output power (W_2) versus (i) % η and (ii) secondary terminal voltage V_2.

Observations and Calculations:

Sr. No	V_1	I_1	W_1*	V_2	$I2$	$W_2 = V_2 \times I_2$	% Efficiency η	% Regulation
1.	230			$V_{2(NL)}$	0	0	0	0
2.								
10.	230			$V_{2(FL)}$	Rated			

* Multiply the observed reading of wattmeter by its multiplying factor. See wattmeter cover.

$$\% \text{ Efficiency } \eta = \frac{W_2}{W_1} \times 100$$

$$\% \text{ Regulation } = \frac{V_{2(NL)} - V_2}{V_{2(NL)}} \times 100$$

Graph:

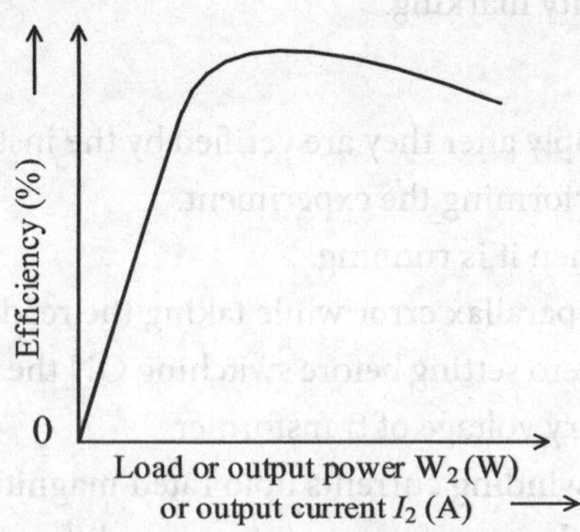

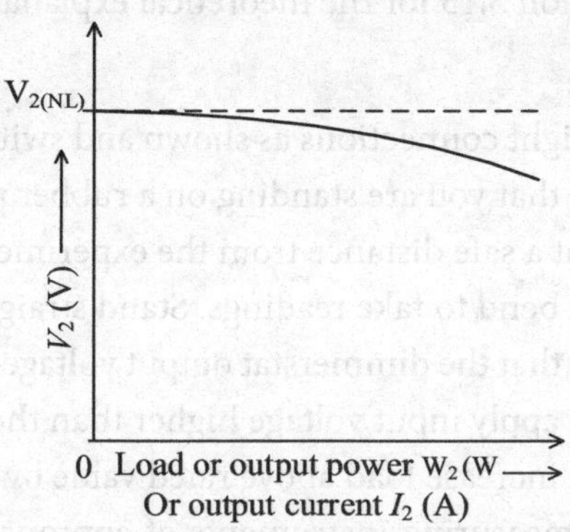

Output power (W_2) versus % efficiency η   Output power (W_2) versus terminal voltage V_2.

Results:

(1) Full-load efficiency = ____________

(2) Full-load regulation = ____________

Conclusion:

(Comment on the nature of graphs)

Discussion:

1. *Give a real-life experience of voltage regulation.*

   Sometimes in our houses, we receive less voltage from the power distribution company due to which the performance of electric appliances and fittings is not satisfactory. This usually happens during the evening time when most of the electrical appliances, electric lamps, and ventilating devices are in ON state. During summer season, it may happen during the afternoons and nights also. Let us investigate the probable reason. It was informed in Section 3.1 and 3.2 that, distribution transformers are employed by power companies to step-down the input voltage to 230V (phase value) which is then delivered to the residential and commercial consumers. This is the secondary terminal voltage V_2 of the transformer.

   Normally, all residential and commercial consumers have inductive (R-L) loads. When the load is increased, the load current (which is also the secondary winding current *I*2 of transformer) will also increase. Consequently, the two voltage drops as referred to secondary side i.e. $I_2 R_{02}$ and $I_2 X_{02}$ will increase. As a result, there is a fall in secondary terminal voltage V_2 of the transformer which we experience in our houses.

2. *Why wattmeter is not used to measure the output power in this experiment?*

   In this experiment, we have connected resistive load across the transformer. For resistance, the power-factor cos ϕ =1. Therefore, $W_2 = V_2 I_2 \cos\theta = V_2 I_2$ watts. Therefore, a product of voltmeter and ammeter reading can serve the purpose and hence wattmeter is not employed.

3. *What is wattmeter multiplying factor?*

   Wattmeter is an electrodynamic type measuring instrument which has a current coil (M-L) and a potential coil (C-P). Commercially, the wattmeters are provided with multiple selections for current range and voltage range. This is accomplished by shunt and series resistors of appropriate magnitudes inside the instrument. However for all combinations of current and voltage ranges, the scale below the pointer remains same. Hence to distinguish the readings for different voltage and current range selections, multiplying factors (MF) are provided by the manufacturer of the wattmeter.

4. *Can we use R-L load in this experiment?*

   Yes, R-L load can also be used. However, with R-L load you have to connect a wattmeter on the secondary side also to measure W_2 because for R-L load, cos $\phi < 1$.

5. *What is the difference between DPST switch and ICDP switch?*

   ICDP and DPST switches stand for 'Iron Clad Double Pole switch' and 'Double Pole Single Throw switch' respectively. DPST switch is a pair of ON-OFF switches which operate together. It can be used in ON/OFF configuration where its terminal-pair is either connected or not connected. ICDP switch is normally used to switch ON or OFF the main single-phase supply and hence is also called as Main Switch. It also contains two fuses. One is for 'Phase' wire and the other is for 'Neutral'.

6. *What is the function of fuse?*

   Fuse breaks the circuit if a fault or over-load causes too much current to flow. Thus it provides protection against over-load and short-circuit fault.

7. How load increases when more number of parallel resistors of the load bank are connected in the circuit?

   It is known that when more number of resistors are added in parallel, the magnitude resultant resistance decreases. This leads to rise in load current and hence the rise in output power in this experiment.

8. *Can the voltage regulation be negative? Justify your answer.*

   Yes voltage regulation can be negative when the load is capacitive. This is because, with rise in capacitive load, the secondary terminal voltage increases till some value of load.

9. *What is the condition for maximum efficiency?*

   Efficiency of a transformer is maximum for that value of load at which, the iron loss and copper loss of a transformer are equal.

10. *Which is the other method for determining efficiency and regulation?*

    Efficiency and voltage regulation of a transformer can also be determined by performing the open-circuit and short-circuit tests.

Experiment No: 3

Aim: To perform open-circuit and short-circuit tests on a single-phase transformer

Transformer specifications:

Refer to the name plate of transformer and note down its specifications whichever are available.

kVA rating: 1 kVA	HV side rated voltage: 230V	LV side rated voltage: 115V
Frequency: 50Hz	Ambient temperature:	Cooling method:
% Impedance:	IS Standard:	Year of manufacturing:
Make:		

Apparatus:

Device	Make	Range	Quantity
1-phase dimmerstat		Input: 0-240 V, 5A	01
AC Voltmeter		0-75 V	01
AC Voltmeter		0-150 V	01
AC Ammeter		0-5 A	01
AC Ammeter		0-1 A	01
1-Phase Wattmeter		0-5 A, 0-75 V	01
1-Phase Wattmeter		0-1 A, 0-150 V	01
ICDP Switch (including fuse)		10 A	01

Circuit connections:

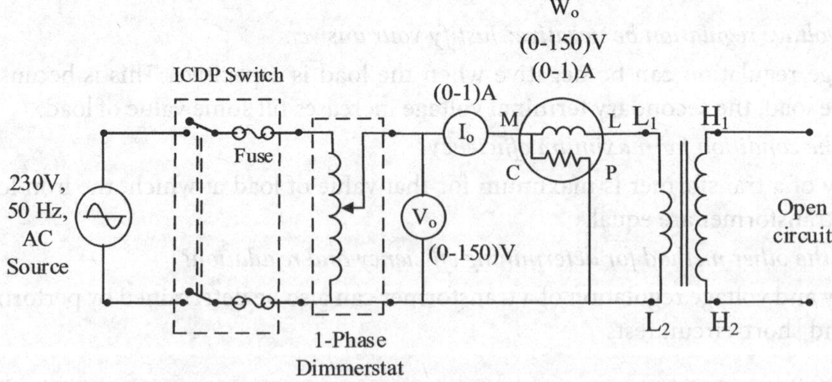

Fig. 3.35: Connection diagram for open-circuit test on single-phase transformer

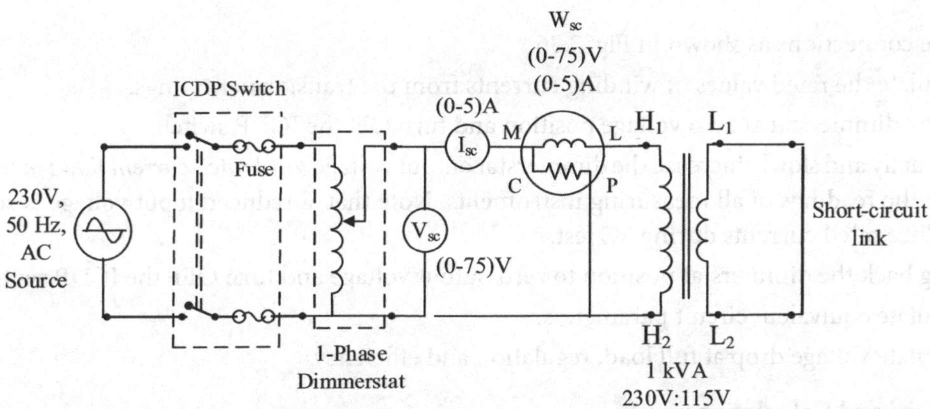

Fig. 3.36: Connection diagram for short-circuit test on single-phase transformer

Theory:

Refer to Section 3.16 for the theoretical explanation on OC and SC tests.

Precautions:

1. Make tight connections as shown and switch on the supply after they are verified by the instructor.
2. Ensure that you are standing on a rubber mat while performing the experiment.
3. Stand at a safe distance from the experimental setup when it is running.
4. Do not bend to take readings. Stand straight and avoid parallax error while taking the readings.
5. Ensure that the dimmerstat output voltage setting is at zero setting before switching ON the supply.
6. Do not apply input voltage higher than the rated primary voltage of transformer.
7. Be careful during short circuit test because a reduced input voltage is sufficient to produce rated currents during SC test. Do not allow the winding currents to exceed rated magnitude.
8. Select measuring instruments of appropriate range if the transformer ratings are different from those given here.

Procedure:

OC Test:

1. Make connections as shown in Fig. 3.35.
2. Set the dimmerstat at zero voltage position and turn ON the ICDP switch.
3. Gradually increase the dimmerstat output voltage *upto the rated voltage of transformer* and keep it undisturbed. Note the readings of all measuring instruments. These are the no-load readings.
4. Bring back the dimmerstat position to zero output voltage and turn OFF the ICDP switch.
5. Calculate no-load parameters of the equivalent circuit as referred to primary
6. At No-Load Condition or during OC Test, the transformer power-factor is very poor (low). Therefore, for more accuracy, the Wattmeter used in OC Test is a *Low Power-factor* (*LPF*) Wattmeter.

SC Test:

1. Make connections as shown in Fig. 3.36.
2. Calculate the rated values of winding currents from the transformer ratings.
3. Set the dimmerstat at zero voltage position and turn ON the ICDP switch.
4. Gradually and slowly increase the dimmerstat output voltage *until rated current flows in the winding.* Note the readings of all measuring instruments. Note that, a reduced input voltage is sufficient to produce rated currents during SC test.
5. Bring back the dimmerstat position to zero output voltage and turn OFF the ICDP switch.
6. Calculate equivalent circuit parameters.
7. Calculate voltage drop at full load, regulation and efficiency.

Observations and Calculations:

OC Test:

V_o	I_o	W_o = Wattmeter reading x MF	$\cos\theta_o = \frac{W_o}{V_o I_o}$	$R_o = \frac{V_o}{I_o \cos\cos\theta_o}$	$X_o = \frac{V_o}{I_o \sin\theta_o}$
Rated value					

* Multiply the observed reading of wattmeter by its multiplying factor (MF). See wattmeter cover.

Note: To obtain the equivalent circuit as referred to HV side (which is considered as primary side in a 230V:115V transformer), all parameters of the equivalent circuit should be as referred to primary. The OC test was performed on LV side therefore the above values of R_o and X_o are as referred to LV side (i.e. secondary side). They should be referred to primary side as:

$$R_o' = \frac{R_o}{K^2} \text{ and } X_o' = \frac{X_o}{K^2} \text{ where, the transformation ratio } K = \frac{V_2}{V_1} = \frac{115}{230}$$

During OC test, since rated voltage (115 V) was applied on input side, the wattmeter reading *Wo* will be equal to the rated iron-loss of transformer. Therefore,

$$P_i = W_o$$

SC Test:

V_{sc}	I_{sc}	W_{sc} = Wattmeter reading x MF	$Z_{sc} = \frac{V_{sc}}{I_{sc}}$	$R_{sc} = \frac{W_{sc}}{I_{sc}^2}$	$X_{sc} = \sqrt{Z_{sc}^2 - R_{sc}^2}$
	Rated value				

* Multiply the observed reading of wattmeter by its multiplying factor (MF). See wattmeter cover.

Note: Since SC test was performed on HV side, the above values of R_{sc} and X_{sc} are as referred to HV side (i.e. primary side). They can also be written as:

$$R_{sc} = R_{o1} \text{ and } X_{sc} = X_{o1}$$

During SC test, since rated current was flowing in the windings, the wattmeter reading *Wsc* will be equal to the rated copper-loss (i.e. full-load copper loss) of transformer. Therefore,

$$P_{cu(\mathrm{FL})} = Wsc$$

Calculate efficiency and voltage regulation:

(i) At Full-load, 0.8 lagging power factor

$$\% \eta_{(FL,0.8)} = \frac{\text{Full load VA capacity} \times 0.8}{\text{Full load VA capacity} \times 0.8 + P_i + P_{\mathrm{cu(FL)}}} \times 100$$

$$\% \text{ Regulation} = \frac{I_{1(FL)}\left(R_{o1} \times 0.8 + X_{o1} \times 0.6\right)}{V_1} \times 100$$

(ii) At Full-load, 0.8 leading power factor

$$\% \eta_{(FL,0.8)} = \frac{\text{Full load VA capacity} \times 0.8}{\text{Full load VA capacity} \times 0.8 + P_i + P_{\mathrm{cu(FL)}}} \times 100$$

$$\% \text{ Regulation} = \frac{I_{1(FL)}\left(R_{o1} \times 0.8 - X_{o1} \times 0.6\right)}{V_1} \times 100$$

(iii) At Full-load, unity power factor

$$\% \eta_{(FL,UPF)} = \frac{\text{Full load VA capacity} \times 1}{\text{Full load VA capacity} \times 1 + P_i + P_{\mathrm{cu(FL)}}} \times 100$$

$$\% \text{ Regulation} = \frac{I_{1(FL)} R_{o1}}{V_1} \times 100$$

Results:

(1) Full-load efficiency at 0.8 lagging, 0.8 leading and UPF = ____________________

(2) Full-load regulation at 0.8 lagging, 0.8 leading and UPF = ____________________

(3) Simplified equivalent circuit as referred to HV side:

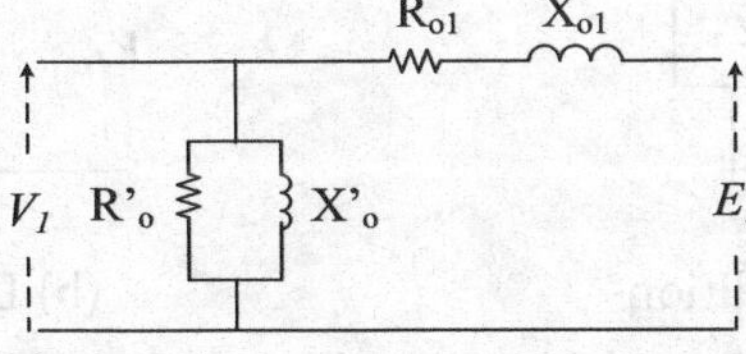

Discussion:

1. *Why is it preferred to perform OC test on LV side and SC test on HV side?*

   It is not mandatory to perform OC test on LV side and SC test on HV side. One can perform both tests on any one side also. Consider a 10kVA, 11kV/230V single-phase transformer. During the OC test, we have to apply rated input voltage. If OC test is performed on HV side then, voltmeter and wattmeter capable of measuring 11kV will be required which are commercially not available. Similarly, during the SC test, rated currents should be passed. The rated current on LV side will be 47.82A. Ammeter and wattmeter of this current range may not be available. Hence, it is preferred to perform OC test on LV side and SC test on HV side.

2. *Why are transformers rated in kVA and not in kW?*

   Refer Section 3.14.

3. *Why iron-loss are neglected in SC test?*

   During SC test, the applied voltage is of reduced magnitude whereas current flowing in the windings are of rated magnitude. Therefore, iron loss taking place in the transformer core is negligibly small as compared to the copper loss take takes place in both windings. Hence with allowable loss of accuracy, it is assumed that the input power (indicated by the wattmeter) is totally consumed by copper loss alone.

4. *Why copper-loss is neglected in OC test?*

   During the OC test, current $I_2 = 0$ and therefore current $I_1 = I_o$ which is just 1 to 5% of rated current. Whereas, the applied voltage has rated magnitude. Therefore, copper loss in secondary winding is zero and that in primary winding is negligibly small as compared to the iron loss. Hence with allowable loss of accuracy, it is assumed that input power (wattmeter reading) is totally consumed by iron loss alone.

5. *What is the output power during OC and SC tests?*

   Output power is given by $P_2 = V_2 I_2 \cos \phi$. During OC test, $I_2 = 0$ and during SC test, $V_2 = 0$. Hence, during OC as well as SC tests, the output power is zero.

6. *Why reduced voltage is applied during SC test?*

   The approximate equivalent circuits as referred to primary side during (a) normal load condition and (b) during SC test are shown below. $Z'_L = Z_L / K^2$ is the load impedance as referred to primary side

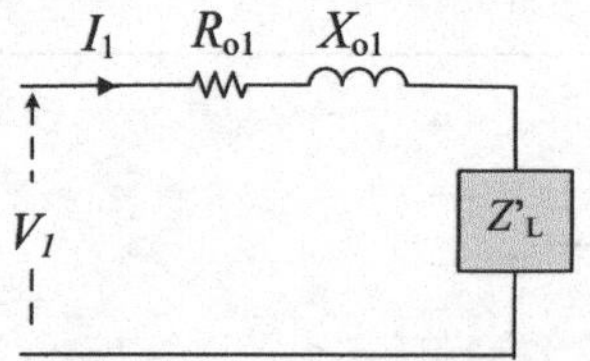

(a) During normal load condition

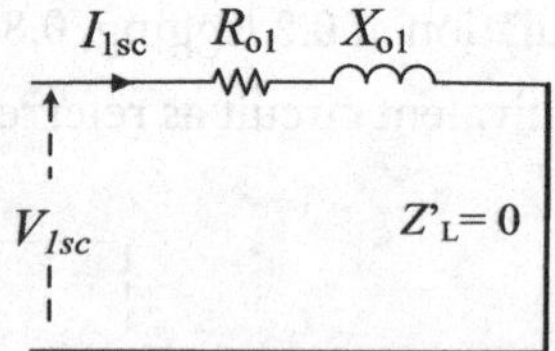

(b) During SC test

Fig. 3.37 Approximate equivalent circuits as referred to primary side

When load is connected across secondary, the input current is, $I_1 = \dfrac{V_1}{Z_{o1} + Z_L^{'}}$

During short circuit test, $Z'_L = 0$ and therefore, the input current is, $I_{1sc} = \dfrac{V_{1sc}}{Z_{o1}}$

Where, $Z_{01} = (R_{01} + jX_{01})$. Since, $Z_{01} < Z'_L$, even a small input voltage V_{1sc} is sufficient to produce rated currents in the windings during SC test. If rated input voltage ($V_{1sc} = V_1$) is applied during SC test than, due to lesser opposition Z_{01}, large current will be produced in the windings which can produce huge amount of copper loss and heat.

Why Low Power Factor (LPF) wattmeter is preferred during OC test?

At no-load (i.e open circuited secondary), the input current Io is mostly of magnetizing nature (i.e inductive) and its active component (I_w) is relatively very small. Therefore, the power factor at no-load is poor (between 0.2 and 0.4). It improves as load is increased due to rise in active component of current. To understand better, refer to the phasor diagrams in Fig. 3.17. At no-load the input phase angle θ_o is the angle between V_1 and I_o whereas after connecting the load, the input phase angle θ_1 is the angle between V_1 and I_1. It can be observed that, $\angle\theta_o > \angle\theta_1$ which means that the input power factor at no-load ($\cos\theta_o$) is smaller than the power factor when load is connected. Hence for more accuracy, Low Power Factor (LPF) wattmeter is preferred during OC test.

In this experiment, OC and SC tests were performed on a 50 Hz supply. What will be the effect on equivalent circuit parameters if the same tests are repeated on a 60 Hz supply?

As inductive reactances are directly proportional to frequency, they will increase in the ratio (60/50). The resistances will remain almost same.

During SC test, $W_{sc} = 75$ W when the current in short-circuit link is 10A. What will be the wattmeter reading if the current in short-circuited secondary is doubled?

Since copper loss is proportional to square of current, it will increase four times if the current is doubled. Hence wattmeter reading will be $Wsc = 300\ W$

Experiment No: 4

Aim: To perform polarity test on single-phase transformer whose polarity markings are masked

Transformer specifications:

Refer to the name plate of transformer and note down its specifications whichever are available.

kVA rating: 1 kVA	HV side rated voltage: 230V	LV side rated voltage: 115V
Frequency: 50Hz	Ambient temperature:	Cooling method:
% Impedance:	IS Standard:	Year of manufacturing:
Make:		

Apparatus:

Device	Make	Range	Quantity
1-phase dimmerstat		Input: 0-240 V, 5A	01
AC Voltmeter		0-300 V	01
AC Voltmeter		0-150 V	01
AC Voltmeter		0-600 V	01
ICDP Switch (including fuse)		10 A	01

Circuit connections:

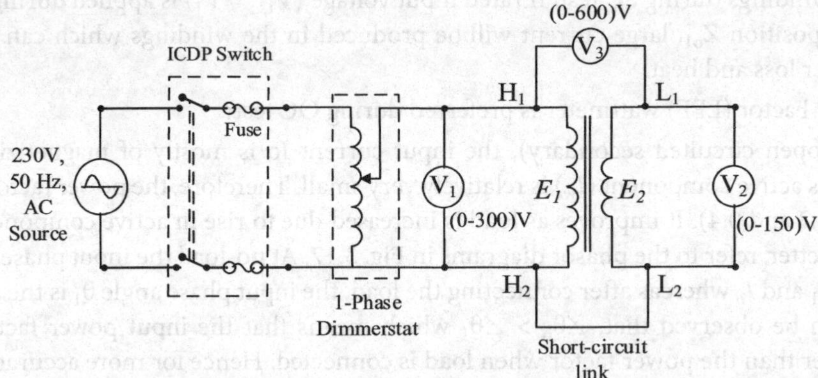

Fig. 3.38: Connection diagram for polarity test on single-phase transformer

Theory:

Refer to Section 3.17 for the theoretical explanation on polarity marking.

Precautions:

1. Make tight connections as shown and switch on the supply after they are verified by the instructor.
2. Ensure that you are standing on a rubber mat while performing the experiment.
3. Stand at a safe distance from the experimental setup when it is running.
4. Do not bend to take readings. Stand straight and avoid parallax error while taking the readings.
5. Ensure that the dimmerstat output voltage setting is at zero setting before switching ON the supply.
6. Do not apply input voltage higher than the rated primary voltage of transformer.
7. Select measuring instruments of appropriate range if the transformer ratings are different from those given here.

Procedure:

1. Make connections as shown in Fig. 3.38.
2. Connect the HV side to input source through a dimmerstat.
3. Short circuit the terminals H_2 and L_2.
4. Connect three voltmeters as shown.
5. Set the dimmerstat at zero voltage position and then turn ON the source.
6. Assume a dot (•) on any one terminal of HV winding (representing either + or – polarity) say at H_1.
7. Increase the input voltage gradually upto rated value by adjusting the dimmerstat position.
8. Note the voltmeter readings and analyse them to identify the position of corresponding dot on LV side.
   - If $V_3 = (V_1 - V_2)$ than, the transformer has subtractive polarities.
   - If $V_3 = (V_1 + V_2)$ than, the transformer has additive polarities.

Observations:

Sr. No	V_1	V_2	V_3	If, $V_3 = (V_1 - V_2)$	If, $V_3 = (V_1 + V_2)$
				The transformer is of subtractive polarities. Terminals H_1, L_1 will have similar polarities. Similarly, the instantaneous voltage polarities of H_2, L_2 will be same.	The transformer is of additive polarities. Terminals H_1, L_2 will have similar polarities. Similarly, the instantaneous voltage polarities of H_2, L_1 will be same.

Conclusion:

Discussion:

1. *What is transformation ratio? What will be its value for a step-up, step-down and isolation transformer?*

   Transformation ratio is the ratio of number of turns of secondary winding N2 to the number of turns of primary winding N1. It can also be expressed as the ratio of voltages or as the ratio of currents.

   $$K = \frac{N_2}{N_1} = \frac{E_2}{E_1} = \frac{V_2}{V_1} = \frac{I_1}{I_2}$$

   Where, V_1 and V_2 are the terminal voltages of primary and secondary windings respectively.

   For a step-up transformer: $V_2 > V_1$, $N_2 > N_1$, $\therefore K > 1$

   For a step-down transformer: $V_2 < V_1$, $N_2 < N_1$, $\therefore K < 1$

   For an isolation transformer: $V_2 = V_1$, $N_2 = N_1$, $\therefore K = 1$

2. *What do you mean by the statement 'transformer provides isolation'?*

   Transformer provides electrical isolation between circuits that are connected to primary and secondary windings. Due to the magnetic mutual coupling, the effect of load on the secondary side gets reflected on primary side. The isolation feature is useful when a pulse-transformer is employed to couple the gate- pulse generator with gate terminal of a power semiconductor based switching device like silicon controlled rectifier (SCR).

3. *Polarity test gives us the continuous polarity or instantaneous polarity?*

   It gives instantaneous polarity only. Transformer being an *ac* machine, the voltage polarities as well as the current direction alternate after every half cycle. Polarity test simply indicates, which two terminals from primary and secondary side have same instantaneous polarities.

4. *The rated copper loss and rated iron loss of a transformer are 400W and 150 W respectively. What will be their values at half-full load?*

   Copper loss varies with load and hence is called as 'variable loss' whereas, iron-loss is independent of load and is therefore called as 'constant loss'.

   $$P_{cu\left(\frac{1}{2}FL\right)} = \left(\frac{1}{2}\right)^2 P_{cu(FL)} = 100\text{W} \text{ and } P_{i\left(\frac{1}{2}FL\right)} = P_{i(FL)} = 150\text{W}$$

5. *Calculate the rated primary current of a 2 kVA, 230/115 volts single-phase transformer. What will be its values at half-full load? Can you obtain the corresponding values on secondary side?*

$$I_{1(FL)} = \frac{\text{Full load voltamperes}}{\text{rated primary voltage}} = \frac{2000}{230} = 8.69\text{A}$$

$$I_{1(\frac{1}{2}FL)} = \left(\frac{1}{2}\right) I_{1(FL)} = \frac{8.69}{2} = 4.345\text{A}$$

$$I_{2(FL)} = \frac{1}{K} I_{1(FL)} = \left(\frac{V_1}{V_2}\right) I_{1(FL)} = \frac{230}{115} \times 8.69 = 17.391\text{A}$$

$$I_{2(\frac{1}{2}FL)} = \left(\frac{1}{2}\right) I_{2(FL)} = \frac{17.391}{2} = 8.69\text{A}$$

6. *Full-load efficiency of a single phase transformer is 96% at 0.8 lagging power factor. What will be the change in efficiency on full-load 0.8 leading power factor?*

   Efficiency does not depend on the lagging or leading nature of power factor. Hence, it will remain same.

7. *Could you think of a situation where incorrect connections at transformer terminals would be harmful?*

   Two or more transformers connected wrongly in parallel with respect to their polarities will result into short circuit. Heavy current will circulate within the transformer windings.

8. *Determine the direction of secondary emf and current in the following example.*

   Hint: Use Right hand thumb rule (Refer appendix of Unit-1: Direction of flux established by a current carrying coil) to find the direction of primary flux ϕ and then show secondary flux ϕ_2 in opposite direction of ϕ. Using Right hand thumb rule again, determine the direction of I_2 from ϕ_2

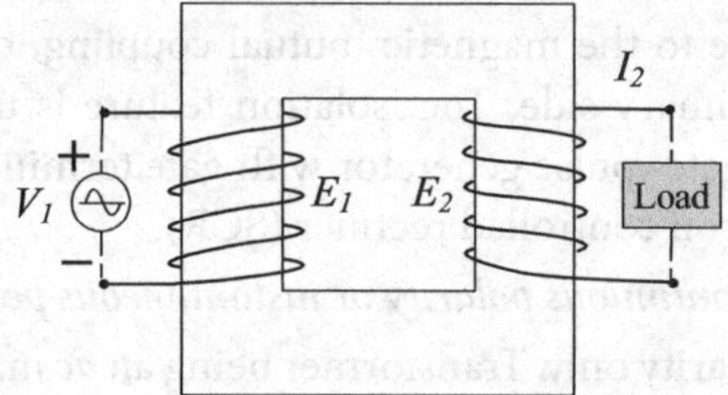

9. *Indicate similar polarity terminals by dot marking on LV and HV sides of Question 7.*

   Hint: Refer Section 3.17 for details.

10. Draw the lines of magnetic flux produced by currents in the conductors shown below.

    Hint: Use Right hand thumb rule (If unable to answer, you may refer to Appendix of Unit-1: Ampere's Right Hand Thumb rule, Dot & Cross marking).

Experiment No: 5

Aim: To perform parallel operation of two single-phase transformers and,

(i) Determine the load current sharing *

(ii) Determine the apparent and real power sharing *

* Teachers may divide this experiment into two separate experiments.

Transformer specifications:

Refer to the name plates of both transformers and note down the specifications whichever are available.

TRANSFORMER: A	Make:	
kVA rating: 1 kVA	HV side rated voltage: 230V	LV side rated voltage: 115V
Frequency: 50Hz	Ambient temperature:	Cooling method:
% Impedance:	IS Standard:	Year of manufacturing:

TRANSFORMER: B	Make:	
kVA rating: 1 kVA	HV side rated voltage: 230V	LV side rated voltage: 115V
Frequency: 50Hz	Ambient temperature:	Cooling method:
% Impedance:	IS Standard:	Year of manufacturing:

Apparatus:

Device	Make	Range	Quantity
1-phase dimmerstat		Input: 0-240 V, 5A	01
AC Voltmeter		0-300 V	01
AC Voltmeter		0-75-150 V	01
AC Voltmeter		0-600 V	01
AC Ammeter		0-5 A	01
AC Ammeter		0-10 A	03
AC Ammeter		0-20 A	01
1-Phase Wattmeter		0-5 A, 0-75 V	01
1-Phase Wattmeter		0-10 A, 0-150 V	02
1-Phase Wattmeter		0-20 A, 0-150 V	01
Variable rheostatic load bank		2 kW, 115V	01
ICDP Switch (including fuse)		10 A	01
DPDT Switch		20 A	01

Circuit connections:

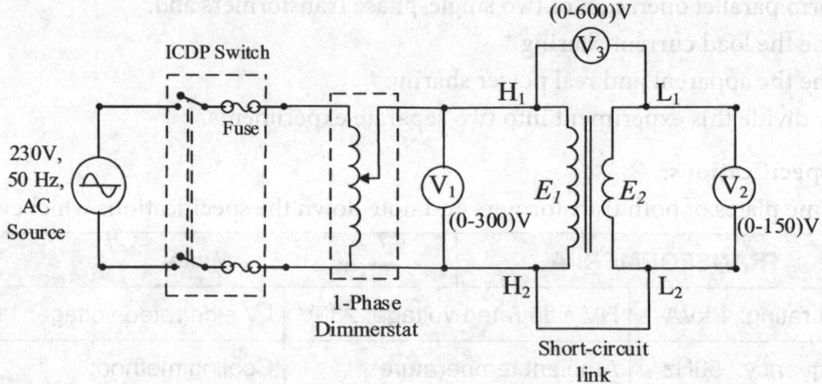

Fig. 3.39: Connection diagram for polarity test

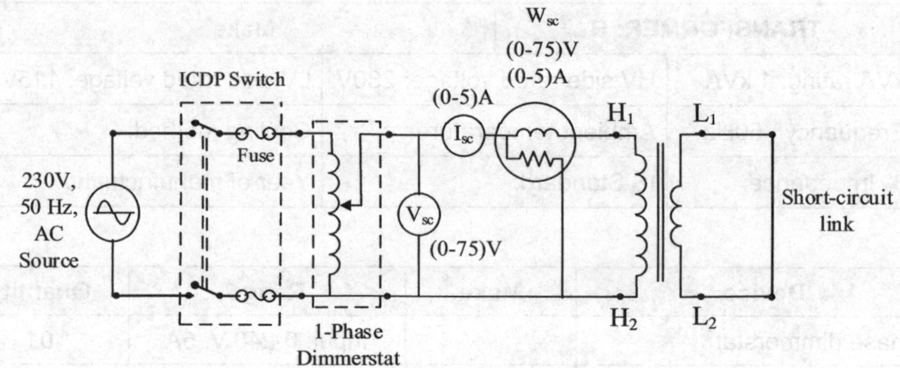

Fig. 3.40: Connection diagram for short-circuit test or impedance test

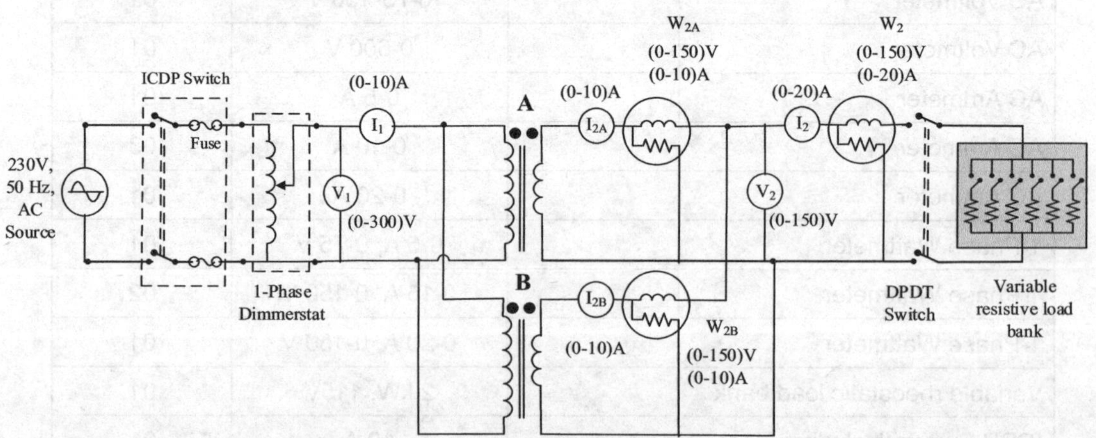

Fig. 3.41: Connection diagram for parallel operation

Theory:

Refer to Sections 3.18 and 3.19 for the theoretical explanation on parallel operation of transformers.

Precautions:

1. Make tight connections as shown and switch on the supply after they are verified by the instructor.
2. Ensure that you are standing on a rubber mat while performing the experiment.
3. Stand at a safe distance from the experimental setup when it is running.
4. Do not bend to take readings. Stand straight and avoid parallax error while taking the readings.
5. Ensure that the dimmerstat output voltage setting is at zero setting before switching ON the supply.
6. Do not apply input voltage higher than the rated primary voltage of transformer.
7. Do not allow the load kVA rating to exceed the sum of kVA ratings of the two transformers.
8. Select measuring instruments of appropriate range if the transformer ratings are different from those given here.

Procedure:

1. For each of the transformers, note down the name plate ratings and calculate the rated currents for both the windings.
2. As explained in Section 3.18, the two transformers must have identical polarities, identical voltage ratios, equal ratios of equivalent leakage reactance to equivalent resistance and the equivalent leakage impedances inversely proportional to the respective kVA ratings. It is assumed that these conditions are already satisfied. If they are not verified than:
3. First perform Polarity Test on both transformers as per the connection diagram shown in Fig. 3.39 and identify their similar polarity terminals. Mark them by 'dot'. For detail procedure on polarity test, refer to Experiment No. 4.
4. Also, confirm that the no-load secondary voltages of both transformers match in magnitude.
5. Then perform Short Circuit Test on both transformers as per the connection diagram shown in Fig. 3.40 and calculate equivalent leakage impedances Z_{o2A}, Z_{o2B}; equivalent leakage reactances X_{o2A}, X_{o2B} and equivalent resistances R_{o2A}, R_{o2B}. Mark them by 'dot'. For detail procedure on Short-circuit test, refer to Experiment No. 3.
6. Now connect the two transformers in parallel across a common variable resistive load as shown in connection diagram of Fig. 3.41.
7. Ensure that initially, the dimmerstat position is at zero. Also ensure all the load switches as well as the DPDT switch in OFF position.
8. Turn ON the input source and increase the dimmerstat output voltage gradually until rated voltage appears across the primaries of each transformer.
9. Increase the load in steps till rated value and note the readings. Ensure that, the load kVA rating does not exceed the sum of kVA ratings of the two transformers.
10. After all sets of reading are noted, turn OFF the load switches, turn OFF the DPDT switch, adjust the dimmerstat position to zero and turn OFF the ICDP switch.

Observations:

POLARITY TEST

Transformer	V_1	V_2	V_3	Verify Voltage ratio $K=\frac{V_2}{V_1}$	If, $V_3 = (V_1 - V_2)$	If, $V_3 = (V_1 + V_2)$
A					The transformer is of subtractive polarities. Similar polarity terminals are: H_1, L_1 and H_2, L_2	The transformer is of additive polarities. Similar polarity terminals are: H_1, L_2 and H_2, L_1
B						

SHORT-CIRCUIT TEST

Transformer	V_{sc}	I_{sc}	W_{sc}	$Z_{o1}=Z_{sc}=\frac{V_{sc}}{I_{sc}}$	$R_{o1}=R_{sc}=\frac{W_{sc}}{I_{sc}^2}$	$X_{o1}=X_{sc}=\sqrt{Z_{sc}^2-R_{sc}^2}$
A						
B						
Considering HV side of both transformers as primary and LV sides as secondary.						

Transformer	$Z_{o2} = K^2 Z_{o1}$	$R_{o2} = K^2 R_{o1}$	$X_{o2} = K^2 X_{o1}$	Verify whether,	Verify whether,
A				$\frac{Z_{o2B}}{Z_{o2A}}=\frac{\text{kVA}_A}{\text{kVA}_B}$	$\frac{X_{o2A}}{R_{o2A}}=\frac{X_{o2B}}{R_{o2B}}$
B					

PARALLEL OPERATION

Sr. No.	V_1	I_1	V_2	I_2	W_2	I_{2A}	W_{2A}	I_{2B}	W_{2B}

Calculations:

(i) To determine the load current sharing

Sr. No.	I_2	I_{2A}	Calculated value of I_{2A} $= I_2 \frac{Z_{o2B}}{(Z_{o2A}+Z_{o2B})}$	I_{2B}	Calculated value of I_{2B} $= I_2 \frac{Z_{o2A}}{(Z_{o2A}+Z_{o2B})}$	Remark

(ii) To determine the apparent power sharing

Sr. No.	$S = V_2 I_2$	$S_A = V_2 I_{2A}$	Calculated value S_A $= S \frac{Z_{o2B}}{(Z_{o2A}+Z_{o2B})}$	$S_B = V_2 I_{2B}$	Calculated value of S_B $= S \frac{Z_{o2A}}{(Z_{o2A}+Z_{o2B})}$	Remark

(iii) To determine the real power sharing

Sr. No.	W_{2A}	W_{2B}	W_2	Calculated value $W_{2A} = V_2 I_{2A} \cos\theta$	Calculated value $W_{2B} = V_2 I_{2B} \cos\theta$	Calculated value $W_2 = W_{2A} + W_{2B}$	Remark
Since the load is resistive, $\cos\theta = 1$							

Conclusion:

____

____

Discussion:

1. *Why do we need to operate transformers in parallel?*
   The main reason of parallel operation is to fulfill the future rise in load demand at optimum cost by increasing the MVA capacity of existing system. Otherwise, the other solution would have been to replace the currently operating low MVA transformer by a higher MVA transformer which is not cost effective. Secondly, running two or more low capacity transformers in parallel offers

additional advantages like, improvement in system reliability by maintaining the continuity of supply, improved economy, increase in overall efficiency and reduction in the cost of stand-by unit.

2. *What are the conditions to be satisfied for parallel operation of transformers?*

   The conditions that should be fulfilled before running two or more transformers in parallel are:

   1. Interconnected terminals of two transformers should have same polarities.
   2. The transformers should have same voltage ratios.
   3. If the transformers of different MVA ratings are to be connected in parallel than, their kVA ratings should be inversely proportional to their respective equivalent leakage impedances.
   4. The transformers should have same ratio of equivalent leakage reactance to equivalent resistance.

   Amongst above four conditions, the first one is mandatory.

3. *If the two secondaries of the transformers in parallel are not connected with proper polarity and a voltmeter is connected in series with the winding then the reading will be?*

   The voltmeter will read the sum of no-load secondary voltages of the two transformers.

4. Why is the rating of the transformer expressed in kVA?

   Transformer rating is expressed in terms of apparent power i.e voltamperes (either kVA or MVA) and not in terms of active power (i.e. *VI* cos θ watts). This is because, the iron loss depends on voltage and the copper loss depends on current. These two losses are responsible for the production of heat which the transformer should be able to sustain safely. The power factor directly does not contribute to any loss or heat.

5. *If a 100kVA transformer has* $Z_{o2A} = 2\Omega$, then the other transformer of 500kVA in parallel must have equivalent leakage impedance of ?

   $$Z_{o2B} = \frac{100}{500} \times 2 = 0.2\Omega$$

6. *At no-load if* E_{2A} *and* E_{2B} *are the secondary voltages of transformers A and B and their equivalent leakage impedances are* Z_{o2A} *and* Z_{o2B} *then circulating current is ?*

   - If the transformers are connected in parallel with correct polarities than the circulating current in secondary loop will be,

     $$I_c = \frac{E_{2A} - E_{2B}}{Z_{o2A} + Z_{o2B}}$$

   - But if they are connected with incorrect polarities on the secondary side than,

     $$I_c = \frac{E_{2A} + E_{2B}}{Z_{o2A} + Z_{o2B}}$$

7. *Can two transformers of different sizes be operated in parallel?*

   Yes transformers of different kVA or MVA ratings can be connected in parallel provided that, the condition 3 in Ques. 2 is satisfied. However, voltage ratios and voltage ratings should be equal.

8. *When is the circulating current produced in the parallel operation of two transformers?*

   It is produced in two cases: (i) when the polarities are incorrect and (ii) when voltage ratios are not equal.

9. *What is the effect of circulating currents when two transformers are connected in parallel?*
   Circulating current results in additional copper loss.
10. *Among transformers and rotating machines, which one has higher efficiency?*
   Transformer has higher efficiency as compared to all rotating machines. This is because, the rotating machines have mechanical losses too (friction and windage) which are absent in transformers.

APPENDIX

Lagging and Leading nature of ac quantities:

All sinusoidally varying ac quantities (current, voltage, magnetic flux etc.) can be represented graphically as well as mathematically. The graphical forms include (i) waveforms and (ii) phasor diagrams. Waveforms represent instantaneous magnitude and are denoted by small-case letters like *i*, *v*, *e* etc. Phasors represent Root Mean Square (RMS) value and are denoted by capital letters like I, V, E etc. Consider a set of *ac* currents i_1, i_2, i_3 and i_4 as shown by waveforms in Fig. 3.42. The x-axis represents 'time' axis (ωt) whereas the y-axis represents the instantaneous magnitude.

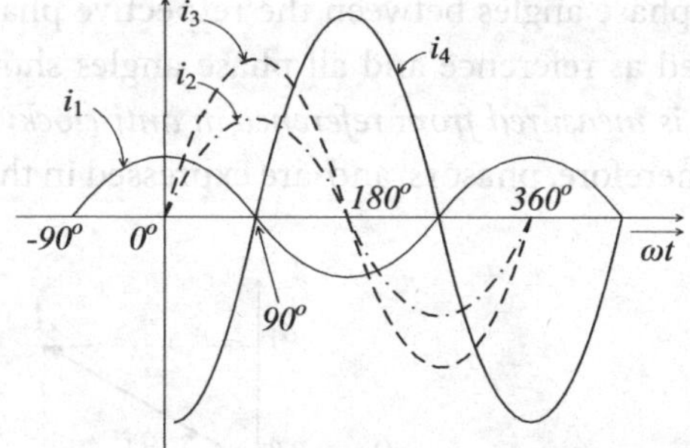

Fig. 3.42

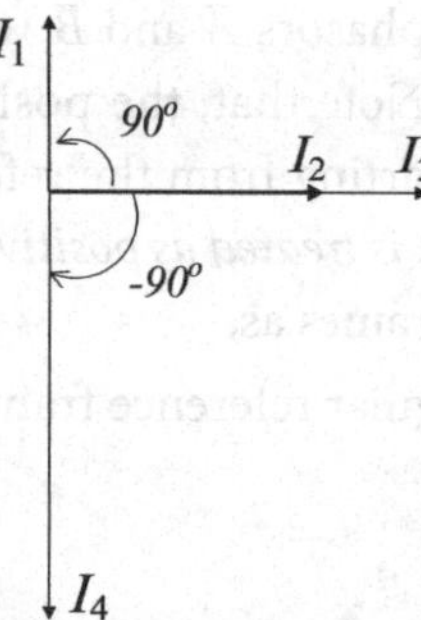

Fig. 3.43

Though any one of the four currents can be treated as reference, it is convenient to consider that waveform as reference whose positive half cycle starts at $\omega t = 0$. Therefore, let i_2 be the reference current. You will notice that the changes in magnitude and direction of i_2 and i_3 are identical and simultaneous. Hence i_2 and i_3 are said to be in-phase with each other. In other words, the phase angle between them is zero. On the time-axis (ωt), the positive peak of i_4 comes 90° later than the positive peaks of i_2 and i_3. Hence, i4 is said to be lagging i2 and i3 by 90o. Similarly you will find on the time axis that, the positive peak of i1 comes 90o earlier than the positive peaks of i_2 and i_3. Hence, i_1 is said to be leading i_2 and i_3 by 90°.

The above finding can also be represented in the form of phasors as shown in Fig. 3.43. Remember that, the *phasors always rotate in counter-clockwise direction* at a speed governed by frequency. Therefore, if the acute angle between phasors and their rotation are considered, you will observe that (i) I_2 and I_3 are *in-phase,* (ii) I_4 is lagging I_2 and I_3 by 90o and (iii) I_1 is leading *I*2 and I_3 by 90°.

Mathematically, the instantaneous values will be,

$$i_2 = I_{m2} \sin \omega t$$
$$i_3 = I_{m3} \sin \omega t$$
$$i_4 = I_{m4} \sin (\omega t - 90°)$$
$$i_1 = I_{m1} \sin(\omega t + 90°)$$

Negative phase angle indicates lagging nature and positive phase angle indicates leading nature of the quantity with respect to reference. I_{m1}, I_{m2}, I_{m3} and I_{m4} are the peak values of four currents.

Calculations in complex frames:

Phasor calculations are normally carried out in (*i*) Rectangular reference frame and (*ii*) Polar reference frame. The *Rectangular reference frame is preferred for addition and subtraction of phasors* while *Polar reference frame is preferred for multiplication and division of phasors.*

In Rectangular reference frame, the *ac* quantity is represented in terms of its real component and imaginary component. Consider two phasors $\bar{A}$ and $\bar{B}$ in a complex frame as shown in Fig. 3.44. The X-axis is called as real axis and Y-axis as imaginary axis. and are the real and imaginary components of $\bar{A}$. Similarly, $-b_r$ and $-b_i$ are the real and imaginary components of $\bar{B}$. For example: When we write impedance as $Z = (R \pm jX)\Omega$, the resistance R is real component of Z and reactance $\pm X$ is its imaginary component.

In Polar reference frame, phasors are represented in terms of their magnitude (RMS value) and the phase angle between that phasor and the reference. Considering Fig. 3.44 again. $|\bar{A}|$ and $|\bar{B}|$ are the magnitudes of phasors $\bar{A}$ and $\bar{B}$ whereas θ_1 and θ_2 are the phase angles between the respective phasor and reference. Note that, the positive real axis is considered as reference and all phase angles should be measured starting from the reference. *If the phase angle is measured from reference in anti-clockwise direction than it is treated as positive otherwise as negative.* Therefore, phasors and are expressed in these two reference frames as,

In Rectangular reference frame:

$$\bar{A} = a_r + ja_i$$

$$\bar{B} = -b_r + jb_i$$

In Polar reference frame:

$$\bar{A} = |\bar{A}| \angle \theta_1$$

$$\bar{B} = |\bar{B}| \angle -\theta_2$$

Where,

Operator $|j| = \sqrt{-1}$, $j_2 = -1$

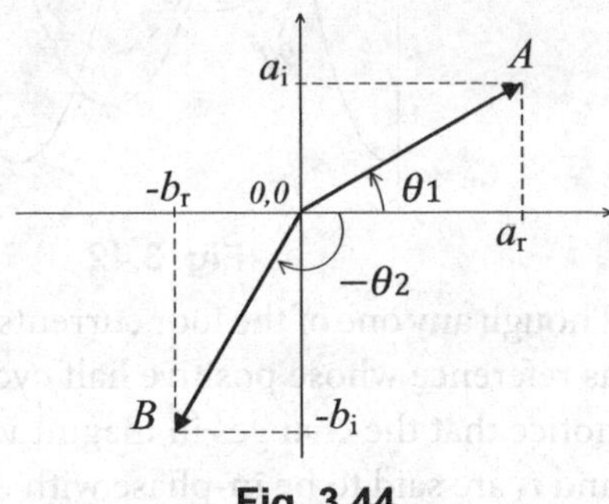

Fig. 3.44

AC quantities can be converted from Rectangular reference frame to Polar reference frame and vice-versa.

Rectangular to Polar → $|\bar{A}| = \sqrt{a_r^2 + a_i^2}$

$$\theta_1 = \tan^{-1} \tan^{-1} \frac{a_i}{a_r}$$

Polar to Rectangular → $a_r = |\bar{A}| \cos \theta_1$

$$a_i = |\bar{A}| \cos \theta_1$$

Addition, Subtraction, Multiplication and Division of phasors

Recall that *addition and subtraction should be preferably done in Rectangular frame* whereas, *multiplication and division in Polar frame. If the values are not available in required reference frame than, first they should converted into the required frame* using above conversion formulae. For the phasors in Fig. 3.44,

Addition: $\bar{A} + \bar{B} = [a_r + (-b_r)] + j[a_i + (-b_i)]$

Subtraction: $\bar{A} - \bar{B} = [a_r - (-b_r)] + j[a_i - (-b_i)]$

Multiplication: $\bar{A} \times \bar{B} = |\bar{A}| \times |\bar{B}| + \angle(\theta_1 + \theta_2)$

Division: $\dfrac{\bar{A}}{\bar{B}} = \dfrac{|\bar{A}|}{|\bar{B}|}\angle(\theta_1 - \theta_2)$

Note: Ready facilities for all above operations in the two reference frames are available on all scientific calculators. Read their manuals carefully to understand and use them.

KNOW MORE

Distinguished Electrical Engineers of India known for their Illustrious Work

Maneklal Sankalchand Thacker was an Indian power engineer and the Director General of Council of Scientific and Industrial Research (CSIR), the largest R & D organization in India. He completed BSc (Engineering) from University of Bristol, UK. and received four honoris causa doctoral degrees; Doctor of Science from Mysore University and BHU, Doctor of Literature from Annamalai University and Doctor of Engineering from Roorkee University. He also worked as Professor, department of Power Engineering and Director IISc, Bangalore; Secretary, Ministry of Education and Scientific Research.

Maneklal S.Thacker (1904–1998)

He did considerable research in power engineering: spanning design problems in electrical plant; thermal and hydropower projects, their economics and instrumentation for power installations.

He received *Padma Bhushan* in 1955 from Government of India and was elected Fellow of American Institute of Electrical Engineers and Indian Academy of Sciences, Bangalore.

Ayyagari Sambasiva Rao was an Indian Scientist and founder of Electronics Corporation of India Ltd. (ECIL) Hyderabad, Telangana. He completed his M.Sc in Physics from Banaras Hindu University where Dr. Sarvepally Radhakrishnan was the Vice-Chancellor. and then Masters in Electrical Engineering from Stanford University.

Ayyagari S Rao (1914–2003)

His greatest achievement was in leading his team in bringing APSRA, India's first nuclear reactor to a critical stage.

In recognition of his services to the nation he received the *Padma Shri Award* in 1960, *Shanti Swaroop Bhatnagar* Award for Engineering Sciences in 1965, Doctor of Science degree, honoris causa, from Andhra University in 1969, *Padma Bhushan* in 1972, and Fellow of Indian Academy of Sciences in 1974. A postal stamp was released on 16 Nov. 2014 on the occasion of his birth centenary.

Bantval J. Baliga
(born 28th April 1948)

Bantval Jayant Baliga received his B.Tech. degree in electrical engineering from IIT Madras in 1969, M.S. and Ph.D degrees from Rensselaer Polytechnic Institute in 1971 and 1974 respectively. He is well-known for his invention of Insulated Gate Bipolar Transistor (IGBT).

From 1974, he worked for 15 years at General Electric Research and Development Center in Schenectady NY, then joined North Carolina State University in 1988 as Full Professor. He introduced the concepts of functional integration of MOS with bipolar physics and merging of PIN with Schottky physics. He has written 19 books and over 500 papers in peer-reviewed journals. He holds 120 U.S. patents.

Amongst many awards received by him, the most prominent ones include highest recognition by the US government, the *National Medal of Technology and Innovation* in 2011 by the hands of President Barack Obama, the Russian top technology award *Global Energy Prize* in 2015 and the *IEEE Medal of Honor* in 2014.

N. R. Narayan Murthy
(born 21st Aug 1946)

N. R. Narayan Murthy, the Founder and Chairman Emeritus of Infosys, completed his Bachelor's degree in electrical engineering from the University of Mysore in 1967 and a Master's degree in Technology from the Indian Institute of Technology, Kanpur, in 1969.

He was awarded Padma Vibhushan, award in India by the Government of India in 2008.

Satya Nadella
(born 19th August 1967)

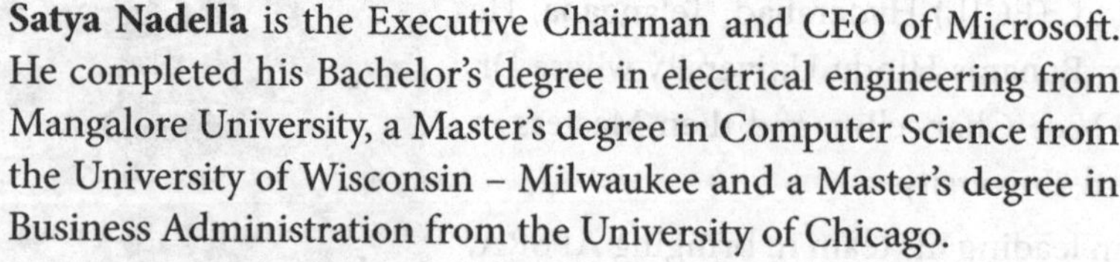

Satya Nadella is the Executive Chairman and CEO of Microsoft. He completed his Bachelor's degree in electrical engineering from Mangalore University, a Master's degree in Computer Science from the University of Wisconsin – Milwaukee and a Master's degree in Business Administration from the University of Chicago.

In 2022, Nadella was awarded Padma Bhushan, the third highest civilian award in India by the Government of India.

REFERENCES AND SUGGESTED READINGS

1. A. E. Fitzgerald, C. Kingsley Jr., A. Kusko, Electric Machinery 3e, McGraw Hill, 1971
2. T. Wildi, Electrical Machines, Drives, and Power Systems, Pearson Education Asia, 2001
3. R. K. Srivastava, Electrical Machines 2e, Cengage Learning, 2013
4. P. S. Bhimbra, Electrical Machinery, Khanna Publishers, 2009.
5. A. K. Sawhney, Electrical Machine Design, Dhanpat Rai & Co., 2012.
6. B. L. Theraja, A Text book of Electrical Technology Vol-II, S. Chand & Co., 2015.
7. S. G. Tarnekar, P. K. Kharbanda, S. B. Bodkhe, Laboratory Courses in Electrical Engineering, S. Chand & Co., 2013.
8. A. Krings, M. Cossale et al, 'Magnetic Materials Used in Electrical Machines', *IEEE Ind. App. Magazine*, November/December 2017 (DOI: 10.1109/MIAS.2016.2600721)

QR Code for Further Reading

Unit 3: Single Phase Transformers

4 Three Phase Transformers

UNIT SPECIFICS

Through this unit the following aspects are discussed:

- Need of 3-phase transformer;
- Comparison between a bank of three 1-phase transformers and a 3-phase transformer
- Construction and functions of different parts of transformer and the cooling methods;
- 3-phase transformer connections as per IS:2026 (Part IV)-1977;
- Selection of transformer as per IS: 10028 (Part I)-1985;
- Power transformer and Distribution transformer;
- Amorphous core type distribution transformer;
- Criteria for selection of distribution transformer and power transformer;
- Specifications of three-phase distribution transformers as per IS:1180 (part I)-1989;
- Three phase to two phase conversion (Scott connection);
- Need of parallel operation of three phase transformers;
- Conditions for parallel operation;
- Phasing out test on 3-phase transformer.

After making the students familiar with 1-phase transformer, a comprehensive discussion on different aspects of three-phase transformer with sufficient number of diagrams is provided in this Unit. As three-phase transformers are widely employed for power transmission and distribution, it is advisable that by the end of this Unit, the students of electrical engineering should attain the expected outcomes which are given below. With due consideration to Bloom's taxonomy, multiple-choice questions, questions for short and long answers and numerical problems are provided for practice and self-assessment.

A laboratory experiment on the phasing-out test is included that will help students to test a given three-phase transformer practically whose phase markings are masked. The "Know More" section supplements additional information about the overall assembly of a commercial three-phase transformer which curious students will enjoy reading. Finally, the list of references and a dynamic QR code are given at the end for suggestive self-learning. Happy reading!

RATIONALE

By this time you must have gained a good level of understanding about the importance and functional aspects of single-phase transformer. Do you know that the electric power that is generated, transmitted and distributed amongst different consumers in all countries across the world is not of single-phase type but it is of three-phase type? Do you know that although a residential consumer needs single-phase power at 230V level and an industrial consumer needs three-phase power at 11kV - 66kV level, it is first stepped-up to very high voltages before transmission (220kV, 400kV, 765kV etc.) and then at the end of transmission line it is stepped down to low voltage levels? Do you know that practically in power systems, the three-phase transformers are extensively employed as compared to 1-phase transformers?

This Unit will satisfy the quest for answers to above questions and much more. Note that although there are differences between the construction of three-phase and single-phase transformers but, the basic operating principle remains same. In this Unit, you shall also journey to the related Indian Standards which all transformer manufacturers in India are required to follow as per law.

PRE-REQUISITES

Basics of Electrical Engineering

UNIT OUTCOMES

By the end of this Unit, the students will be able to,

U4-O1: Compare a three-phase transformer with a bank of three single-phase transformers;

U4-O2: Compare power transformer and distribution transformer on different aspects;

U4-O3: Explain the construction, types and cooling methods of three-phase transformer;

U4-O4: Describe amorphous core type distribution transformer;

U4-O5: Identify the three-phase transformer connection using clock-hour diagram as per IS:2026 (Part-IV) 1977;

U4-O6: Select a three-phase transformer as per IS:10028 (Part-I) 1985 and discuss the criteria for selection;

U4-O7: Explain the specifications of three-phase transformer as per IS:1180 (Part-I) 1989;

U4-O8: State the application of two-phase power and explain the conversion from three-phase power to two-phase using Scott connection;

U4-O9: Explain the need of parallel operation of three-phase transformers and the conditions for parallel operation;

U4-O10: Perform the phasing-out test on three-phase transformer in laboratory and make valid conclusion from the observations.

Unit-4 Outcome	EXPECTED MAPPING WITH COURSE OUTCOMES (1 - Weak Correlation; 2 - Medium Correlation; 3 - Strong Correlation)				
	CO-1	CO-2	CO-3	CO-4	CO-5
U4-O1	-	-	-	3	-
U4-O2	-	-	-	3	-
U4-O3	-	-	-	3	-
U4-O4	-	-	-	3	-
U4-O5	-	-	-	3	-
U4-O6	-	-	-	3	-
U4-O7	-	-	-	3	-
U4-O8	-	-	-	3	-
U4-O9	-	-	-	3	-
U4-O10	-	-	-	3	-

4.1 INTRODUCTION

In section 3.1 it was informed that, after the generation of electricity, power is transferred over long distances and distributed in a wider network. Generation of power takes place at 11kV or slightly higher in *power stations*. It is then stepped-up to higher voltage levels like 132kV, 220kV, 400kV, 765kV at *step-up or transmission substations* for transmission over long distances. After transmission, it is stepped-down to distribution voltage levels (66kV, 33kV, 11kV) at *step-down or distribution substations* and then to utilization voltage level (433V). Finally, it is delivered to the consumer. This generation, transmission and distribution of electrical power is of three-phase type. Transmission of power at high voltage reduces the magnitude of current and ohmic line losses. This makes it possible to use higher gauge (thinner) copper wire, thus reducing both material and labour costs significantly. Each time, the voltage is stepped-up or stepped-down, a three-phase transformer is required. It would be interesting to understand the reasons why the electric grid opts three-phase power for generation, transmission and distribution.

4.1.1 Why three-phase system is used in electrical power system?

The number of phases in an electrical power system is determined by several factors, including efficiency, cost, and ease of use. Three-phase systems are widely used because they offer a good balance between these factors. In a three-phase system, power is transmitted using three conductors, each carrying an alternating current that is displaced in phase from that in other conductors by 120 degrees. Following are some of the advantages of 3-phase system.

- *Cost-effective:* A single-phase system requires two conductors whereas a three-phase system requires three conductors when the load is balanced. With only one extra conductor, a three-phase system can transmit three times the power of single phase.
- *Load balancing:* In a three-phase system, the power is evenly divided between the three phases. Thus it allows for a balanced distribution of power and prevents overloading of any single phase.

- *Higher efficiency:* For the same magnitude of power, conductor current in a three-phase system is less than that in a single-phase system. This reduces the amount of energy lost in ohmic losses, increases the transmission efficiency and allows for more power to be transmitted over longer distances. Thus, three-phase systems can handle higher magnitudes of power in Megawatts and are more suited for heavy industrial and commercial loads.
- *Improved motor performance:* A single phase motor provides marginal performance. A two-phase motor offers slightly improved performance but it requires three conductors. With same number of three conductors, a three-phase motor is able to produce an uniformly rotating magnetic field and offers various advantages like self-starting nature, smooth operation, silent operation, reduction in size and cost, superior performance and reduced maintenance. This is the reason why more than 70% motors used in industries are three-phase induction motors.
- Whenever required, a single-phase can be easily obtained from the three-phase supply.

In view of the above discussion, one may question why higher number of phases such as 6, 9, 12 etc. are not preferred for transmission and distribution of power. Higher number of phases would require additional conductors and make the system more complex, more expensive. Also, increasing the number of phases does not result in proportional rise in the efficiency of the system. It will complicate the generator, transformer, switch gear, and transmission line construction, and would increase the cost significantly. These reasons make three-phase system a preferred choice.

However, transformers used in rectification from three-phase *ac* to *dc* have multiphase transformers for reducing the ripples content in output *dc*.

4.2 COMPARISON WITH A BANK OF THREE 1-PHASE TRANSFORMERS

For every voltage transformation from one voltage level to another, a three-phase transformer or a bank of three interconnected single-phase transformers can be used. Years ago, the common practice was to employ a bank of three single-phase transformers and interconnect their windings in star or delta on both primary and secondary sides. This practice has become obsolete in current times. Due to the improvement in design, manufacturing process, better acquaintance of operating persons and cost reduction, three-phase transformers of equivalent rating have become a favoured choice. However, in hilly regions or mines where the transportation of huge three-phase transformer could be a challenge, a bank of three single-phase transformers can be preferred. Both practices have their relative merits and demerits. They are presented in Table 4.1.

Table 4.1: Comparison between three-phase transformers and a bank of three single-phase transformers

	Three-Phase Transformer		Bank of three Single-Phase Transformers
1.	Occupies less floor space for the same rating.	1.	Occupies more floor space.
2.	Weighs less for the same rating.	2.	Weighs more.
3.	Costs about 15% less for the same rating.	3.	Costs more because the three separate magnetic cores require more iron and more auxiliary apparatus.
4.	Because of relatively less iron in the common magnetic core, the iron loss P_i is less and hence efficiency is more.	4.	More amount of iron in three separate cores results in high iron loss and low efficiency.

	Three-Phase Transformer		Bank of three Single-Phase Transformers
5.	On both LV and HV sides, the three windings can be connected in star or delta inside the transformer tank and therefore on HV side, only three high voltage leads are brought out and three bushings are required.	5.	On both LV and HV sides, the star or delta connections are made outside the transformer tank and therefore on the HV side, six high voltage leads must be brought out which require six bushings that increases the cost.
6.	In case of fault on any winding, the complete unit needs to be disconnected for repairs thus disconnecting the complete power supply to the load.	6.	In case of fault on any one unit, the remaining two single-phase transformers can be operated in V-V or open-delta mode and continuity of supply at reduced power level (57.7%) can be maintained.
7.	The cost of stand-by unit is more because its ratings will be same as the three-phase transformer.	7.	The cost of stand-by unit is less because its ratings will be same as single-phase transformer.
8.	Repair work is relatively difficult.	8.	The three units being separate, their repair work is easy.

4.3 POWER TRANSFORMER AND DISTRIBUTION TRANSFORMER

The transformers used in step-up and step-down substations are power transformer, distribution transformer, current transformer (CT) and potential transformer (PT). CT and PT are the instrument transformers which are meant for measurement purpose and shall be discussed in Unit-5. The power and distribution transformers are employed for the transformation of voltage levels of electrical power which is finally delivered to the consumers. They differ in terms of power handling capacity, duty of operation and maximum efficiency condition.

Though these two transformers work on the same principle of operation which was discussed in Unit-3 and have identical construction, there exists a few differences with respect to their service conditions and hence their design aspects. This is because during the 24 hours load cycle, the power demanded by loads is not constant but keeps on varying between no-load (light load) and full-load (i.e. rated load) conditions. In Indian regions located on the northern part of Deccan plateau, during the winter season, the load demand is maximum in the evening and minimum at night. Typical views of a power transformer and pole-mounted distribution transformer are shown in Fig. 4.1 (a) and (b) followed by Table 4.2 highlighting major differences between the two.

(a) Power Transformer

(b) Pole-mounted Distribution Transformer

Fig. 4.1: Three-Phase Transformer glimpses

Table 4.2: Differences between Power Transformer and Distribution Transformer

	Power Transformer		Distribution Transformer
1.	Power transformers come in ratings above 500 kVA. They are employed in substations to step-up or step-down the voltages (765kV, 400kV, 220kV, 132kV, 66kV).	1.	Distribution transformers are upto 500 kVA rating, relatively smaller in size and are used as step-down transformer for the distribution of electrical energy at low voltages which are suitable for utilization at consumer premises. ($\leq$ 33kV for industrial consumers and 230V (phase voltage) for domestic consumers).
2.	Power transformers are normally connected in parallel with each other. They are controlled to operate almost always at or near their rated capacity (kVA rating). Therefore, during the light load periods, some of the power transformers are disconnected and the overall load is shifted on those which are kept operating. Thus, whenever a power transformer is in energized condition, it is normally near the full-load condition.	2.	Distribution transformers are always kept energized for 24 hours a day. The load across their secondary varies over a wide range from little or no-load to rated value. A distribution transformer installed in a residential colony will be under-loaded for a considerable period of the day but in the evenings it may get loaded to rated value. Thus for most of the time, distribution transformers work at or near light-load condition.
3.	This saves the iron loss during light load periods because, iron loss is a constant loss which is independent of connected load.	3.	The copper loss which is directly proportional to the square of load current varies with load condition but, the wastage of energy in iron loss keeps on taking place throughout the day and night.
4.	Hence, power transformers are designed to have minimum copper loss because at higher loads, the copper loss is much more than the iron loss.	4.	Hence, distribution transformers are designed to have minimum iron loss.
5.	In view of above, a power transformer is designed to have maximum efficiency at or near its full-load kVA (i.e. rated kVA) condition.	5.	In view of above, a distribution transformer is designed to have maximum efficiency at about one-half of its rated kVA (i.e. full-load kVA).
6.	Hence the choice of the power transformer out of a large number of competing transformers should be based on full-load efficiency.	6.	Consequently, the full-load efficiency is less than maximum efficiency.
7.	Power transformer does not come in direct contact with consumer terminals.	7.	Distribution transformer may have its secondary side directly connected to the consumer terminals.
8.	Large leakage reactance and short-circuit protection in Power transformers.	8.	Less leakage reactance.

4.4 CONSTRUCTION OF THREE-PHASE TRANSFORMER

As informed in Section 3.3, the single-phase and three-phase transformers have two main parts. They are: (i) A magnetic core comprising of limbs and yokes and (ii) one winding on the primary side and one winding on the secondary side both of which are wound on the same limbs of magnetic core. Based on the shape of magnetic core, three-phase transformers are classified as

(i) Core type transformer and

(ii) Shell type transformer.

Recall that, in a single-phase transformer, there is only one single-phase winding on the primary side and one single-phase winding on the secondary side. However in a three-phase transformer, as the name implies, there should be a three-phase winding on primary side and a three-phase winding on the secondary side. In other words, the primary side has a set of three, single-phase windings and similarly, a set of three, single-phase windings is present on the secondary side. These single-phase windings are interconnected either in star, delta or zig-zag configuration. Thus altogether, there are six single-phase windings in a three phase transformer when connected in star or delta. In case of zigzag configuration, each single-phase zigzag winding is split into two-parts. Therefore, if zigzag connections are made on the secondary side then there are total nine segments of single-phase windings.

Detailed discussion on the construction of magnetic core by using CRGO Silicon-steel alloy for the reduction of hysteresis loss, use of laminated core for the reduction of eddy-current loss, types of core cross-section, types of joints and interleaving of joints are presented in Sections 3.3.1 and 3.4.3 of Unit-3. The same is valid and applicable to three-phase transformers also. Students are advised to revisit the above sections of Unit-3 for more understanding.

4.4.1 Core Type Construction

In Fig. 3.4 of Unit-3, we saw that the core-type construction of magnetic core of a single-phase transformer has two limbs and two yokes. The construction of three-phase core-type transformer has evolved from the construction of single-phase transformer and is very interesting to know. Refer to Fig. 4.2(a). It shows three single-phase, core-type transformers (T1, T2 and T3) physically arranged such that, their un-wound limbs are brought in close contact. Whereas the other limb of each transformer is wound with both primary and secondary windings. When the three primary windings are connected across a balanced three-phase *ac* supply, three alternating fluxes (ϕ_1, ϕ_2 and ϕ_3) will be established by them in their respective cores. The three un-wound limbs which are in close contact with each other will carry the total flux ($\phi_1 + \phi_2 + \phi_3$). The three alternating fluxes can be expressed as,

$$\phi_1 = \phi_m \sin \omega t$$

$$\phi_2 = \phi_m \sin (\omega t - 120°)$$

$$\phi_3 = \phi_m \sin (\omega t + 120°)$$

$$\therefore \quad (\phi_1 + \phi_2 + \phi_3) = 0 \qquad (4.1)$$

Since, these three fluxes are equal in amplitude and time-displaced from each other by 120°, their phasor sum will always be zero. Thus the central limbs will carry no flux and hence can be eliminated as shown in Fig. 4.2(b). However, such type of construction can be very expensive and it is therefore simplified further. The three-phase core type transformer shown in Fig. 4.2(c) is the final form adopted commercially. However in this form, the central phase has relatively less reluctance as compared to those on outer limbs due to which the magnetizing current in central phase windings is slightly less than the magnetizing current in two outer limb phase windings. But this difference is negligibly small. Fig. 4.2 (c) also shows the HV winding of each phase wound over the LV winding of that respective phase. This is called as *concentric type winding* and is used in core type transformers.

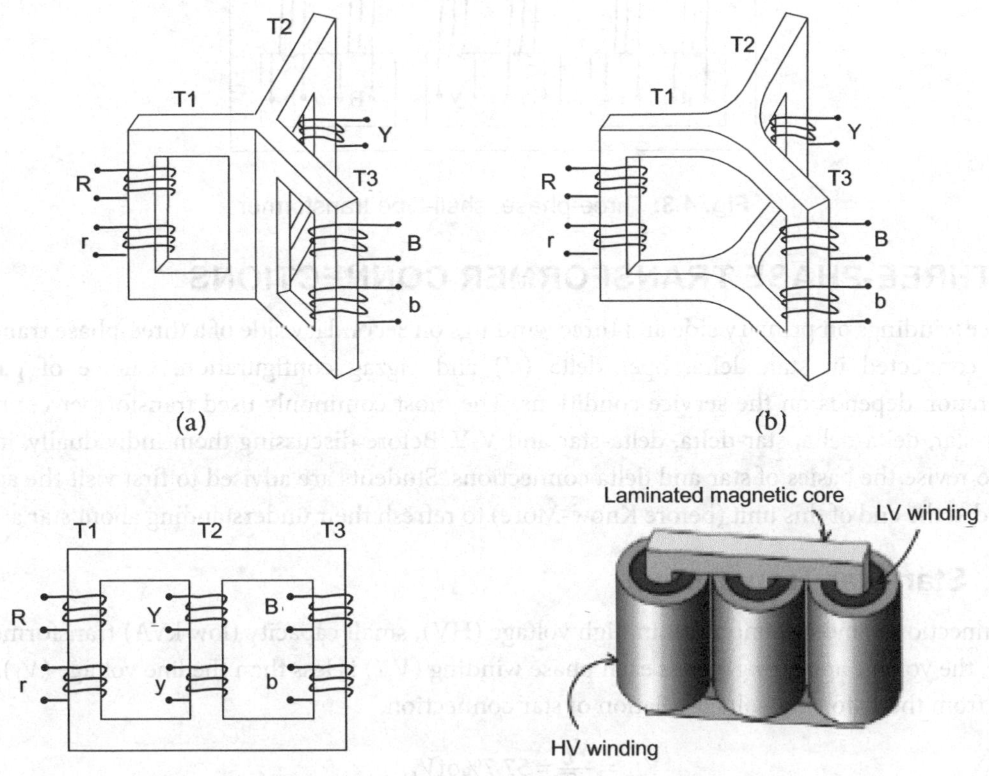

(c) Final shape of three-phase core type transformer

Fig. 4.2: Evolution of three-phase, core-type transformer

4.4.2 Shell Type Construction

A three-phase shell-type transformer is shown in Fig. 4.3. Its shape looks as if three single-phase shell-type transformers are placed side by side. The primary and secondary windings of each phase are placed on the central limb. To allow equal width and equal cross-section area of all unwound limbs (i.e. the limbs on which no windings are wound), the direction in which the windings of the central phase are wound is reversed as compared to the outer two phases. This helps in saving the core material and hence reduces the cost.

In shell type transformers, *sandwich type HV and LV windings* are almost exclusively used. In this type of winding, both HV and LV windings are sub-divided into many sections. The HV sections lie in between the LV winding sections. Consequently, most of the flux produced by one winding of a phase links with the other winding of that phase. In other words, the leakage flux gets reduced and therefore, there is a reduction in leakage reactances thus reducing the voltage regulation.

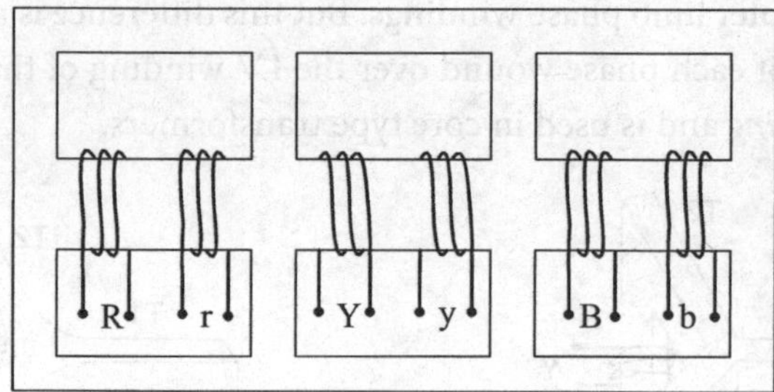

Fig. 4.3: Three-phase, shell-type transformer

4.5 THREE-PHASE TRANSFORMER CONNECTIONS

The three windings on primary side and three windings on secondary side of a three-phase transformer can be connected in star, delta, open-delta (V) and zigzag configuration. Choice of particular configuration depends on the service conditions. The most commonly used transformer connections are: star-star, delta-delta, star-delta, delta-star and V-V. Before discussing them individually, it will be useful to revise the basics of star and delta connections. Students are advised to first visit the appendix provided at the end of this unit (before Know-More) to refresh their understanding about star and delta.

4.5.1 Star-Star Transformer

This connection is most economical in high voltage (HV), small capacity (low kVA) transformers. It is because, the voltage appearing across each phase winding (V_{ph}) is less than the line voltage (V_L). This is evident from the following voltage relation of star connection,

$$V_{ph} = \frac{V_L}{\sqrt{3}} = 57.7\% \text{of} V_L \tag{4.2}$$

As a result, the insulation required by phase-windings is stressed by only 57.7% of line voltage. This reduces the required insulation of phase-windings and also reduces the number of turns. Hence as compared to delta, star-star connection is preferred in cases where the input and output line voltages are high. However in star connection, the phase current is equal to the line current ($I_L = I_{ph}$) therefore thick copper conductors will be necessary to carry the large phase currents. Hence, the use of this connection is restricted to small kVA rating transformers. The drawback of star-star transformer is that, it performs satisfactorily with balanced loads but not with unbalanced loads. The phase shift between primary and secondary line voltages can be either 0° or 180°.

Note that, for the output voltages to be sinusoidal, it is necessary that the resultant flux in the core should be sinusoidal. Because of nonlinear shape of B-H curve of the magnetic core, the flux can be sinusoidal only when the third harmonic component is present in the phase currents. In star connection, the third harmonic currents can exist only when the neutral point is grounded. But if none of the neutral points in a star-star transformer are grounded, then there will be no path available for the flow of *third harmonic currents*. Therefore, they cannot exist in phase windings. This may result in non-sinusoidal wave shape of the output voltage waveform.

Consider a star-star transformer as shown in Fig. 4.4. Let N_1 and N_2 represent the number of turns per phase on the two sides. Suffix '1' indicates primary side and suffix '2' indicates secondary side.

$$\text{Transformation ratio K} = \frac{N_2}{N_1} = \frac{V_{ph2}}{V_{ph1}} = \frac{I_{ph1}}{I_{ph2}}$$

$$\text{Input voltamperes} = 3V_{ph1}I_{ph1} = 3\frac{V_{L1}}{\sqrt{3}}I_{L1} = \sqrt{3}V_{L1}I_{L1}$$

$$\text{Output voltamperes} = 3V_{ph2}I_{ph2} = 3\left(KV_{ph1}\right)\left(\frac{I_{ph1}}{K}\right) = 3\left(\frac{V_{L1}}{\sqrt{3}}\right)I_{L1} = \sqrt{3}V_{L1}I_{L1} \tag{4.3}$$

∴ Three-phase input voltamperes = three-phase output voltamperes

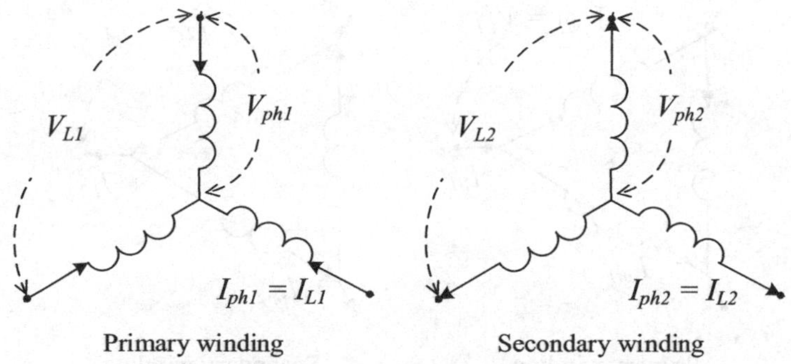

Fig. 4.4: Three-phase, star-star transformer

4.5.2 Delta-Delta Transformer

This connection is suitable for low voltage (LV), large capacity (high kVA) transformers. It is because, the voltage appearing across each phase winding is equal to the line voltage (*VL*).

$$V_L = V_{ph}$$

As a result, the insulation of all phase windings are stressed by whole of the line voltage. This increases the insulation requirement of phase-windings and also increases the required number of turns. Hence as compared to star, delta-delta connection is preferred in applications where the input and output line voltages are low. However, in delta connection, the phase current is less than line current because,

$$I_{ph} = \frac{I_L}{\sqrt{3}} = 57.7\% \text{ of } I_L \tag{4.4}$$

Therefore relatively, thin copper conductors will be sufficient to carry the phase currents. Hence, the use of this connection can be allowed to large kVA rating transformers. Secondly, the close delta loops provide paths for the flow of third harmonic currents in phase windings. Consequently, the magnetic flux is sinusoidal and hence, the output voltages are also sinusoidal. The phase shift between primary and secondary line voltages can be either 0o or 180o.

Consider a delta-delta transformer as shown in Fig. 4.5. Let N1 and N2 represent the number of turns per phase on the primary and secondary sides respectively. Suffix 1 and 2 represent the primary and secondary side respectively.

$$\text{Transformation ratio } K = \frac{N_2}{N_1} = \frac{V_{ph2}}{V_{ph1}} = \frac{I_{ph1}}{I_{ph2}}$$

$$\text{Input voltamperes} = 3V_{ph1}I_{ph1} = 3\frac{V_{L1}}{\sqrt{3}}I_{L1} = \sqrt{3}V_{L1}I_{L1}$$

$$\text{Output voltamperes} = 3V_{ph2}I_{ph2} = 3\left(KV_{ph1}\right)\left(\frac{I_{ph1}}{K}\right) = 3\left(\frac{V_{L1}}{\sqrt{3}}\right)I_{L1} = \sqrt{3}V_{L1}I_{L1} \quad (4.5)$$

∴ Three-phase input voltamperes = three-phase output voltamperes

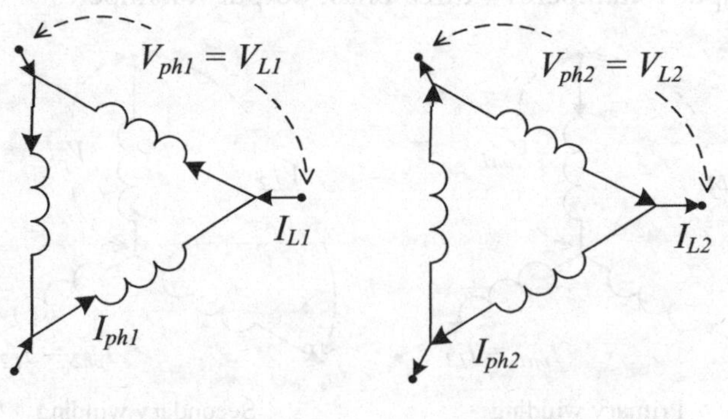

Fig. 4.5: Three-phase, delta-delta transformer

4.5.3 Star-Delta Transformer

This connection is commonly used in step-down transformers for stepping down the voltage from high level to a medium or low level. It is employed in substations after the completion of transmission line. A star-delta connected transformer is shown in Fig. 4.6. In these transformers, there is a 30o phase shift between the primary and secondary line voltages due to which, it cannot be connected in parallel with a star-star or delta-delta transformer.

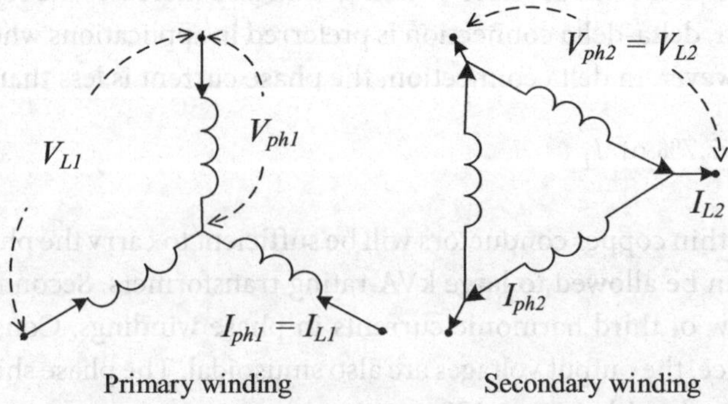

Fig. 4.6: Three-phase, star-delta transformer

Transformation ratio $K = \frac{N_2}{N_1} = \frac{V_{ph2}}{V_{ph1}} = \frac{I_{ph1}}{I_{ph2}}$

The ratio between secondary line voltage and primary line voltage is $\frac{1}{\sqrt{3}}$ times the transformation ratio of each transformer.

$$\frac{V_{L2}}{V_{L1}} = \frac{K}{\sqrt{3}}$$

On the input side due to star connection,

Input voltamperes = $3V_{ph1}I_{ph1} = 3\frac{V_{L1}}{\sqrt{3}}I_{L1} = \sqrt{3}V_{L1}I_{L1}$

On output side due to the delta connection,

$$\text{Output voltamperes} = 3V_{ph2}I_{ph2} = 3\left(KV_{ph1}\right)\left(\frac{I_{ph1}}{K}\right) = 3V_{L1}\left(\frac{I_{L1}}{\sqrt{3}}\right) = \sqrt{3}V_{L1}I_{L1} \quad (4.6)$$

∴ Three-phase input voltamperes = three-phase output voltamperes

4.5.4 Delta-Star Transformer

This connection is commonly used in step-up transformers for stepping up the voltage from low or medium level to a high level. It is employed in substations at the beginning of transmission line. A delta-star connected transformer is shown in Fig. 4.7. There is a 30o phase shift between the primary and secondary line voltages in a delta-star transformer due to which, it cannot be connected in parallel with a star-star or delta-delta transformer.

Transformation ratio $K = \frac{N_2}{N_1} = \frac{V_{ph2}}{V_{ph1}} = \frac{I_{ph1}}{I_{ph2}}$

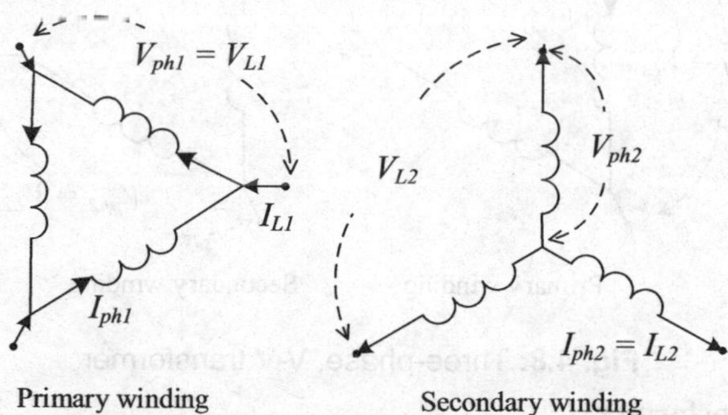

Fig. 4.7: Three-phase, delta-star transformer

The ratio between secondary line voltage and primary line voltage is $\sqrt{3}$ times the transformation ratio of each transformer.

$$\frac{V_{L2}}{V_{L1}} = \sqrt{3}K$$

On the input side due to delta connection,

Input voltamperes $= 3V_{ph1}I_{ph1} = 3\frac{V_{L1}}{\sqrt{3}}I_{L1} = \sqrt{3}V_{L1}I_{L1}$

On output side due to the star connection,

Output voltamperes= $3V_{ph2}I_{ph2} = 3\left(KV_{ph1}\right)\left(\frac{I_{ph1}}{K}\right) = 3V_{L1}\left(\frac{I_{L1}}{\sqrt{3}}\right) = \sqrt{3}V_{L1}I_{L1}$ (4.7)

∴ Three-phase input voltamperes = three-phase output voltamperes

4.5.5 V-V or Open-Delta Transformer

The V-V or Open-delta transformer results when one of the single-phase transformer or one phase winding (from both, primary and secondary sides) of a delta-delta transformer is removed. The input three-phase supply is connected across remaining two windings. V–V connected transformer is employed in following situations:

- When one of the single phase transformer of a delta-delta bank is disconnected due to the fault developed on it and the supply is to be continued to the load at reduced power level.
- When it is anticipated that in future, there may be a rise in load demand and would require higher kVA rating of the transformer then in that case initially, it is operated in V-V mode. After the rise in load demand, the third phase is closed and it is operated in delta-delta mode at higher power level.

It should be noted that, the total load that can be carried by a V-V transformer is not two-third of the total kVA capacity in delta-delta mode but, it is only 57.7% of it. This is proved as given below. The V-V connected transformer is shown in Fig. 4.8.

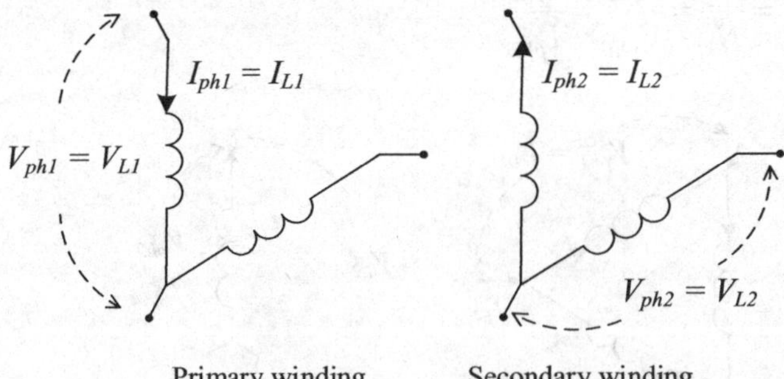

Fig. 4.8: Three-phase, V-V transformer

For a delta-delta transformer:

$$V_L = V_{ph}$$

$$I_L = \sqrt{3}I_{ph}$$

Output voltamperes $= \sqrt{3}V_{L2}I_{L2} = 3V_{ph2}I_{ph2}$

When connected in V-V mode:

The line voltage and phase voltage are equal, $V_L = V_{ph}$

The line current and phase current are equal, $I_L = I_{ph}$

$\therefore$ Output voltamperes $= \sqrt{3}V_{L2}I_{L2} = \sqrt{3}V_{ph2}I_{ph2}$

$$\therefore \frac{\text{voltampere rating in V}-\text{V mode}}{\text{voltampere rating in delta}-\text{delta mode}} = \frac{\sqrt{3}V_{ph2}I_{ph2}}{3V_{ph2}I_{ph2}} = 0.577 \quad (4.8)$$

4.5.6 Zigzag Connection

The zigzag connection is also called as 'inter-star' connection. It is employed on the secondary side and is primarily used to suppress the *third harmonic voltages* between phase and neutral so that the output voltage of the transformer becomes sinusoidal.

In zigzag connection, each single-phase winding is made of two halves and both halves are wound in the same direction. For example, the single-phase winding of phase-a will have two halves as: a_1-a_2 and a_3-a_4. Similarly, the other two phases will have b_1-b_2, b_3-b_4, c_1-c_2 and c_3-c_4 as shown in Fig. 4.9 (a). The polarity of a_1 and a_3 are same. Similarly, the polarity of b_1 and b_3 are same. Also, the polarity of c_1 and c_3 are same.

Let the voltage across each half of the zigzag winding be V_2 then the phase voltage will be and the line voltage will be . This can be proved by analysing the circuit connection in Fig. 4.9 (a) and phasor diagrams shown in Fig. 4.9 (b) and (c).

$$|V_{a1a2}| = |V_{a3a4}| = V_2$$
$$|V_{b1b2}| = |V_{b3b4}| = V_2$$
$$|V_{c1c2}| = |V_{c3c4}| = V_2$$

From Fig. 4.9 (a) the phase voltages will be,

$$V_{ph1} = \bar{V}_{a1a2} - \bar{V}_{c3c4}$$
$$V_{ph2} = \bar{V}_{b1b2} - \bar{V}_{c3c4}$$
$$V_{ph3} = \bar{V}_{a1a2} - \bar{V}_{c3c4}$$

These equations can be plotted to obtain the phasor diagram as shown in Fig. 4.9 (b). After simple graphical simplification and analysis on this phasor diagram, we get the phase voltages as,

$$|V_{ph1}| = |V_{ph2}| = |V_{ph3}| = \sqrt{3}V_2$$

Fig. 4.9 (a) : Three-phase zigzag winding

These phase voltages can be used to obtain the line voltages by equations,

$$V_{L1} = \bar{V}_{ph1} - \bar{V}_{ph3}$$
$$V_{L2} = \bar{V}_{ph2} - \bar{V}_{ph2}$$
$$V_{L3} = \bar{V}_{ph3} - \bar{V}_{ph1}$$

Another phasor diagram can be drawn using above relations as shown in Fig. 4.9 (c). After simplification. the line voltages will be obtained as,

$$|V_{L1}| = |V_{L2}| = |V_{L3}| = 3V_2$$

If I_2 is the current in each single-phase winding then the three–phase volt-amperes of zigzag connection will be,

$$\text{Voltamperes } = 3V_{ph}I_{ph} = 3\left(\sqrt{3}V_2\right)I_2 = 3\sqrt{3}V_2I_2 \quad (4.9)$$

But if we consider the individual capacities of both the halves of all three single-phase windings than, their arithmetic sum will be $6V_2I_2$ voltamperes. Thus it means that, when connected in zigzag configuration, the three-phase capacity decreases from $6V_2I_2$ to $3\sqrt{3}V_2I_2$ voltamperes. Thus only 86.66% of the combined individual capacity is available in zigzag mode. With zigzag connection, two types of three-phase transformer connections are possible:

- **Delta - Zigzag transformer:** This transformer is useful for supply of power to rectifiers (i.e. *ac* to *dc* converters).
- **Star - Zigzag transformer:** This connection permits unbalanced loading even when the neutral of star side (primary) is not grounded. Recall here that, in star-star transformers, without grounding of neutral, the unbalanced loads gives poor performance and therefore are not permitted.

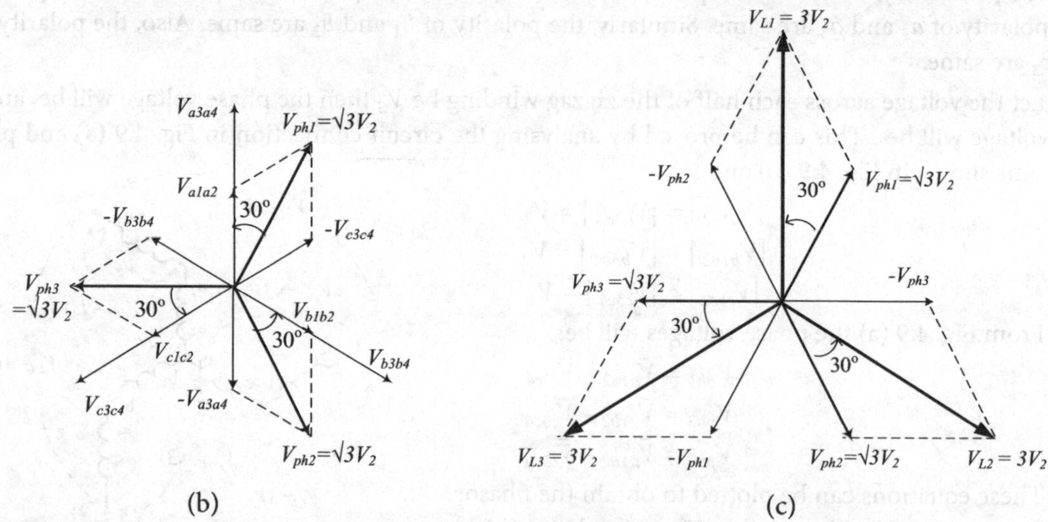

Fig. 4.9: Phasor diagrams of three-phase zigzag connection

Example 4.1: A 3-phase 11kV/433V transformer supplies power to a 100 kVA resistive load. Calculate the line and phase currents carried by the HV and LV windings when the transformer is connected as,

(i) HV in star and LV in delta

(ii) HV in delta and LV in star

(iii) Both HV and LV sides in star

Data: V_{L1} = 11000 volts, V_{L2} = 433 volts, apparent power = 100 kVA

Solution:

Since HV voltage (11 kV) is mentioned first, treat HV side as primary (suffix 1) and LV side as secondary (suffix 2).

(i) When HV winding is in star and LV is in delta:

$$V_{ph1} = \frac{V_{L1}}{\sqrt{3}} = \frac{11000}{\sqrt{3}} = 6350.85V \text{ .... Since HV is in star}$$

$$V_{ph2} = V_{L2} = 433V \text{ .... Since LV is in delta}$$

$$\text{Transformation ratio } K = \frac{V_{ph2}}{V_{ph1}} = \frac{433}{6350.85} = 0.06817$$

Since the three-phase apparent power is written as, $S_{3ph} = \sqrt{3}V_L I_L$

∴ Secondary side currents will be,

$$I_{L2} = \frac{S_{3ph}}{\sqrt{3}V_{L2}} = \frac{100 \times 10^3}{\sqrt{3} \times 433} = 133.33A$$

$$I_{ph2} = \frac{I_{L2}}{\sqrt{3}} = 76.98A$$

Transformation ratio is also expressed as $K = \dfrac{I_{ph1}}{I_{ph2}}$

∴ $$I_{ph1} = KI_{ph2} = 0.06817 \times 76.98 = 5.2477A$$

In star connection; $I_L = I_{ph}$

∴ $$IL_1 = I_{ph1} = 5.2477A$$

(ii) When HV winding is in delta and LV is in star:

$$V_{ph1} = V_{L1} = 11000V \text{........ Since HV is in delta}$$

$$V_{ph2} = \frac{V_{L2}}{\sqrt{3}} = \frac{433}{\sqrt{3}} = 250V \text{ .... Since LV is in star}$$

$$\text{Transformation ratio } K = \frac{V_{ph2}}{V_{ph1}} = \frac{250}{11000} = 0.0227$$

Secondary side currents will be,

$$I_{L2} = \frac{S_{3ph}}{\sqrt{3}V_{L2}} = \frac{100 \times 10^3}{\sqrt{3} \times 433} = 133.33A$$

$$I_{ph2} = I_{L2} = 133.33A$$

Primary side currents are,

$$I_{ph1} = KI_{ph2} = 0.0227 \times 133.33 = 3.03A$$

$$\therefore I_{L1} = \sqrt{3}I_{ph1} = \sqrt{3} \times 3.03 = 5.248A$$

(iii) When both HV and LV windings are in star:

$$V_{ph1} = \frac{V_{L1}}{\sqrt{3}} = \frac{11000}{\sqrt{3}} = 6350.85V$$

$$V_{ph2} = \frac{V_{L2}}{\sqrt{3}} = \frac{433}{\sqrt{3}} = 250V$$

$$\text{Transformation ratio } K = \frac{V_{ph2}}{V_{ph1}} = \frac{250}{6350.85} = 0.0393$$

Secondary side currents will be,

$$I_{L2} = \frac{S_{3ph}}{\sqrt{3}V_{L2}} = \frac{100\times10^3}{\sqrt{3}\times433} = 133.33A$$

$$I_{ph2} = I_{L2} = 133.33\text{A}$$

Primary side currents are,

$$I_{ph1} = KI_{ph2} = 0.0393 \times 133.33 = 5.248\text{A}$$

$$\therefore \quad I_{L1} = I_{ph1} = 5.248\text{A}$$

4.5.7 Clock Hour Method

For a given three-phase transformer, in addition to the awareness of connection (star, delta or zigzag) used in high-voltage (HV) and low-voltage (LV) windings, it is equally important to have information about the phase-angle between the HV and LV line voltages because they may or may not be in-phase with each other. This knowledge is necessary mainly when two or more number of three-phase transformers are to be operated in parallel. The phase angle between HV and LV line voltages depends upon the direction in which the windings are wound and the type of connections employed (star, delta or zigzag) on HV and LV sides. This phase angle can be expressed either in degrees or by a much more convenient method of angle designation called as '*Clock hour method*'.

While representing a three-phase transformer, the HV winding connection is indicated by a capital letter (*Y* for star connection, *D* for delta and *Z* for zigzag). The LV winding connection is indicated by a lower case letter (y for star connection, *d* for delta and *z* for zigzag).

According to Clock method, the HV line voltage phasor is considered as minute-hand of the clock and it is always fixed at the 12 O'clock position (i.e. zero hour position). The LV line voltage phasor is represented by hour-hand and its position is decided by the phase-angle between HV and LV line voltages. For example, a three-phase transformer '*Yd11*' represents HV winding connected in star, LV winding in delta and the position of LV line phasor (hour-hand) at 11 O'clock. This is shown in Fig. 4.10 (a).

We know that, all phasors on a phasor diagram always rotate in anti-clockwise direction at a common speed. With this understanding, you will observe that the LV line phasor (hour-hand) is moving ahead of HV line phasor (minute-hand) by 30°. Hence it means that, in a *Yd*11 transformer, the LV line voltage $V_{L(LV)}$ leads the HV line voltage $V_{L(HV)}$ by 30o. Similarly, *Dy*1 represents HV winding in delta, LV winding is star and $V_{L(LV)}$ lagging behind $V_{L(HV)}$ by 30°. *Yy*0 will mean both windings in star and the line voltages on two sides as in-phase with each other. This is shown in Fig. 4.8 (b) and (c) respectively.

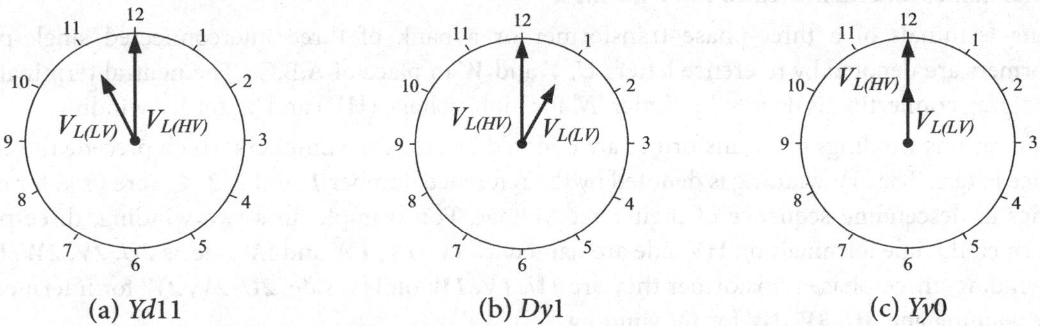

Fig. 4.10: Examples demonstrating clock hour method

The clock hour method can also be applied to three-phase, three-winding (primary, secondary, tertiary) transformers. The first letter apply to highest voltage winding and the next two letters pertain to the other two windings in order of decreasing voltage. Example: *Yy0d1* represents a three-phase, three-winding transformer with HV winding in star, medium-voltage winding in star and LV winding in delta. '0' indicates that the line-voltage of medium-voltage winding is in-phase with $V_{L(HV)}$. Whereas, '1' indicates that $V_{L(LV)}$ lags $V_{L(HV)}$ by 30°.

4.5.8 Connections as Per Indian Standard IS: 2026 (Part –IV) 1977

Indian Standards are the guiding documents published by Bureau of Indian Standards (BIS). They are developed through consensus of experts. BIS is the National Standard Body which functions under the Ministry of Consumer Affairs, Food & Public Distribution, Government of India. It is responsible for the activities of standardization, marking and quality certification of goods. IS: 2026 (Part –IV) 1977 is the Indian Standard titled as "Specification for Power Transformers: Part IV, Terminal Markings, Tappings and Connections". It provides guidelines on transformer connections, tappings and markings. Details of connection related guidelines are as discussed below.

Symbols of three-phase connection:

As per this standard, in a three-phase transformer or in a bank of three interconnected single-phase transformers, the connection between phase-windings is indicated by symbols shown in Table 4.3. For an auto-transformer in which the two windings have a common part, the winding of the pair which has the lower rated voltage is indicated by letter ***a***.

Table 4.3 : Symbols for three-phase connections

Type of connection	High voltage winding	Intermediate voltage and low voltage windings
Star without neutral terminal	Y	y
Delta	D	d
Zigzag	Z	z
Star connection with neutral terminal brought out	YN	yn
Zigzag connection with neutral terminal brought out	ZN	zn

Terminal names and identification of windings:

The line terminals of a three-phase transformer or a bank of three interconnected single-phase transformers are denoted by reference letters ***U, V*** and ***W*** in place of A,B,C. The neutral terminal of a star or zigzag connection is denoted by letter '***N***' for high-voltage (HV) and '***n***' for LV winding.

The various windings of a transformer are denoted by reference numbers which precede the above reference letters. The HV winding is denoted by the reference number ***1***, and ***2, 3, 4***,... are used for other windings in descending sequence of their rated voltage. For example, in a two-winding, three-phase transformer, the line terminals on HV side are named as, ***1U, 1V, 1W*** and LV side as ***2U, 2V, 2W***. For a three winding, three-phase transformer they are ***1U, 1V, 1W*** on HV side, ***2U, 2V, 2W*** for intermediate voltage winding and ***3U, 3V, 3W*** for LV winding.

Series/Parallel connections:

If phase-windings consist of several parts which may either be connected in series or in parallel, then the end points of these parts are denoted by reference numbers 1, 2, 3, 4,... Wherever clarification is required about the voltage level of the winding, the above reference letters precede these reference numbers as, U1, U2, U3, U4, V1, V2 etc.

Phase Displacement between windings:

To represent a three-phase transformer, the vector relating to the high-voltage (HV) winding is taken as the vector of origin. In other words, it is fixed at 12 O' clock on the clock-hour figure. This is already explained in section 4.5.7.

Examples:

1. Consider a three-phase, three-winding (primary, secondary and tertiary) transformer having voltage ratings of 150 kV (in delta), 60 kV (in star with neutral point not brought out) and 10 kV (in star with neutral point not brought out). Let the two star winding voltages be in-phase with each other and lag behind the delta winding voltagc by 30°. This transformer will be represented as ***Dy1y1***.
2. Consider a three-phase, three-winding (primary, secondary and tertiary) transformer having voltage ratings of 6 kV (in star with neutral point brought out), 380 V (in star with neutral point not brought out) and 220 V (in zigzag with neutral point not brought out). Suppose, the two star winding voltages are in-phase with one another and the zigzag winding voltage lags them by 30°. The representation of this transformer will be ***YNy0z1***

Table 4.4 gives some more examples of different connections and their representation.

Table 4.4: Three-phase transformer connections and their representation

Connection	Representation	Interpretation
1U 1V 1W 2U 2V 2W HV winding LV winding	***Yy0***	Both windings are in star. Their voltages are in-phase with each other.

Connection	Representation	Interpretation
1U 1V 1W LV winding HV winding 2U 2V 2W	***Yy*6**	Both windings are in star. Their voltages are out of phase with each other.
1U 1V 1W 2U 2V 2W HV winding LV winding	***Dd*0**	Both windings are in delta. Their voltages are in-phase with each other.
1U 1V 1W LV winding HV winding 2U 2V 2W	***Dd*6**	Both windings are in delta. Their voltages are out of phase with each other.
1U 1V 1W 2U 2V 2W HV winding LV winding	Yd1	HV winding is in star, LV winding is in delta. The voltage of LV winding lags the voltage of HV winding by 30°.
1U 1V 1W 2U 2V 2W HV winding LV winding	Yd11	HV winding is in star, LV winding is in delta. The voltage of LV winding leads the voltage of HV winding by 30°.
1U 1V 1W 2U 2V 2W HV winding LV winding	***Dy*1**	HV winding is in delta, LV winding is in star. The voltage of LV winding lags the voltage of HV winding by 30°.

Connection	Representation	Interpretation
1U 1V 1W 2U 2V 2W HV winding LV winding	*Dy*11	HV winding is in delta, LV winding is in star. The voltage of LV winding leads the voltage of HV winding by 30°.

4.6 COOLING OF TRANSFORMER

The currents flowing in the windings of a transformer causes the absorption of some amount of active power in the winding resistances and results in generation of heat in the windings. This absorbed power which is not transferred to the load is called as *copper loss or variable loss or load loss*. Similarly, the alternating flux in magnetic core causes some loss of active power in the form of hysteresis loss and eddy current loss resulting into generation of heat in the core. If the generated heat is not properly removed and dissipated, it may result in the damage of insulation and even melting of conducting material. The provision of cooling ducts in the core enhances the dissipation of heat upto some extent. Following are some of the cooling methods commonly employed in transformers.

(a) Air Natural cooling (AN)
(b) Oil-immersed Air Natural cooling (ONAN)
(c) Oil-immersed, Forced oil circulation with Natural cooling (OFAN)
(d) Oil-immersed, Forced oil circulation with Forced air cooling (OFAF)
(e) Oil-immersed, Forced oil circulation with Forced water cooling (OFWF)
(f) Dry-type transformer
(g) Hydrogen cooling (For large power rotating machines).

4.6.1 Oil-Immersed, Natural Cooled Transformer (ONAN)

It is also called as oil-immersed self-cooled transformer. In this arrangement, the transformer is submerged in a tank filled with high-grade transformer oil (mineral oil) for cooling purpose. The oil has high dielectric strength and good thermal conductivity. It transfers the heat from windings and core to the surrounding metal tank where it is dissipated by radiation and convection to the outside air. This method is normally preferred in out-door installations. The oil must be kept as free as possible from moisture and oxygen, dissolved combustible gases, and particulates.

As the power rating increases, external cooling tubes or radiators are connected to the transformer tank to increase the effective cooling surface. The hot oil expands in volume, circulates through the radiators and gets cooled due to natural air flow. The Oil-immersed Natural cooling (ON) of a Distribution transformer is shown in Fig. 4.11 (a).

4.6.2 Oil-Immersed, Forced air Cooled Transformer (ONAF)

For higher rating transformers, external fans are employed to force the air through radiators. As the hot oil circulates through radiators by gravity, it gets cooled. It is also called as oil-immersed air-blast transformer. This is shown in Fig. 4.11 (b).

4.6.3 Oil-immersed, Forced Oil/ Forced Water Cooled Transformer (OFOW)

For transformers in mega voltampere range, cooling is enhanced by an oil-water heat exchanger. Hot oil drawn from the transformer tank is pumped to a heat exchanger where it flows through pipes that are in contact with cool water on exterior side. Such heat exchangers are very effective, but they are costly because water itself has to be continuously cooled and circulated. This is shown in Fig. 4.12.

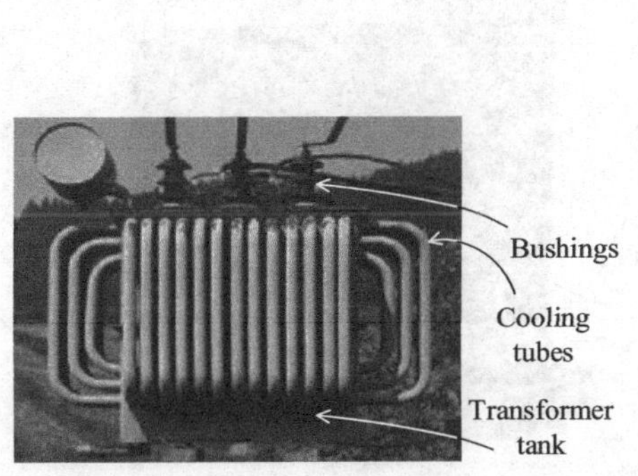

Fig. 4.11 (a): Oil immersed, natural air cooling of distribution transformer

Fig. 4.11 (b): Oil immersed, forced-air cooling of transformer

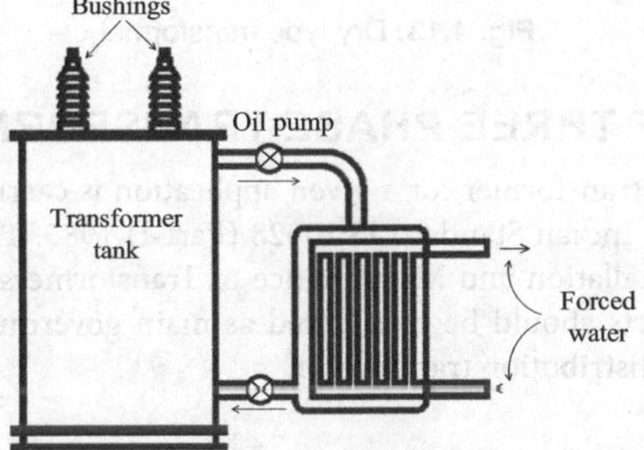

Fig. 4.12: Oil immersed, forced-oil, forced-water cooled transformer

4.6.4 Dry Type Transformer

In dry type transformer, the magnetic core and windings are not immersed in oil. The windings may be encapsulated (encased or surrounded) with solid insulation like Nomex paper or the transformer may be of gas-immersed type or air-immersed type. Nomex is inherently a flame-resistant and high-temperature resistant material that does not melt, drip, or support combustion in air. Cast-resin dry type transformers are used in high moisture areas. In these transformers the primary and secondary windings are encapsulated with epoxy resin which does not retain moisture. The encapsulation prevents moisture from penetrating the winding. Fig. 4.13 shows a cast-resin, dry type distribution transformer.

Indoor transformers below 200 kVA can be directly cooled by the natural flow of surrounding air. In crowded areas and inside the buildings, the installation of oil-immersed transformers is normally avoided and dry type transformers are recommended. The paper insulation of dry transformers have better heat-carrying capacity, apart from the insulating properties. As compared to ONAN type distribution transformers, the Dry type transformers are designed at lower current density. This results in larger weight of core, larger weight of copper and larger surface area for cooling. These transformers are bulky in size and are costly. They have longer life as compared to that of ONAN type transformer.

(a) Epoxy resin insulation enclosures

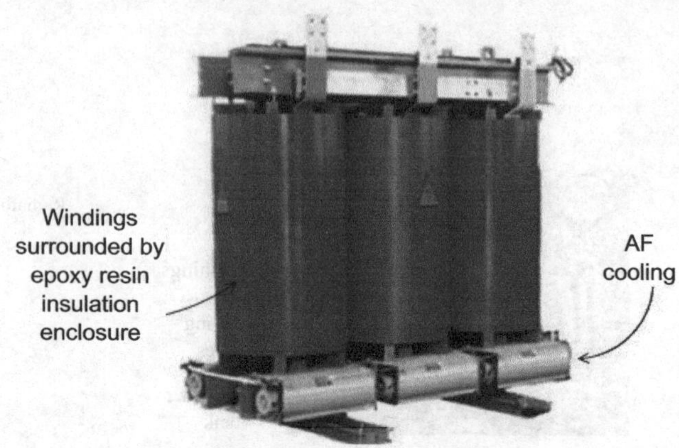

(b) 3-Phase, 1250 kVA, 11kV/433 V, cast-resin, dry-type, indoor distribution transformer,

Fig. 4.13: Dry type transformer

4.7 SELECTION OF THREE-PHASE TRANSFORMERS

The selection of three-phase transformer for a given application is carried out in accordance with the selection criteria specified in Indian Standard IS:10028 (Part-I) 1985. This standard is titled as "Code of Practice for Selection, Installation and Maintenance of Transformers Part-1: Selection". As per this standard, the following aspects should be considered as main governing features for the purpose of selection of both power and distribution transformer.

1. Ratings,
2. Taps,
3. Connection symbol,
4. Impedance,
5. Termination arrangement,
6. Cooling, and
7. Fittings and accessories.

4.7.1 Ratings

This IS Standard gives guidelines about kVA ratings and Voltage ratings of the transformer. It states that, the kVA rating (i.e. apparent power rating) should comply with IS: 2026 (Part-I)-1977. Accordingly for both distribution and power transformers,

- The rated power should be for continuous loading.
- If different values of apparent power are specified for different methods of cooling, the highest of these values is the rated power.
- For a two-winding transformer (primary and secondary) the same value of rated power will be applicable to both windings.
- When rated voltage is applied across primary winding of the transformer, and rated current flows through the terminals of its secondary winding then, it mean that the transformer is receiving rated power for that pair of windings.

Voltage Ratings:

Criteria for Selection of Distribution Transformers (Upto and equal to 1600 kVA)	Criteria for Selection of Power Transformers (More than 1600 kVA)
The no-load secondary voltage should be 433 volts for transformers to be used in 415 V system.	The no-load secondary voltage should be specified as 5% more than the nominal voltage so that the transformer voltage regulation gets compensated partly.

In case if the transformers are to be operated in parallel then, the voltage ratio should be selected in accordance with guidelines given in IS: 10028 (Part 2)-1981. This is applicable to both power transformers and distribution transformers.

4.7.2 Connection Symbol

Criteria for Selection of Distribution Transformers (Upto and equal to 1600 kVA)	Criteria for Selection of Power Transformers (More than 1600 kVA)
The two-winding transformers (primary and secondary) should be preferably connected in delta-star and the exact connection symbol (***Dyn***11 or ***Dyn***1) should be specified. This is in accordance with IS: 2026 (Part 4)-1977.	The two-winding transformers of HV voltage rating up to 66 kV are delta-star (***Dyn***) and star-star (***YNy***). For higher voltages, star-star (***YNyn***) or star/delta (***YNd***) connections may be preferred. For parallel operation with other transformers, selection should be as per IS: 10028 (Part 2)-1981.

4.7.3 Impedance

For both, distribution and power transformers, the impedance values should confirm the guidelines given in IS: 2026 (Part-1) 1977. Also, if these transformers are required for parallel operation then impedance values should be decided with due consideration to IS: 10028 (Part-2) 1981.

Criteria for Selection of Distribution Transformers (Upto and equal to 1600 kVA)	Criteria for Selection of Power Transformers (More than 1600 kVA)
The transformer impedance value should be selected considering the standard available rating of switchgear (circuit breaker) that may be required to be connected on the secondary side of transformer and the associated voltage drops.	The transformer impedance should be decided taking into consideration the secondary fault levels and the voltage dip.

4.7.4 Taps

The output voltage (secondary terminal voltage) of transformers vary in accordance with their input voltage and the load. Normally, the nature of load to whom power is supplied by the transformer is of inductive and resistive type. Therefore during loaded conditions, the output voltage decreases, whereas during off-load conditions, the output voltage increases. In order to balance the voltage variations, *taps* (also called as tappings) are provided on HV winding. To change the tappings as per requirement, a device called as *Tap-changer* is used.

Tap changers can be either ON-load tap changers or OFF-load (OFF-circuit) tap changers. In an ON-load tap changer, the tapping can be changed without isolating the transformer from the supply. In an OFF-load tap changer, it is done after disconnecting the transformer.

Criteria for Selection of Distribution Transformers (Upto and equal to 1600 kVA)	Criteria for Selection of Power Transformers (More than 1600 kVA)
Distribution transformers of this rating are normally provided with OFF-circuit taps on HV side. Only in special cases ON-load tap changer (OLTC) may be required.	Power transformers have ON-load tap changers (OLTC) on HV side. The total number of taps should be 16 in steps of 1.25 %.

The standard range for OFF-circuit taps which are provided on HV side should be + 2.5% and + 5.0%. This is also applicable to power transformers whenever, Off-circuit taps are required.

4.7.5 ermination Agreement

In both, distribution and power transformers, the HV and LV terminals may be bare *outdoor bushings*, cable boxes or bus trunking depending upon the method of installation. However in power transformers of up to 33 kV, porcelain bushings and for transformers of 66 kV and above, oil-filled condenser type bushings should be specified. For installation in polluted atmospheres, the bushings should be specified with extra creepage distances as given in IS: 2099-1973. Fig. 4.14 shows the HV and LV bushings of a three-phase transformer.

4.7.6 Cooling

Criteria for Selection of Distribution Transformers (Upto and equal to 1600 kVA)	Criteria for Selection of Power Transformers (More than 1600 kVA)
Generally ONAN, AN or ANAN cooling methods are employed. Details of different types of cooling methods are explained in Section 4.6.	Generally ONAN, ONAN/ONAF, ONAN /ONAF/ OFAF or OFWF cooling methods are employed.

Fig. 4.14: HV and LV bushings of transformer

4.7.7 Fittings and Accessories

Criteria for Selection of Distribution Transformers (Upto and equal to 1600 kVA)	Criteria for Selection of Power Transformers (More than 1600 kVA)
Explosion vent should be provided for all 11 kV transformers above 400 kVA rating and all 33 kV transformers. When radiators are required to be detached for transportation, isolating valves should be specified. Alarm and trip contacts for oil temperature indicator, winding temperature indicator and gas actuated relay may be specified when monitoring and control devices are available for the transformer protection.	In addition, the following provisions should be made: (a) Skids and hauling lugs, (b) Magnetic oil level gauge with low level alarm contacts; (c) Rollers; (d) Winding temperature indicator with electrical contacts for alarm/trip and controlling fans and pumps; (e) Gas and oil actuated relay with alarm and trip contacts.; (f) Explosion vent; (g) Accessories for cooling; (h) Oil temperature indicator with alarm and trip contacts; (i) Two sampling valves at top and bottom of main tank.

4.8 SPECIFICATIONS OF THREE-PHASE DISTRIBUTION TRANSFORMER

The specifications of a three-phase, oil-immersed, naturally air-cooled, two-winding (primary and secondary), non-sealed type, outdoor mounted, distribution transformer are governed by the Indian Standard IS: 1180 (Part-I) 1989. It is titled as "Outdoor type three-phase Distribution Transformers up to and including 100 kVA, 11 kV- Specification, Part-1: Non-Sealed type". This standard provides guidelines on following specifications:

1. Ratings:
   - kVA rating,
   - Rated frequency,
   - Nominal system voltage
2. No-Load voltage ratios
3. Winding connections and vectors
4. Tappings and tapping methods
5. Transformer tank
6. Transformer oil
7. Fittings:
   - Standard Fittings,
   - Conservator,
8. Terminal arrangement:
   - Bushings and their dimensions,
   - Clearances between the bushings
9. Mounting arrangement
10. Insulation levels
11. Limits of temperature rise
12. Losses and impedance values:
    - Losses,
    - Impedance
13. Tolerances
14. Ability of transformers to with-stand external short-circuit
15. Efficiency and Regulation
16. Marking:
    - Rating plate,
    - Terminal Marking plate

4.8.1 Ratings

kVA Rating: The transformer kVA rating shall be any amongst the standard values i.e. 16 kVA, 25 kVA, 63 kVA and 100 kVA.

Rated Frequency: The rated frequency shall be 50 Hz.

Nominal system voltage: Nominal system voltage shall be chosen from 3.3 kV, 6.6 kV and 11kV.

4.8.2 No-Load Voltage Ratios

The No-load voltage ratios shall be as follows: 3300 volts/433-250 volts; 6600 volts/433-250 volts; 11000 volts/433-250 volts; 10450/433-250 volts.

4.8.3 Winding Connections and Vectors

The primary winding shall be connected in Delta and the secondary winding in Star with neutral terminal brought out such that the vector symbol is ***Dyn*11**. This will make a positive phase displacement of 30° from the primary to secondary vectors of the same phase. In Distribution transformers upto 100kVA rating, the secondary winding is connected in Star so that, the neutral connection could be made available to the consumers. Hence, neutral should be brought out to a separate insulated terminal.

4.8.4 Taps

Taps (also called as tapings) are normally not required on distribution transformers of ratings upto 100 kVA and 11 kV. However, they may be provided if specifically desired by the purchaser. In that case, tap changing is with transformer in OFF-condition (OFF-load tap changers). For more information on taps, revisit Section 4.7.4.

4.8.5 Transformer Oil

The transformer oil shall comply with the requirements of IS 335: 1983. Primarily, it should be suitable for use as an insulating and heat transfer medium.

4.8.6 Transformer Tank

The transformer tank which will house the oil-immersed transformer inside it should be of adequate strength to withstand any pressure built-up inside the tank. The exterior of the tank and all other ferrous fittings shall be given a triple coat of weather-resistant paint or enamel. Also, all steel screws, nuts and fasteners exposed to atmosphere shall be either galvanized or cadmium-plated.

4.8.7 Fittings

Standard Fittings: Transformer tank should be fitted with the following:

- Two earthing terminals with ⏚ symbol
- Oil level guage for indication of oil level inside the transformer tank
- Lifting lugs
- Ratings Plate and Terminals Marking Plate

- Plain breathing device which can prevent entry of rain water or insects inside the transformer tank
- Drain-cum-sampling value at the bottom of transformer tank to drain out the oil whenever necessary during preventive maintenance or repairs.
- Thermometer pocket for transformers above 25 kVA rating
- Oil-filling hole

Conservator: On transformers of rating 63kVA and 100kVA, the provision of conservator tank is mandatory. This tank is provided on the top of the main tank and both are connected by a hollow pipe between them. This provides additional space for the gases evolved from transformer oil when it becomes hot. For more details see section 'Know More' at the end.

4.8.8 Terminal Arrangement

Bushings: These transformers should be fitted with three high-voltage (HV) and four low-voltage (LV) outdoor type bushings as shown in Fig. 4.14 above. Their electrical characteristics shall be in accordance with IS 2099:1986 for HV bushings and IS 7421: 1974 for LV bushings.

Bushing clearances: The minimum phase-to-phase and phase-to-earth external clearances (gaps) between the adjacent bushings shall be as follows:

For LV bushings upto 1.1 kV: 75 mm and 40 mm respectively

For HV bushings: 255 mm and 140 mm respectively.

4.8.9 Insulation Levels

The insulation levels should be such that the transformer shall be capable of withstanding the power frequency (50 Hz) and impulse test voltage as given in IS 2026 (Part 3): 1981.

4.8.10 Limits of Temperature-Rise

For transformer winding: The permissible temperature-rise shall not exceed 55°C (when measured by resistance method).

For top oil: Temperature-rise shall not exceed 45°C when measured by thermometer.

4.8.11 Losses and Impedance Values

Losses: For transformers without taps, losses should not exceed the following.

Transformer kVA Rating	No-Load loss (Fixed loss or constant loss)	Load loss at 75°C (Variable loss)
16 kVA	80 W	475 W
25 kVA	100 W	685 W
63 kVA	180 W	1235 W
100 kVA	260 W	1760 W

Impedance: The recommended value of impedance at 75oC is 4.5%.

4.8.12 Efficiency and Voltage Regulation

The values of efficiency and regulation shall be based on loading at rated kVA and Unity power-factor. However, if desired by the purchaser, they may be calculated at other power-factors also.

4.8.13 Marking

Each transformer should be provided with non-detachable, weather-proof Rating Plate and Terminal Marking Plate. They may also be combined into a single plate. Fig. 4.15 (a) and (b) shows specimen terminal marking plate and rating plate.

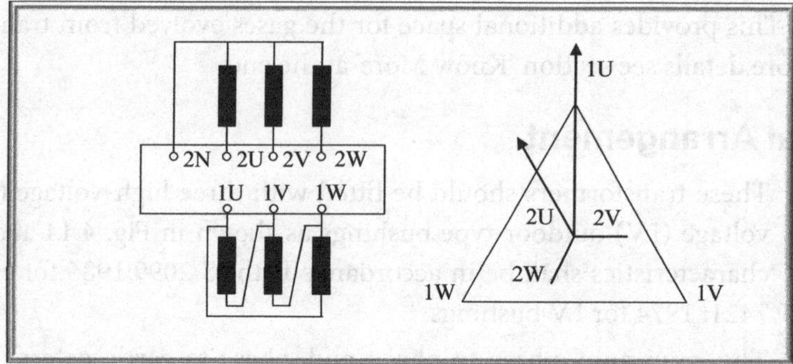

(a) Terminal Marking Plate

MADE IN INDIA

3 PHASE TRANSFORMER			Sr. No.		
STANDARD			FREQUENCY	Hz	50
kVA			TYPE OF COOLING		ONAN
VOLTS AT NO LOAD	HV		VECTOR GROUP		Dyn11
	LV		MASS OF OIL	kg	
AMPERES	HV		TOTAL MASS	kg	
	LV		VOL OF OIL	l	
IMPED VOLT	%		MONTH & YEAR OF MFG.		
CUSTOMER					
ORDER NUMBER					

MANUFACTURER'S NAME

(b) Rating Plate

Fig. 4.15: Marking Plates

4.9 PHASOR GROUPS

Depending on the phase displacement between lines-voltages of HV and LV winding, three-phase transformers are classified into four phasor groups (Also called as *vector groups*). This is shown in Table 4.5.

Table 4.5: Phasor groups of three-phase transformers

Group Number	Phase angle between HV and LV Line-voltages		Types of connection	Examples
Group No. 1	0°	$V_{L(LV)}$ is in-phase with $V_{L(HV)}$	star-star, delta-delta	*Yy0*, *Dd0*
Group No. 2	180°	$V_{L(LV)}$ is out of phase with $V_{L(HV)}$	star-star, delta-delta	*Yy6*, *Dd6*
Group No. 3	-30°	$V_{L(LV)}$ lags behind $V_{L(HV)}$ by 30°	star-delta, delta-star, star-zigzag	*Yd1*, *Dy1*, *Yz1*
Group No. 4	+30°	$V_{L(LV)}$ leads $V_{L(HV)}$ by 30°	star-delta, delta-star, star-zigzag	*Yd11*, *Dy11*, *Yz11*

4.10 THREE-PHASE TO TWO-PHASE CONVERSION OR SCOTT CONNECTION OR T-CONNECTION

Two-phase supply is usually required in special cases like, two-phase electric arc furnaces or for interlinking a three-phase system with two-phase system. Note that, the voltages in a two-phase balanced system have equal magnitudes but they are displaced from each other by a phase angle of 90o. Conventionally, power stations produce three-phase voltages which have a mutual phase displacement of 120o. One may think of taking any two phases and the neutral wire from a three-phase source to supply power to a two-phase load. But with this type of arrangement, the three-phase supply system will become unbalanced and the 120o phase displacement will not result into satisfactory performance of the load.

In 1890s, an electrical engineer and Professor from Yale University, Charles F. Scott invented a circuit connection for the conversion of three-phase supply to two-phase supply and vice-versa. In his honour, it is called as Scott connection. It requires two single-phase transformers. One of them should have a centre tap both on its primary side and is called as *Main Transformer*. The second one should have a tapping at 86.6% of its turns on primary side and is called as *Teaser Transformer*. The circuit is connected as shown in Fig. 4.16. Let the primary and secondary winding terminals of teaser transformer be denoted by **P, Q** and **p, q** respectively. Similarly, let the terminals of main transformer be **R, S** and **r, s** respectively. As shown in figure, **Q** is connected to the centre tap of main transformer. A three-phase balanced *ac* source *A-B-C* is connected across the terminals **P, R** and **S.** A two-phase load is connected across the terminals **p-q** and **r-s.** Consider the input line-voltage *VBC* as reference phasor. If *VL* is the magnitude of input line-voltage than,

$$\therefore \quad V_{BC} = V_{RS} = V_L \angle 0°$$

$$V_{CA} = V_{SP} = V_L \angle -120°$$

$$V_{AB} = V_{PR} = V_L \angle 120° \tag{4.10}$$

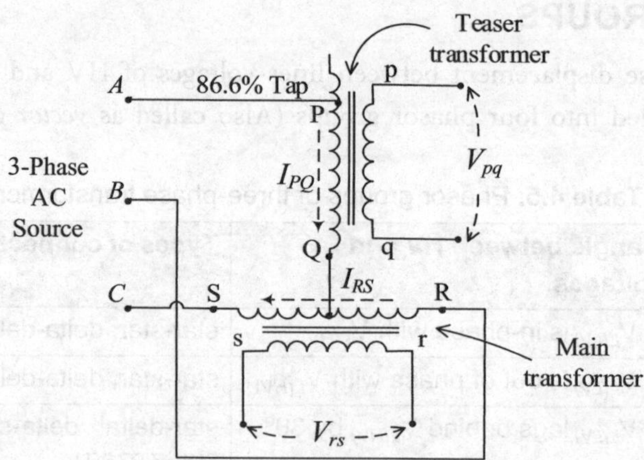

Fig. 4.16: Scott connection for three-phase to two-phase conversion

$$\therefore \quad V_{BC} = V_{RS} = V_L \angle 0°$$

$$V_{CA} = V_{SP} = V_L \angle -120°$$

$$V_{AB} = V_{PR} = V_L \angle 120° \tag{4.11}$$

These relations can be represented graphically as shown in Fig. 4.17(a). Let the positive directions of current and voltage in the two primary windings of teaser and main transformers be taken along I_{PQ} and I_{RS} respectively as shown in Fig. 4.16. Traversing the circuit along **P-Q-R**, the voltage equation will be,

$$\bar{V}_{PR} = \bar{V}_{PQ} + \bar{V}_{QR}$$

Or,

$$\bar{V}_{PR} = \bar{V}_{PQ} + \bar{V}_{RQ}$$

Since **Q** is connected to the centre tap between **R** and **S**, it can be rewritten as,

$$\bar{V}_{PR} = \bar{V}_{PQ} - \frac{1}{2}\bar{V}_{RS}$$

$$\bar{V}_{PQ} = \bar{V}_{PR} + \frac{1}{2}\bar{V}_{RS}$$

Substituting the values from equations (4.1),

$$\bar{V}_{PQ} = V_L 120^o + \frac{1}{2} V_L 0^o$$

$$\bar{V}_{PQ} = V_L \left(1 \angle 120^o + \frac{1}{2} \angle 0^o \right)$$

By vector addition in rectangular reference frame (refer Appendix in Unit-3), we get,

$$\bar{V}_{PQ} = 0.866\ V_L \angle 90° \tag{4.12}$$

Equation (4.12) is the voltage across primary winding of teaser transformer. Note that, it is only 86.6% of full voltage. It can also be proved by phasor diagram shown in Fig. 4.17(b). Since the primary winding of main transformer is connected across the supply terminals B and C, therefore from equation (4.11) again, the voltage across primary side of main transformer is,

$$\therefore \quad V_{RS} = V_{BC} = V_L \angle 0° \qquad (4.13)$$

Thus the two primary voltages V_{PQ} and V_{RS} are now phase shifted by 90°. Assuming that all windings are wound in same direction, the voltage across secondary windings V_{pq} and V_{rs} should be drawn in parallel with their respective primary voltages V_{PQ} and V_{RS} as shown in Fig. 4.17(c). Thus the three-phase voltages are converted into two-phase voltages V_{pq} and V_{rs}.

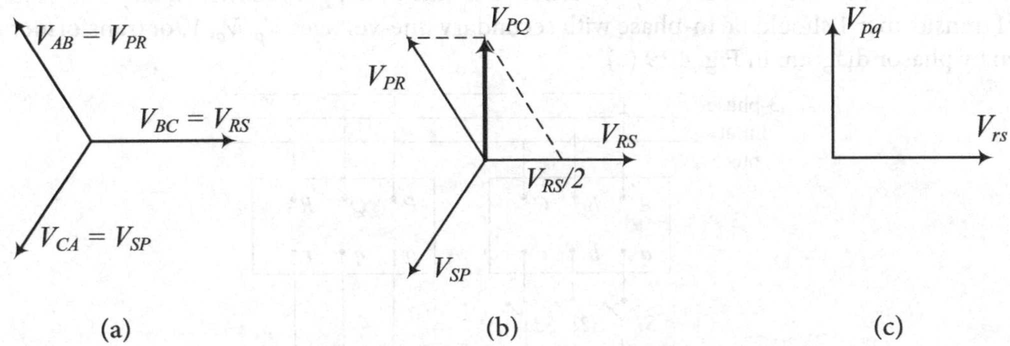

Fig. 4.17: Phasor diagrams related with Scott connection

Refer to equations (4.12) and (4.13) again. The voltages appearing across the primary windings of teaser and main transformers are not equal i.e. *0.866* V_L and V_L respectively. To have equal volts per turn in the two primary windings either, the number of turns of teaser primary should be 0.866 times the number of turns of main primary or the teaser transformer should have a tap at 86.6% turns on its primary side.

4.11 PARALLEL OPERATION OF THREE-PHASE TRANSFORMERS

In Section 3.18 of Unit-3, we have studied the parallel operation of single-phase transformers. The conditions that should be fulfilled before parallel operation were discussed in Section 3.18.1. They were:

1. Both transformers must have similar polarities.
2. Both transformers should have same voltage-ratios.
3. Their kVA ratings should be inversely proportional to respective equivalent impedances. In other words, they should have equal per-unit leakage reactances.
4. The ratios of equivalent leakage reactance to equivalent resistance should be equal.

All the above conditions are applicable for parallel operation of three-phase transformers also and they have same consequences when not fulfilled. Therefore, readers are expected to revisit Section 3.18 first. In addition to above, for the parallel operation of three-phase transformers, fulfilment of following two conditions is also necessary.

5. Both transformers should belong to the same Phasor Group (*Vector Group*). In other words, the phase-displacement between the secondary line-voltages corresponding to respective phases should be zero.
6. The phase-sequence should be same for both transformers.

4.11.1 Same Phasor Groups

Both, ***Yy*0** and ***Dd*0** belong to Phasor group No. 1. Therefore, they can be connected in parallel. Similarly, ***Yy*6** and ***Dd*6** belong to Phasor group No. 2 and can be connected in parallel. Let us now analyse what will happen when the transformers belonging to different phasor groups are connected in parallel. Obviously their secondary line-voltages corresponding to respective phases will not be in phase.

Consider that, two three-phase transformers 1 and 2 shown in Fig. 4.18 are to be connected in parallel by means of switches S_1, S_2 and S_3. Before closing these switches, if the voltage across each of them is zero then after closing them, there won't be any current circulating within the two secondary windings and the switch. This is treated as healthy condition. For this to happen, the secondary line-voltages V_a, V_b, V_c of transformer 1 should be in-phase with secondary line-voltages V_p, V_q, V_r of transformer 2. This is shown by phasor diagram in Fig. 4.19 (a).

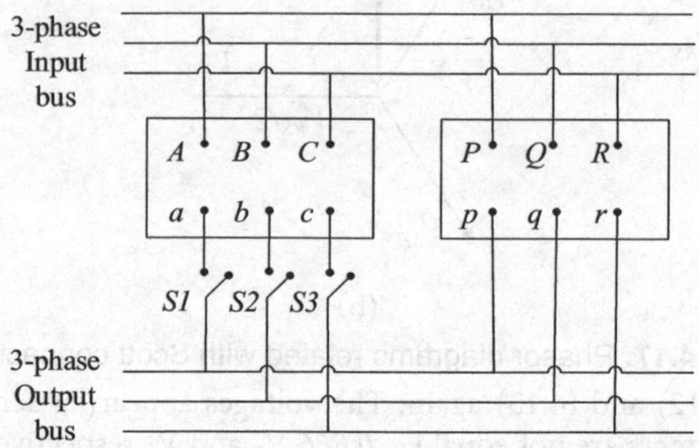

Fig. 4.18: Parallel operation of three-phase transformers

Now imagine that, these line voltages are not in-phase with each other. In other words, the two transformers belong to different phasor groups. The voltage phasor diagram for such a situation is shown in Fig. 4.19 (b). It can be clearly seen that, following voltages will appear across the switches before their closure.

$$\bar{V}_{S1} = \bar{V}_a - \bar{V}_p$$
$$\bar{V}_{S2} = \bar{V}_b - \bar{V}_q$$
$$\bar{V}_{S3} = \bar{V}_c - \bar{V}_r \tag{4.14}$$

After closing the switches, these voltages will result in large currents that will circulate within the two secondary windings even at No-load and cause additional losses and temperature-rise in the windings. They will not contribute to the load current.

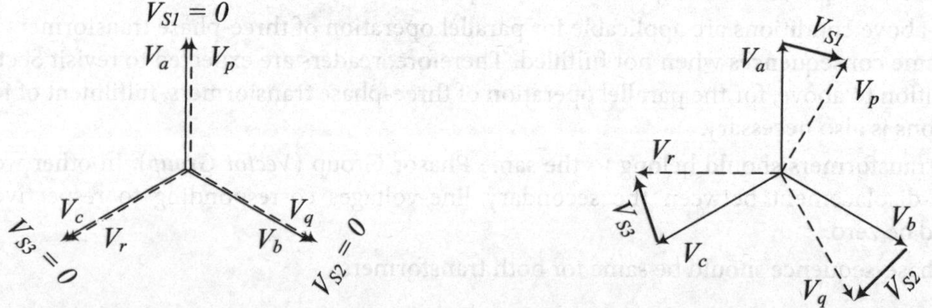

(a) When secondary voltages of two transformers are in-phase

(b) When secondary voltages of two transformers are not in-phase

Fig. 4.19: Voltage phasor diagrams

4.11.2 Same Phase Sequence

Phase sequence is the order in which the voltage waveforms of a poly-phase *ac* source reach their respective peaks. For a three-phase system, there are only two possible phase sequences: *U-V-W* and *U-W-V* corresponding to the two different possible directions of *ac* generator rotation.

Consider again the two three-phase transformers to be connected in parallel by switches S_1, S_2 and S_3 as shown in Fig. 4.20. Note the change in connection on secondary sides. Imagine that, the switch S_2 is now wrongly connected across terminals **b** and **r** in place of **b** and **q**. Similar error is with switch S_3. It means that on one side of the switches, the phase sequence is a-b-c. Whereas on its other side, it is p-r-q in place of p-q-r.

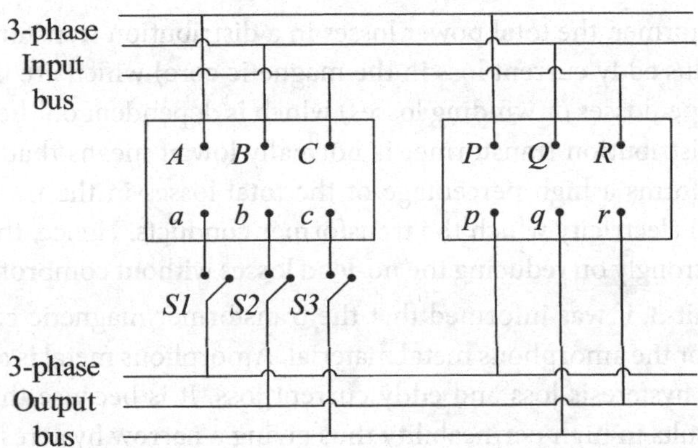

Fig. 4.20: Parallel operation of three-phase transformers with incorrect phase sequence

When they are connected with correct phase sequence like Fig 4.18, the secondary voltage phasor diagram will be as shown in Fig. 4.21 (a) which is similar to that shown in Fig. 4.19 (a). But when they are connected with different phase sequences then the resulting voltage phasor diagram will be as shown in Fig. 4.21 (b).

(a) When phase sequences are same (b) When phase sequences are different

Fig. 4.21: Voltage phasor diagrams

It clearly shows that the voltage across switch S_1 will be zero but the voltages across S_2 and S_3 will have large magnitudes at the time of closing. They will be as,

$$\bar{V}_{S1} = \bar{V}_a - \bar{V}_p = 0$$

$$\bar{V}_{S2} = \bar{V}_b - \bar{V}_r$$

$$\bar{V}_{S3} = \bar{V}_c - \bar{V}_q \tag{4.15}$$

Thus large currents will circulate at No-load within the two secondary windings causing additional losses and significant temperature-rise in the windings. They will not contribute to the load current.

4.12 AMORPHOUS CORE DISTRIBUTION TRANSFORMER

In Section 4.3, it was informed that, distribution transformers are maintained in energized state continuously for all the 24 hours every day. They carry a load which varies from time to time during the day & night and the maximum load occurs only for few hours in a day. Despite this fact, the kVA capacity of distribution transformer has to be decided to cater the maximum load.

Like any other transformer, the total power losses in a distribution transformer consists of no-load loss (i.e. hysteresis loss plus eddy current loss in the magnetic core) which are independent of the load, and the load loss (i.e. copper losses or winding losses) which is dependent on the loading of transformer. As the load-factor of a distribution transformer is normally low, it means that, the no-load loss which take place continuously forms a high percentage of the total losses in the transformer. It amounts to more than 2% of the total electricity which the transformer conducts. Hence, the design of distribution transformers is focused strongly on reducing the no-load losses without compromising the performance.

In Section 3.4 of Unit-3, it was informed that the transformer magnetic core is made from either CRGO silicon steel alloy or the amorphous metal material. Amorphous metal is a ferromagnetic material that offers both reduced hysteresis loss and eddy current loss. It is because this material has random grain structure which results in high permeability thus giving a narrow hysteresis loop compared to the conventional CRGO silicon steel material. It also increases the resistivity of the material. This makes the amorphous metal more popular for distribution transformers. Typical grain structures are shown in Fig. 4.22.

(a) Regular crystalline structure of grains in CRGO Silicon steel

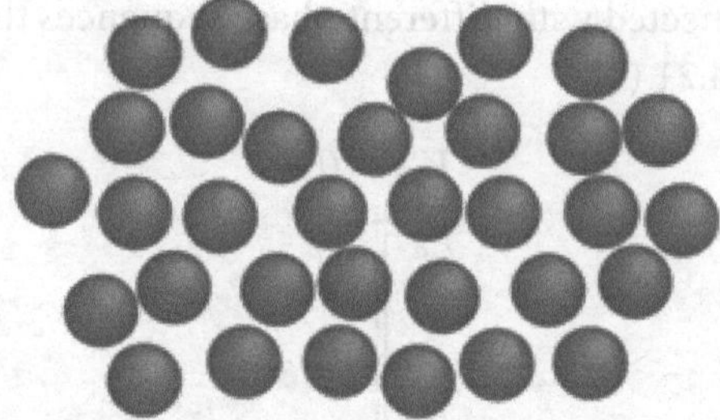

(b) Random structure of grains in Amorphous metal

Fig. 4.22: Grain structures

The CRGO Silicon steel magnetic core is made from thin laminations of standard shapes (like E, I, C) and the joints between them are interleaved and mitred. This was explained in detail in Section 3.3.1 of Unit-3. The core of an amorphous metal is made from a thin ribbon or foil of amorphous material which is wound spirally to get the desired dimension of the core. A typical five legged, three-phase amorphous core without and with windings is shown in Fig. 4.23 (a) and (b) respectively.

Some of the key differences between a CRGO Silicon steel core transformer and an amorphous core transformer are listed in Table 4.6.

(a) Unwound amorphous core (5 legs)

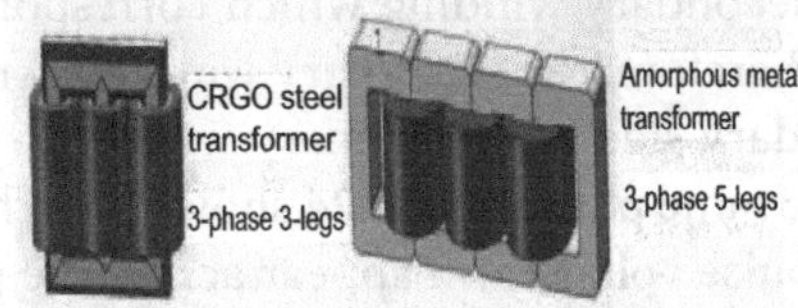

(b) Amorphous core with windings (shell type construction)

Fig. 4.23: Three-phase, amorphous core transformer

Table 4.6 : Comparison between CRGO core transformer and amorphous core transformer

Sr. No.	Parameter	CRGO core transformer	Amorphous core transformer
1.	No-load losses	Higher	Lower. Hence, the temperature rise is also less.
2.	Maximum efficiency loading point	At around 35% - 40% of rated load	Maximum efficiency point shifts more towards lighter loads
3.	Efficiency at full-load	High	Low
4.	Weight of transformer	Heavy	Lighter in weight
5.	Size of transformer for the same kVA rating	Comparable	Slightly higher
6.	Ease of repair	Easy	Relatively difficult
7.	Initial cost	Comparatively low	Comparatively high
8.	Repair cost	Comparable	High

Amorphous core transformer gives better harmonic wave tolerances. However, it has the following drawbacks:

- Degradation of efficiency over time
- High failure rates
- Low ability to withstand the short circuit impact and Poor overload capacity

4.13 PHASING OUT TEST

This test is performed on three-phase transformers to identify the primary and secondary windings belonging to the same phase. It is relevant when the terminals of all the primary and all the secondary windings are either masked or not marked.

In this test, all the windings of a three-phase transformer are short circuited except any one primary and any one secondary. A low voltage *dc* supply is applied across the selected primary and a galvanometer or ammeter is connected to the terminals of secondary winding which is not shorted. The galvanometer reading is noted. Now this secondary winding is short-circuited and the galvanometer is connected across another secondary winding and the procedure is repeated. However, the primary winding across which the *dc* source was connected is not changed. After testing all the secondary windings, conclusion is drawn from the galvanometer readings.

At the instant when *dc* supply is switched ON, the rate of change of flux-linkages will induced an *emf* in that secondary winding which corresponds to the primary winding across which the *dc* voltage is applied. Therefore, the secondary winding across which maximum deflection has occurred is marked as the secondary side of the primary winding to which the source was connected. In other words, both belong to the same phase. Fig. 4.24 shows the schematic diagram of phasing-out test. When the key 'K' is pressed, *dc* pulse voltage will appear across the primary winding. The winding across which maximum deflection occurs is the secondary phase winding that corresponds to primary winding to which dc source is connected. To understand the detail procedure, refer to the experiment in the later part of this Unit.

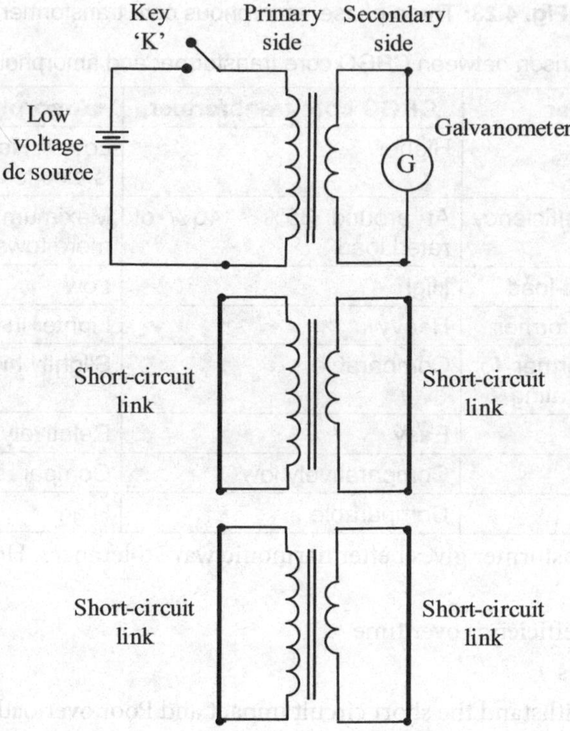

Fig. 4.24: Phasing-out test on three-phase transformer

UNIT SUMMARY

- Power transmission is carried out at high voltage levels to reduce the magnitude of current. This allows the use of thin wires for current and thus reduces the cost.
- Reduction in the magnitude of current also reduces the copper loss and voltage drop in transmission line.
- Three-phase power transformation from one voltage level to another can be carried out by using a three-phase transformer as well as by employing a bank of three single-phase transformers.
- Distribution transformers are kept in energized state continuously for 24 hours a day whereas some of the power transformers are disconnected during light load periods.

- Distribution transformers are designed such that their iron-losses will be minimum whereas power transformers are designed to have minimum copper loss.
- The horizontal parts of magnetic core are called as yoke while the vertical legs are called as limb.
- Third harmonic currents have frequency three times more than the fundamental frequency.
- The load that can be carried by an open-delta (V-V) transformer is only 57.7% of the total load that it can carry when connected in delta-delta mode.
- Clock hour method is used to represent the connections of transformer HV and LV windings and the phasor relationship between their line voltages.
- In clock hour method, the minute hand of the clock represents HV side line voltage whereas the hour hand represents LV side line voltage.
- kVA rating of power and distribution transformers is always selected for continuous loading.
- Tap-changers are of two types: ON-load tap changer and OFF-load tap changer.
- Distribution transformers are preferably connected in delta-star with neutral terminal of the secondary side being grounded.
- Taps are provided on the HV winding. They are used to balance the transformer output voltage between No-load and Full-load conditions. Taps can be changed by using a device called as 'Tap-Changer'.
- As per IS: 1180 (Part-I) 1989, the value of efficiency and voltage regulation to be specified by the manufacturer should be based on loading at rated kVA and Unity power-factor. However, if desired by the purchaser, they may be calculated at other power-factors also.
- Transformers are classified in four phasor groups on the basis of the phase displacement between line voltages on HV and LV sides as Group No.1, 2, 3 and 4.
- Scott connection also called as T-connection is used to convert a balanced three-phase supply into two-phase supply.
- For parallel operation of three-phase transformers, it is necessary that they should have same polarities and they should belong to the same phasor group. Also their phase sequence should be same.
- Amorphous core are normally preferred in distribution transformers because they produce less iron loss.
- The method of cooling of a three-phase transformer is selected on the basis of voltampere rating.
- Dry type transformers are normally preferred for indoor installations and crowded areas.
- The phasing-out test is carried out to identify the HV-LV winding pairs of each phase in a three-phase transformer.

EXERCISES

Multiple Choice Questions

1. A three-phase transformer has delta connected HV winding and star connected LV. It can be represented as:

(a) Dy0 or Dy6 (b) Dy11 or Dy1
(c) Dy0 only (d) Dy11 only

2. A three-phase transformer having star connected HV winding and delta connected LV can be represented as:
(a) Yd1 or Yd11 (b) Yd0 or Yd6
(c) Yd1 only (d) Yd6 only

3. A three-phase star-star transformer can be represented as:
(a) Yy1 or Yy11 (b) Yy0 or Yy6
(c) Yy1 only (d) Yy0 only

4. A three-phase delta-delta transformer can be represented as:
(a) Dd1 or Dd11 (b) Dd0 or Dd6
(c) Dd1 only (d) Dd0 only

5. YNd1 represents a transformer with
(a) Neutral terminal of HV side brought out
(b) Neutral terminal of HV side not brought out
(c) Neutral terminal of LV side brought out
(d) Neutral terminal of LV side not brought out

6. A Dy11 transformer can work in parallel with: (i) Yd11, (ii) Dy1, (iii) Yz11, (iv) Dy11
(a) All of the above (b) (i) and (ii) only
(c) (i) only (d) (i), (iii) and (iv)

7. As compared to delta-delta transformer, the capacity of V-V transformer is,
(a) 50% (b) 66.66%
(c) 57.7% (d) 100%

8. When a delta-delta transformer is delivering rated load, suddenly one of the phase gets disconnected and the transformer now operates in V-V mode. The overload on transformer will be
(a) 73.2% (b) 57.7%
(c) 66.66%

9. When a V-V transformer is converted into delta-delta, its capacity will increase by,
(a) 73.2% (b) 57.7%
(c) 66.66%

10. Which of the following connections should be selected for the conversion of three-phase into two-phase supply?
(a) star-delta (b) delta-star
(c) V-V (d) Scott

11. A three-phase, delta-star connected, 11kV/433 V transformer has a primary side line current of 5A. The corresponding line current on secondary side will be:
(Hint: the voltage rating so given are always the line voltages)
(a) 142A (b) 127A
(c) 86A

12. The input power taken by a 3-phase, star-delta transformer is 10 kVA. The output power will be,
(a) 5.77 kVA (b) 10 kVA
(c) 17.32 kVA

13. Consider a 50 kVA, 11kV/433V, delta-star transformer. Its rated primary and secondary line currents will be, (Hint: the voltage rating so given are always the line voltages)
(a) 4.7521A, 56.3332A (b) 3.798A, 64.468A
(c) 2.6243A, 66.6686A

14. Consider a 50 kVA, 11kV/433V, delta-star transformer. Its rated primary and secondary phase currents will be, (Hint: the voltage rating so given are always the line voltages)
(a) 4.7521A, 56.3332A (b) 3.798A, 64.468A
(c) 1.5151A, 66.6686A

15. Consider a 50 kVA, 11kV/433V, star-star transformer. Its rated primary and secondary line currents will be, (Hint: the voltage rating so given are always the line voltages)
(a) 2.6243A, 66.6686A (b) 4.7521A, 56.3332A
(c) 3.798A, 64.468A

16. Consider a 50 kVA, 11kV/433V, star-star transformer. Its rated primary and secondary phase currents will be, (Hint: the voltage rating so given are always the line voltages)
(a) 3.798A, 64.468A (b) 2.6243A, 66.6686A
(c) 4.7521A, 56.3332A

17. Consider a 50 kVA, 11kV/433V, delta-delta transformer. Its rated primary and secondary line currents will be, (Hint: the voltage rating so given are always the line voltages)
(a) 4.7521A, 56.3332A (b) 2.6243A, 66.6686A
(c) 3.798A, 64.468A

18. Consider a 50 kVA, 11kV/433V, delta-delta transformer. Its rated primary and secondary phase currents will be, (Hint: the voltage rating so given are always the line voltages)
(a) 4.7521A, 56.3332A (b) 3.798A, 64.468A
(c) 2.15A, 46.823A (d) 1.5151A, 38.4911A

19. Consider a 50 kVA, 11kV/433V, star-delta transformer. Its rated primary and secondary line currents will be, (Hint: the voltage rating so given are always the line voltages)
(a) 2.6243A, 66.6686A (b) 4.7521A, 56.3332A
(c) 3.798A, 64.468A

20. Consider a 50 kVA, 11kV/433V, star-delta transformer. Its rated primary and secondary phase currents will be, (Hint: the voltage rating so given are always the line voltages)
(a) 3.798A, 64.468A (b) 4.7521A, 56.3332A
(c) 2.6243A, 38.4911A

ANSWERS															
1.	(b)	2.	(a)	3.	(b)	4.	(b)	5.	(a)	6.	(d)	7.	(c)	8.	(a)
9.	(a)	10.	(d)	11.	(b)	12.	(b)	13.	(c)	14.	(c)	15.	(b)	16.	(b)
17.	(d)	18.	(d)	19.	(a)	20.	(c)								

Short Answer Questions

1. What are the different winding connections used in a three-phase transformer?
2. Name the types of windings used in core type and shell type three-phase transformers.
3. How sandwich type winding reduces the leakage flux?
4. Why star connection is preferred for high voltage, low kVA transformer?
5. Why delta connection is preferred for high line currents?
6. Why a star-delta transformer cannot be connected in parallel with star-star or delta-delta transformer?
7. When is the zigzag connection used in three-phase transformers?
8. Describe the following transformers: Dd6, Dyn1, YNd11.
9. Name the symbols suggested by Indian Standard to represent the line terminals of a three-phase transformer.
10. Name the governing features for the selection of three-phase transformer.
11. List the standard kVA ratings and nominal system voltages of a three-phase distribution transformer as specified by Indian Standard IS: 1180 (Part-I) 1989.
12. Why in distribution transformers, the secondary winding is connected in star?
13. Explain Scott connection.
14. Why is amorphous core popular in distribution transformers?
15. Why cooling of transformer is important? List the different methods of cooling in a three-phase transformer.
16. When is phasing-out test performed?

Long Answer Questions

1. Why three-phase system is preferred for power generation, transmission and distribution?
2. Compare a three-phase transformer with a bank of three single-phase transformers.
3. Compare power transformer and distribution transformer.
4. Compare core type and shell type three-phase transformers.
5. When is the V-V connection used in a three-phase transformer?
6. Explain zigzag connection.
7. Describe clock hour method.
8. State the considerations for selection of kVA rating of a transformer?
9. Explain the need of tappings in a three-phase transformer.
10. Classify three-phase transformer on the basis of phasor groups. Explain the type of connection and phase displacement in each of them with examples.
11. Prove that the two phase voltages obtained from Scott connection are displaced from each other by 90°.
12. Explain parallel operation of three-phase transformers.
13. What are the conditions for parallel operation of three-phase transformer? Which of them are common with the parallel operation of single-phase transformers?
14. Compare CRGO core transformer with amorphous core transformer.
15. Explain dry type transformer. Where is it preferred?
16. Explain the functions of conservator and breather.

PRACTICAL

Experiment No: 1

Aim: To perform phasing-out test on a three-phase transformer.

Equipment and Apparatus:

Device	Make	Range	Quantity
3-Phase transformer with all 12 terminals of six windings brought out		2 kVA	01
1-phase dimmerstat		Input: 0-240 V, 5A	01
Rectifier		12V	01
Multimeter/ DC voltmeter		0 – 10 V	01
Galvanometer		0 – 30 mA	01
ICDP Switch			01

Circuit connections:

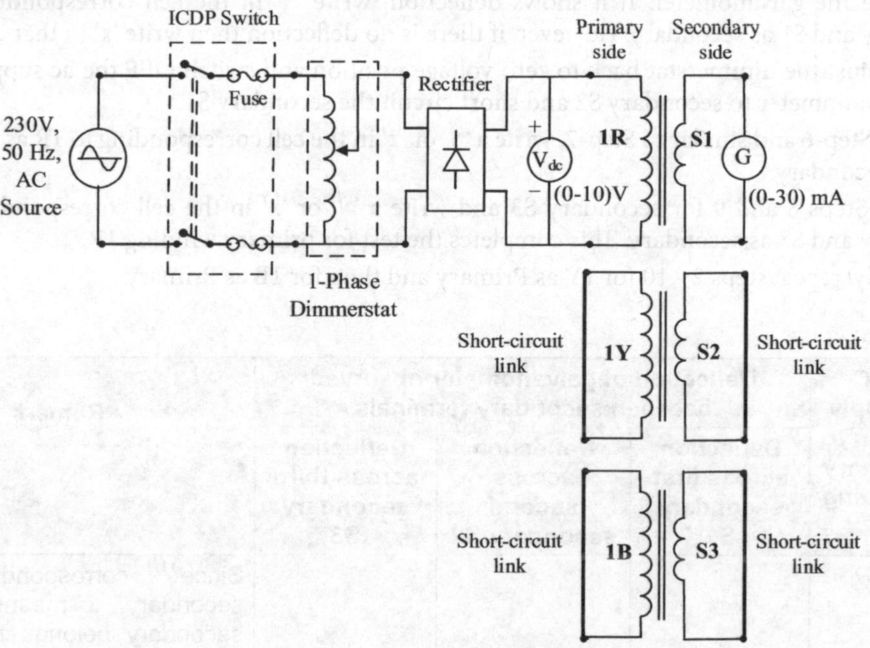

Fig. 4.25: Connection diagram for phasing-out test on 3-phase transformer

Theory:

Refer to Section 4.13.

Precautions:

1. Make tight connections as shown and switch on the supply after they are verified by the instructor.
2. Ensure that you are standing on a rubber mat while performing the experiment.
3. Stand at a safe distance from the experimental setup when it is running.
4. Do not bend to take readings. Stand straight and avoid parallax error while taking the readings.
5. Ensure that the dimmerstat output voltage setting is at zero setting before switching ON the supply.
6. Do not apply more than 1V *dc* voltage across primary windings.

Procedure:

1. Name the three primary windings as 1R, 1Y and 1B and the three secondaries as S1, S2 and S3.
2. Connect an input ac source through a dimmerstat and rectifier across the primary winding 1R and a galvanometer across any one secondary say 'S1'.
3. Short-circuit the primaries 1Y, 1B and the secondaries S2, S3.
4. After connecting the circuit, ensure that the dimmerstat is at zero output voltage position.
5. Switch on the 1-phase ac supply.
6. Adjust the dimmerstat position very gradually so as to get a low output voltage of say *1V* dc on the output side of the rectifier. This step should be executed with due care.
7. Observe the galvanometer. If it shows deflection, write '√' in the cell corresponding to 1R as primary and S1 as secondary. However, if there is no deflection then write 'x' in that cell.
8. Now adjust the dimmerstat back to zero voltage position and switch OFF the ac supply. Connect the galvanometer to secondary S2 and short circuit the secondary S1.
9. Repeat Step-6 and similar to Step-7, write a '√' or 'x' in the cell corresponding to 1R as primary and S2 as secondary.
10. Repeat Steps 8 and 9 for secondary S3 and write a '√' or 'x' in the cell corresponding to 1R as primary and S3 as secondary. This completes the test for primary winding 1R.
11. Similarly, repeat steps 2 - 10 for 1Y as Primary and then for 1B as Primary.

Observations:

Sr. No.	DC Supply given to Primary winding	Deflection of galvanometer observed between secondary terminals			Remark
		Deflection across first secondary S1	Deflection across second secondary S2	Deflection across third secondary S3	
1.	1R				Since '√' corresponds to _____ secondary, it means that this secondary belongs to Phase-R and should be renamed as 2R.
2.	1Y				Since '√' corresponds to _____ secondary, it means that this secondary belongs to Phase-Y and should be renamed as 2Y.
3.	1B				Since '√' corresponds to _____ secondary, it means that this secondary belongs to Phase-B and should be renamed as 2B.

Conclusion:

Discussion:

1. *Why the remaining windings are short-circuited during phasing-out test?*

   All windings in a three-phase transformer have large inductance. During the switching-ON and switching-OFF instants of a highly inductive circuit, there is a possibility of high induced *emf* in the mutually coupled remaining windings. To avoid any risk to the persons and equipment, all the remaining windings are short-circuited.

2. *Why the low voltage dc supply is connected across primary winding and not ac supply in this test?*

   The objective of phasing-out test is to identify the primary-secondary pairs of different phases and this is possible even by momentary induction of *emf* in the corresponding secondary. In this test, this momentary induction of *emf* takes place at the switch-ON and OFF instants. If an *ac* source is connected across primary in place of *dc* than there will be a continuous induction of *emf* in the secondary winding which is not required.

   Secondly, since *dc* supply is connected across primary winding, the frequency of winding currents is zero and that results in zero inductive reactances ($X_L = 2\pi f L\ \Omega$) of the windings. Consequently, the opposition to winding currents will be due to winding resistances alone. Since the winding resistances are normally very small it means that, the total winding impedance (in the absence of X_L, it will be equal to the winding resistance, Z = R) will be very less. Hence to limit the winding currents at low value, a small dc voltage is applied.

3. *Can we perform phasing-out test on single-phase transformer?*

   A single-phase transformer has only one primary winding and the winding on secondary side is mutually coupled with the primary. In other words it is already known that, they correspond to each other. Hence phasing-out test is not required.

4. *What is the difference between galvanometer and ammeter?*

   Ammeter indicates the magnitude of the current flowing through it. But, a galvanometer indicates both, the direction and the magnitude of the current being measured. This is because, the galvanometer has a centre-zero scale. With reversal in direction of current, the direction of deflection of pointer also reverses. However, galvanometers come in very low current range (Example: 30 mA) whereas, ammeters are available in higher ranges also (Example: 0 – 20A).

APPENDIX

Star and Delta connections:

Consider three identical, phase-windings each having two terminals. One of them can be named as starting terminal while the other one as finishing terminal. Let V_{ph} and I_{ph} represent phase voltage and phase current whereas *VL* and *IL* represent the line voltage and line current respectively. Assume that the three-phase system is balanced. Therefore,

$$|V_L| = |V_{AB}| = |V_{BC}| = |V_{CA}|$$

$$|I_L| = |I_{L1}| = |I_{L2}| = |I_{L3}|$$

$$|V_{ph}| = |V_{AA'}| = |V_{BB'}| = |V_{CC'}|$$

$$|I_{ph}| = |I_{AA'}| = |I_{BB'}| = |I_{CC'}|$$

Star connection:

If similar terminals of the three phase-windings are joined together to form a neutral point (also called as star-point) then, the resulting configuration is called star-connection as shown in Fig.4.26. The line voltages in a star connection will lead their respective phase voltages by a phase angle of 30°. Therefore as shown in phasor diagram of Fig. 4.27,

V_{AB} will lead $V_{AA'}$ by 30°

V_{BC} will lead $V_{BB'}$ by 30°

V_{CA} will lead $V_{CC'}$ by 30°

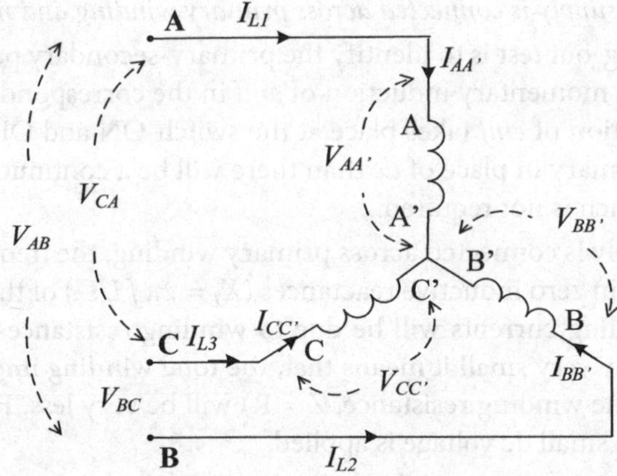

Fig. 4. 26 Star connection

Relationships in star connection:

Current relation: $I_L = \sqrt{3}I_{ph}$

Voltage relation $V_L = V_{ph}$

Apparent power:

$$S_{3Ph} = 3V_{ph}I_{ph} = \sqrt{3}V_L I_L \text{ VA}$$

Fig. 4.27: Voltage phasor diagram of star connection

Delta connection:

If dissimilar terminals of the three phase-windings are joined together to form a close circuit (loop) then, the resulting configuration is called delta-connection as shown in Fig.4.28.

The line currents in a delta connection lag behind their respective phase currents by a phase angle of 30°. Therefore as shown in phasor diagram of Fig. 4.29.

I_{L1} will lag $I_{AA'}$ by 30°

I_{L2} will lag $I_{BB'}$ by 30°

I_{L3} will lag $I_{CC'}$ by 30°

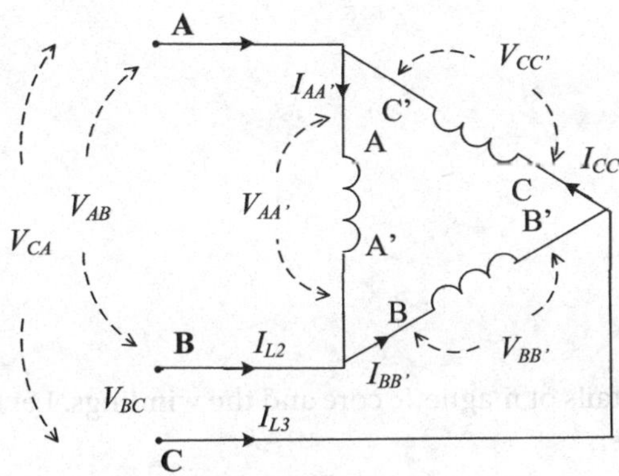

Fig. 4.28: Delta connection

Relationships in delta connection:

Current relation: $I_L = I_{ph}$

Voltage relation $V_L = \sqrt{3}V_{ph}$

Apparent power:

$S_{3Ph} = 3V_{ph}I_{ph} = \sqrt{3}V_L I_L$ VA

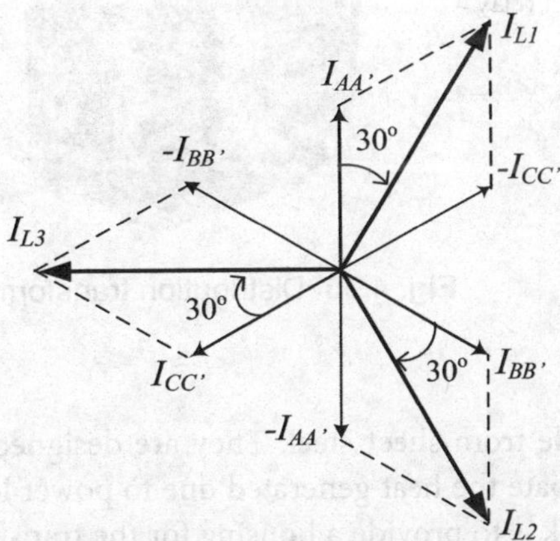

Fig. 4.29: Current phasor diagram of delta connection

KNOW MORE

Parts of Oil-immersed Transformer

Majority of the three-phase transformers installed at outdoor locations are of oil-immersed type. They are immersed in a tank filled with oil. This oil serves two functions. It acts as an insulating medium as well as a cooling medium. The basic parts of an oil-immersed transformer are:

1. Laminated magnetic core
2. 3-Phase HV and LV windings
3. Transformer tank
4. Transformer oil
5. Cooling tubes/Radiators
6. Oil Conservator
7. Breather
8. Buchholz relay
9. Explosion vent
10. Bushings

We have already studied the details of magnetic core and the windings. Let us now have a quick overview of the other parts of transformer.

Fig. 4.30 Distribution transformer

Transformer tank

Transformer tanks are made from sheet steel. They are designed such that sufficient cooling surface should be available to dissipate the heat generated due to power losses in the core and windings. Main function of transformer tank is to provide a housing for the transformer oil and the transformer. It also provides protection from external physical objects.

Transformer oil

The oil which is extensively used is called transformer oil and is a mineral oil obtained from fractional distillation of crude petroleum. To avoid the risk of fire, synthetic oils like *Askarel* are also developed for transformers. Since the whole transformer is immersed in oil, it is necessary that the oil should have good electrical insulting property (high dielectric breakdown strength). At the same time, it should also be able to transfer the generated heat from winding and core to the surrounding tank.

Cooling tubes and Radiators

Cooling tubes and Radiators provide additional surface area for the effective cooling of transformer oil. The transformer oil is circulated through the cooling tubes/ radiators. The circulation may either be natural or forced. In natural circulation, when the temperature of the oil increases, the hot oil naturally rises to the top, and the cold oil sinks downward. Thus the oil naturally circulates through the tubes. In forced circulation, an external pump is used to circulate the oil.

Oil conservator

Oil conservator is an airtight, metallic, cylindrical drum that is fitted above the transformer. It is connected to the main transformer tank which is completely filled with oil by means of a pipe. The normal oil level in the conservator is approximately in the middle of the conservator. During the transformer operation, the oil expands due to heating and some oil enters into the conservator from the transformer tank. Similarly, during light load periods, when oil temperature is less, it contracts and some oil goes back into the transformer tank from the conservator.

Breather

During the transformer operation ON-load condition when the oil in transformer tank expands due to heating and enters into the conservator, it forces some air above the oil surface to move out of the conservator into the atmosphere. Similarly, when the temperature is less, the oil contracts and some of the oil moves from conservator to the transformer tank. During this process, the volume above the oil surface in conservator increases and air enters into the conservator from atmosphere. Thus the transformer breathes like all living things. As the atmospheric air may contain moisture, it tends to contaminate the oil and reduces its dielectric strength. To avoid the pollution of transformer oil, a device called as *Breather* is fitted on the mouth of the pipe that connects the conservator to the atmosphere. This device consists of *Silica Gel* beads made from amorphous and porous form of silicon dioxide. Silica Gel absorbs moisture from atmospheric air before allowing it to enter inside the conservator. Thus the breather acts like an air filter.

The colour of Silica Gel beads used in transformer is blue. After absorbing moisture these beads turn pink in colour. One of the advantages of Silica Gel is that it is fully reversible and can be reused multiple times after drying or baking until it is fully saturated.

Buchholz relay

Buchholz relay is a protective device that is fitted in the pipe which connects the transformer tank to the oil conservator. The operation of this relay depends on the fact that most internal faults within the transformer generates gases in the oil. During the movement of oil from transformer tank towards the conservator, the gas bubbles in the oil gets trapped in the Buchholz relay. After sensing the presence of gases in the oil, the Buchholz relay activates the alarm circuit as well as the tripping circuit which leads to disconnecting the transformer from grid.

Explosion vent

During heavy internal faults, the oil temperature increases significantly. The explosion vent allows the boiling oil to expel outside the transformer to avoid any explosion like accident. The level of explosion vent is normally maintained above the level of conservator.

Bushings

In power and distribution transformers, the out-coming terminals of transformer windings have to pass over the transformer tank which is normally earthed for safety. To ensure complete separation from the transformer tank, these out-coming live terminals of a transformer are provided by bushings. For systems upto 66 kV, porcelain or oil filled bushings also called as non-condenser bushings are used. The oil-filled bushing consists of a hollow porcelain cylinder of special shape with a conductor, preferably a hollow tube through its centre. The space between the conductor and porcelain is filled with insulating oil.

REFERENCES AND SUGGESTED READINGS

1. T. Wildi, Electrical Machines, Drives, and Power Systems, Pearson Education Asia, 2001
2. P. K. Mukherjee, S. Chakravorti, Electrical Machines, Dhanpat Rai Publications, 2010.
3. P. S. Bhimbra, Electrical Machinery, Khanna Publishers, 2009.
4. R. K. Srivastava, Electrical Machines 2e, Cengage Learning, 2013
5. B. L. Theraja, A Text book of Electrical Technology Vol-II, S. Chand & Co., 2015
6. IS 2026, Part-4, 1977, Power-transformers, Terminal marking, Tapping and Connections
7. IS 10028, Part-1, 1985, Selection Installation and Maintenance of transformers
8. IS 1180, Part-1, 1989, Specifications of 3-phase distribution transformers

QR Code for Further Reading

5 Special Purpose Transformers

UNIT SPECIFICS

Through this unit, the different aspects of following special transformers are discussed:

- Single-phase and three-phase auto transformers;
- Current transformer;
- Potential transformer;
- Pulse transformer;
- Isolation transformer;
- Welding transformer;
- 'K' factor of transformers and;
- Overheating due to non-linear loads and harmonics.

Although the basic construction and operating principle of all transformers remain same, special considerations are necessary while designing special purpose transformers for dedicated applications. Autotransformers, instrument transformers, isolation transformer, pulse transformer, welding transformer, traction transformer, furnace transformer are some of the examples.

This unit will present an overview of special purpose transformers with detail explanation and diagrams. Laboratory experiments are also included to validate the theoretical understanding. The "Know More" section supplements additional information on the diagnostic method of detection of incipient faults and health estimation of transformer. This method is called as Dissolved Gas Analysis (DGA) and is popularly used by all major power utilities. Finally, the list of references and a dynamic QR code are given at the end for suggestive self-learning. Happy reading!

RATIONALE

Transformers designed for special purpose applications are different in terms of design aspects although their basic construction and working remains the same. For example, a transformer designed to operate at 400 Hz frequency will be smaller in size as compared to 50 Hz transformer of identical voltampere (VA) rating. Obviously, such low-weight, high-frequency transformers are preferred in aircrafts, spaceships and airborne equipment. In power system, the transformers are employed to either increase or decrease the voltage. However, there are applications where the voltage level need not be changed but the transformers are used to protect the load from voltage surges and dc transients. Some high precision transformers are simply used to measure the current and voltage of high magnitudes. The primary and secondary side of some transformers consist of a single, common winding resulting into the absence of electrical isolation.

This Unit will take the students on a journey to visit different members of the transformer family, each one of which is very special in itself.

PRE-REQUISITES

Basics of Electrical Engineering

UNIT OUTCOMES

By the end of this Unit, the students will be able to,

U5-O1: Compare an autotransformer with two winding transformer and discuss its construction, working, merits/demerits and applications;

U5-O2: Distinguish between CT and PT and discuss their operational aspects;

U5-O3: Explain the need and functioning of pulse transformer and isolation transformer;

U5-O4: Describe welding transformer and explain its V-I characteristics;

U5-O5: Discuss the overheating due to non-linear loads and K-factor rating of a transformer;

U5-O6: Conduct laboratory experiments on special purpose transformers and write reports.

Unit-5 Outcome	EXPECTED MAPPING WITH COURSE OUTCOMES (1 - Weak Correlation; 2 - Medium Correlation; 3 - Strong Correlation)				
	CO-1	CO-2	CO-3	CO-4	CO-5
U5-O1	-	-	-	-	3
U5-O2	-	-	-	-	3
U5-O3	-	-	-	-	3
U5-O4	-	-	-	-	3
U5-O5	-	-	-	-	3
U5-O6	-	-	-	-	3

5.1 AUTOTRANSFORMER

The single-phase and three-phase transformers discussed in Unit 3 and 4 are the two-winding transformers which mean that there are two separate windings per phase (primary and secondary) and they are *electrically isolated* from each other. The transfer of power from primary to secondary winding in a two-winding transformers takes place by mutual magnetic induction.

Autotransformers are single-winding transformer and the same winding is used for both, the primary and secondary sides. Obviously, such arrangement cannot provide an electrical isolation between the two sides. The transfer of power from primary to secondary side in an autotransformer takes place by both methods; *direct conduction* and *mutual induction*. Autotransformers are classified as

- Step-up type,
- Step-down type and
- Variable voltage type

As the names imply, in a step-up autotransformer, the output voltage (V_2) is greater than the input voltage (V_1). In a step-down autotransformer, the output voltage is less than the input voltage. Variable voltage autotransformers can be of both step-up and step-down types and they provide a wide range of variable ac voltage at the output from a fixed ac voltage at the input. They are commonly used in laboratories and industries. The variable voltage autotransformers are also called as Dimmerstat and Variac. Autotransformers can also be classified on the basis of number of phases as,

- Single-phase autotransformer and
- Three-phase autotransformer

Autotransformers have following advantages:

- They have low leakage reactance because predefined portions of the same coil act as primary as well as secondary.
- They require copper less than that of a two-winding transformer of same kVA capacity and hence costs less.
- They have efficiency higher than the two-winding transformer of same kVA capacity because of low ohmic loss and low core loss due to the reduction in material.
- Power rating of an autotransformer is higher than that of an equivalent two-winding transformer because the power is transferred by both mutual induction as well as by conduction.

However, an autotransformer has some demerits as well:

- The equivalent impedance of an autotransformer is less than the equivalent impedance of a two-winding transformer of same rating. This results in high current when a short-circuit fault occurs.
- Autotransformers do not provide an electrical isolation between the primary and secondary sides and as a result, they are not able to block the flow of *dc* transients throughout a three-phase power system.
- During the fault condition, if there is an open-circuit in the common part of the winding then, the secondary side is subjected to primary voltage minus a small voltage drop.

5.1.1 Single-Phase Autotransformer

Single-phase autotransformers have only one winding that is common to both primary and secondary.

A. Construction

Constructionally, it consists of a laminated Silicon-steel core and one single-phase copper winding wound on the core. This is shown in Fig. 5.1. At least three tappings A, B, and C are taken out from the winding. One of them (C) acts as a common neutral point for both primary and secondary sides.

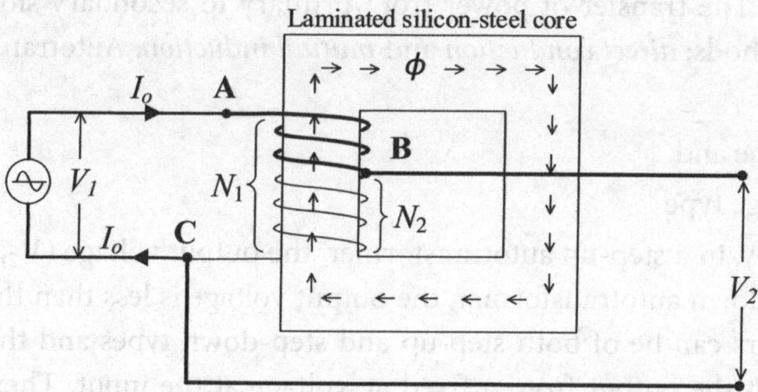

Fig. 5.1: Single-phase, step-down type autotransformer on NO-load

For step-down application, the input ac source is connected across the entire coil between the terminals A and C whereas the load is connected across a portion between the terminals B and C. This portion of the primary coil that is connected to the load also acts as a secondary. In case of step-up application, the input ac source is connected across a portion between B and C and which acts a primary whereas, the load is connected across the entire coil. Symbolic representations of step-down and step-up autotransformer with a fixed tap (B) on the output side are shown in Fig. 5.2(a) and (b). However, if it is desired to obtain a variable output voltage V_2 then the tap (B) is connected to a carbon brush (housed in a brush holder) that can be moved over the coil. As a result, the number of turns on the primary and secondary side will change and cause a smooth variation in the output voltage from zero to slightly more than the maximum input voltage V_1. Such variable voltage autotransformers are also called as Dimmerstat or Variac. They consist of a single layer of a high conductivity insulated copper wire, wound over an insulated toroidal core made of ribbon of thin insulated Silicon-steel. Fig. 5.2 (c) and (d) shows the symbolic representation of single-phase dimmerstat and its commercial view.

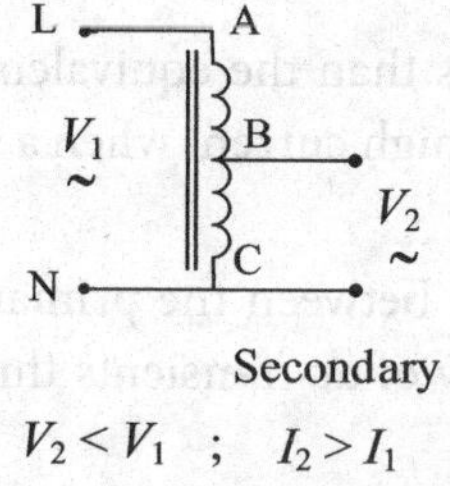

Fig. 5.2 (a): Step-down type autotransformer

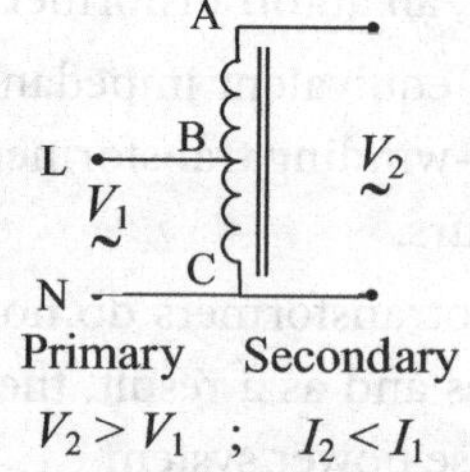

Fig. 5.2 (b): Step-up type autotransformer

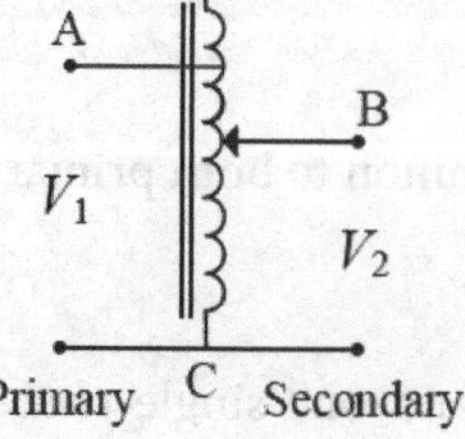

Fig. 5.2 (c): Schematic of single-phase dimmerstat

Fig. 5.2 (d): Single-phase dimmerstat

B. Working

Consider an autotransformer having a winding A-C of N1 turns that is wound on the magnetic core as shown in Fig. 5.1. Let this winding be connected across an ac voltage source V1 which will circulate a current Io and establish a flux in the magnetic core. Let there be a tapping 'B' on the winding such that, the number of turns between terminals B and C are N2. By transformation ratio K, (Refer to Section 3.5.2 of Unit-3), the output voltage V2 across the terminals B and C can be computed as,

$$V_2 = KV_1 = \left(\frac{N_2}{N_1}\right)V_1 \tag{5.1}$$

Thus it acts as a two-winding transformer with primary voltage V_1 connected across A and C while the secondary voltage V_2 appearing across terminals B and C. However there is absence of electrical isolation between the two sides. Now if a load is connected across the secondary terminals B and C then both primary and secondary circuits will carry the currents I_1 and I_2 respectively. However, by Kirchoff's current law, the portion B-C will carry a current $(I_2 - I_1)$. This is shown in Fig. 5.3.

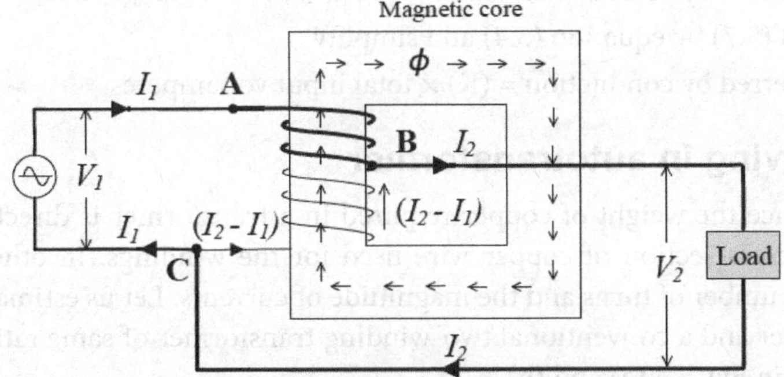

Fig. 5.3: Single-phase, step-down type autotransformer ON-load condition

By transformer action, the *mmf* due to current *I1* should be equal and opposite to the *mmf* produced by current I_2.

$$\therefore \quad I_1 (N_1 - N_2) = (I_2 - I_1) N_2$$

After simplification we get,

$$I_1 N_1 = I_2 N_2 \tag{5.2}$$

Assuming that both the transformer losses and the no-load current *Io* are negligible, then the output apparent power drawn by the load should be equal to the input apparent power supplied by the source.

$$V_1 I_1 = V_2 I_2 \tag{5.3}$$

Equations (5.1), (5.2) and (5.3) of single-phase autotransformer are identical to those of a two-winding transformer having a transformation ratio $K = N_2/N_1$.

C. Power transfer in autotransformer

In Section 5.2 above, it was informed that, the transfer of power from primary side to secondary in an autotransformer takes place by both methods (i) *direct conduction* and (ii) *mutual induction.*

The total voltamperes supplied by input voltage source is $V_1 I_1$ (5.4)

Refer to Fig. 5.3 again.

The *mmf* of portion A-B is $= I_1 (N_1 - N_2) = I_1 N_1 - I_1 N_2$

Substituting the value of from equation (5.2)

mmf of portion A-B will be $= I_2 N_2 - I_1 N_2 = (I_2 - I_1)N_2$

But $(I_2 - I_1) N_2$ is the *mmf* of portion B-C. It means that, the portion A-B acts as primary winding and portion B-C acts as secondary winding of the autotransformer because in any transformer, the primary *mmf* and secondary *mmf* are equal. Therefore as per transformer action, voltamperes from portion A-B should be transferred to portion B-C by mutual induction.

$\therefore$ Power transferred by mutual induction will be $= V_{AB} = (V_1 - V_2) I_1$ (5.5)

Divide equation (5.5) by equation (5.4) and simplify

$\therefore$ Power transferred by mutual induction = (1 – K) × total input voltamperes (5.6)

Where $K = V_2/V_1$ = transformation ratio.

The remaining power is transferred by direct conduction. Therefore equation (5.5) from equation (5.4),

$\therefore$ Power transferred by conduction will be $= [V_1 I_1 - (V_1 - V_2) I_1] = V_2 I_1$ (5.7)

Divide equation (5.7) by equation (5.4) and simplify

$\therefore$ Power transferred by conduction = (K) × total input voltampere s (5.8)

D. Copper saving in autotransformer

The volume and hence the weight of copper required in a transformer is directly proportional to the length and area of cross-section of copper wire used for the windings. In other words, it is directly proportional to the number of turns and the magnitude of currents. Let us estimate the copper required by an autotransformer and a conventional two-winding transformer of same ratings. Consider the two transformers shown in Fig. 5.4 (a) and (b).

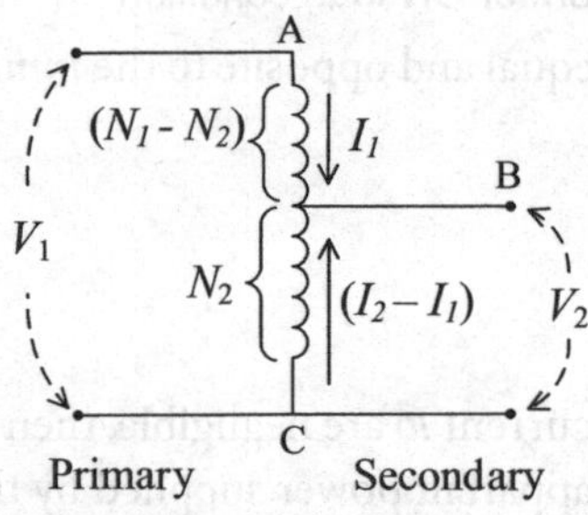

Fig. 5.4 (a): Autotransformer

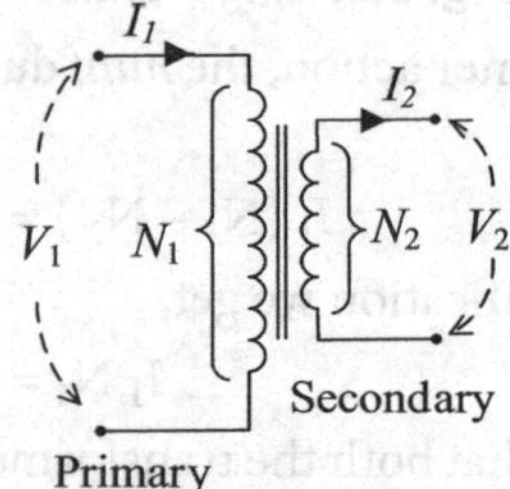

Fig. 5.4 (b): Two-winding transformer of same rating

In an autotransformer:

Weight of copper in portion AB $\propto (N_1 - N_2) I_1$

Weight of copper in portion BC $\propto N_2 (I_2 - I_1)$

Total copper weight in autotransformer (Wa) $= (N_1 - N_2) I_1 + N_2 (I_2 - I_1)$ (5.9)

In a two-winding transformer:

Weight of copper in primary winding $\propto N_1 I_1$

Weight of copper in secondary winding $\propto N_2 I_2$

Total copper weight in two-winding transformer (W) = $N_1 I_1 + N_2 I_2$ (5.10)

Divide equation (5.9) by equation (5.10) and simplify,

Copper weight in autotransformer, $W_a = (1 - K)W$ (5.11)

Where, $K = N_2/(N_1 = V_2/V_1)$ = transformation ratio

W = total copper weight in a two-winding transformer

Saving in copper will be,

$$\text{Copper saving} = (W - W_a) = K \times W \quad (5.12)$$

Thus a considerable amount of copper can be saved with V_2 approaching V_1. For a voltage reduction with transformation ratio K = 0.5, approximately 50% copper can be saved when autotransformer is employed in place of a two-winding transformer.

5.1.2 Three-Phase Autotransformer

In a three-phase autotransformer, for each phase, an individual single-phase winding (coil) is provided. In fact, a three-phase autotransformer can also be constructed by suitably interconnecting three single-phase autotransformers of identical characteristics and ratings, in star or delta. There is no magnetic coupling between the three phases.

In a three-phase autotransformer, three voltage taps are provided on the three coils. They may be either fixed on the coils to provide fixed output voltages or they may be attached to carbon brushes so that they could be moved on the coils thus providing variable output voltages. In the latter case, it is also called as *dimmerstat* or *variac*. Schematic diagrams of three-phase autotransformers in star and delta are shown in Fig. 5.5 (a) and (b). Normally, the three windings of autotransformer are star connected. However, it is important to connect the neutral to common star point. There is no autotransformer action in star connected autotransformer if the common neutral point making star is not connected and almost full input voltage may appear on the output side. Although this a demerit of autotransformer, but it is effectively used in autotransformer starting of three-phase induction motors without momentary disconnection of supply to the induction motor.

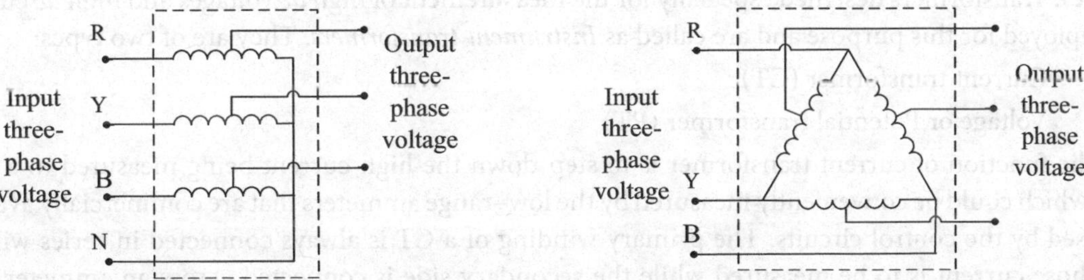

Fig. 5.5 (a): Three-phase autotransformer in star **Fig. 5.5 (b):** Three-phase autotransformer in delta

5.1.3 Applications of Autotransformers

- They serve as variable voltage *ac* source in laboratories so as to test various equipment whose rated voltage is less than the supply voltage available in the laboratory.
- In laboratories, the three-phase dimmerstat is used for reduced voltage starting of three-phase induction motor so that the motor starting current could be restricted within safe limits. This also

helps in the starting of large three-phase induction motors without momentary disconnection of supply to the induction motor.

- In substations, autotransformers are used to transfer the incoming high voltage to the next lower voltage level. They are preferred to transform voltages when the primary to secondary voltage ratio is close to 1. Autotransformers are generally equipped with ON-load Tap Changers.
- They are used for controlling the *dc* output voltage of rectifier by adjusting its *ac* input voltage.
- They are also required for interconnecting two power grids operating at different voltages.

5.2 INSTRUMENT TRANSFORMERS

In *dc* circuits, when high-voltage and high-current are to be measured by the commercially available *dc* voltmeter and *dc* ammeter of lower range, we make use of a series resistor R_{se} and shunt resistor R_{sh} respectively as shown in Fig. 5.6. However, this technique cannot be adopted for the measurement of high *ac* voltages and high *ac* currents.

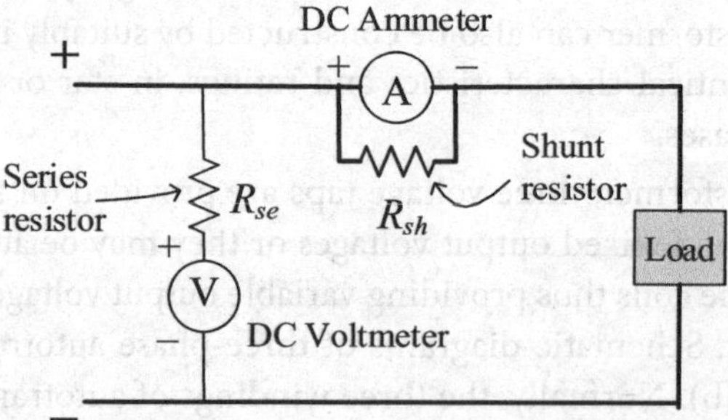

Fig. 5.6: Measurement of dc current and dc voltage (high magnitudes)

In substations, to ensure the protection of human and equipment during fault conditions, suitable protection systems comprising of relays and circuit-breakers are in place. These protection systems involve continuous monitoring of high *ac* voltages (in kilovolts) and high *ac* currents (in hundreds of ampere). Transformers designed especially for the measurement of high *ac* voltages and high *ac* currents are employed for this purpose and are called as *Instrument transformers*. They are of two types:

- Current transformer (CT)
- Voltage or Potential transformer (PT)

The function of current transformer is to step-down the high current being measured to a lower range which could be conveniently measured by the low–range ammeters that are commercially available or sensed by the control circuits. The primary winding of a CT is always connected in series with the line whose current is to be measured while the secondary side is connected across an ammeter or an actuating coil of the relay. Similarly, the function of potential transformer is to step-down the high voltage being measured to a range which could be conveniently measured by the low–range voltmeters that are commercially available or sensed by the control circuits. The primary winding of a PT is connected across the lines whose voltage is to be measured while the secondary side is connected across a voltmeter or an actuating coil of the relay equipment. General connections of CT and PT are shown in Fig. 5.7.

In case of current and voltage transformers, the load appearing across their secondary terminals is called as *burden*.

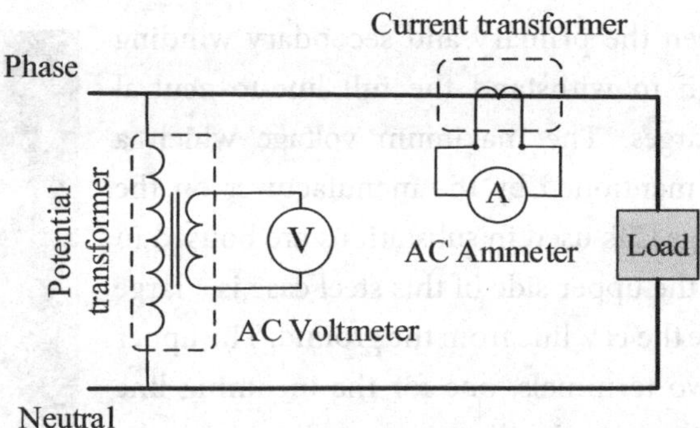

Fig. 5.7: Measurement of ac current and ac voltage (high magnitudes) using instrument transformers

5.2.1 Current Transformer (CT)

Current transformers are the high precision transformers in which the ratio of primary current I1 to secondary current I2 is a known constant and is provided by the manufacturer as CT ratio. For example, a CT with a ratio of 300:5 is rated for 300A primary current at full-load and will produce 5A of secondary current when 300A will flow through the primary. This CT ratio changes very little with the burden and the phase angle between I_1 and I_2 is very small, usually much less than one degree. The primary winding is made from thick copper wire and has a very few number of turns. It is always connected in series with the line. The secondary winding is made from thin wire and has a large number of turns. The nominal secondary current is usually rated at 5A, irrespective of the primary current rating so that standard instruments and relays could be connected on the secondary side.

When the line current exceeds 100A, CTs are made of a laminated torroidal core or a torroidal ferromagnetic core on which several turns of secondary winding alone are wound. The line whose current is to be measured is made to pass through the core and itself acts as a single-turn primary winding. Considering the transformation ratio, $K = I_1 / I_2 = N_2 / N_1$, a CT having 150:5 ratio will have 30 times more turns on the secondary than on the primary. Torroidal CTs are simple, inexpensive and are widely used in medium-voltage (MV) and low-voltage (LV) indoor applications. Fig. 5.8 shows a torroidal core CT and its symbol.

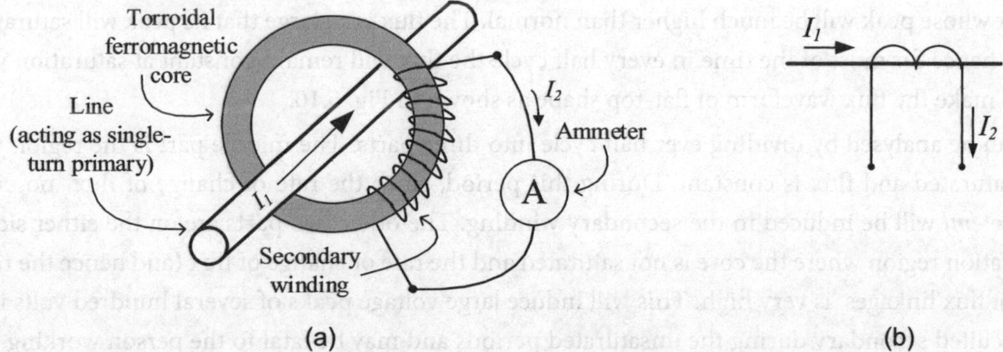

Fig. 5.8: Current Transformer (a) Torroidal core type CT, (b) Symbol of CT

The insulation between the primary and secondary winding must be sufficient enough to withstand the full line-to-neutral voltage, including line surges. The maximum voltage which a CT can sustain is always mentioned by the manufacturer on the nameplate. The high-voltage CTs used in substations are housed in a grounded steel case. On the upper side of this steel case is a large porcelain bushing to isolate the HV line from the ground. The upper end of the bushing has two terminals, one for the incoming line current and the other for the outgoing line current. Therefore, the line current enters into one terminal of CT, flows down the bushing, the primary winding, then up in the bushing and finally leaves the second terminal of the CT. To eliminate the danger of fatal shock when touching the secondary side, one of the secondary winding terminals of a CT installed in substations is always connected to ground. Fig. 5.9 shows the view of a medium voltage CT connected in substation.

Fig. 5.9: Medium voltage CTs

5.2.2 Why the Secondary of a CT Should not be Open-Circuited?

Open-circuiting the secondary winding of a CT while its primary is carrying current is dangerous and may cause serious hazard in terms of electric shock to the operator. In other words, either an ammeter or low-impedance actuating coil of a relay should be connected across the secondary or otherwise the secondary winding of a CT should be closed else, large voltage will appear on the secondary of CT causing insulation failure and resulting in fire. The reason is based on the fundamentals of transformer that were discussed in Section 3.5.1 and 3.7 of Unit 3.

It is known that, the current flowing in the primary of a CT is the line current I_1 of high magnitude. If the secondary winding is open then, the secondary current I_2 and hence both will become zero. As a result, whole of *I1* will act as excitation current *Io* (i.e. No-load current) and consequently, produce a large flux whose peak will be much higher than normal. The flux is so large that its peak will saturate the core and hence for most of the time in every half cycle the flux will remain constant at saturation value. This will make the flux waveform of flat-top shape as shown in Fig. 5.10.

It can be analysed by dividing ever half cycle into three parts. The middle part is the region when core is saturated and flux is constant. During this period, since the rate of change of flux no *emf* or negligible *emf* will be induced in the secondary winding. The other two parts are on the either sides of the saturation region where the core is not saturated and the rate of change of flux (and hence the rate of change of flux linkages is very high. This will induce large voltage peaks of several hundred volts in the open-circuited secondary during the unsaturated periods and may be fatal to the person working on it. Also, failure of insulation of CT secondary causes fire.

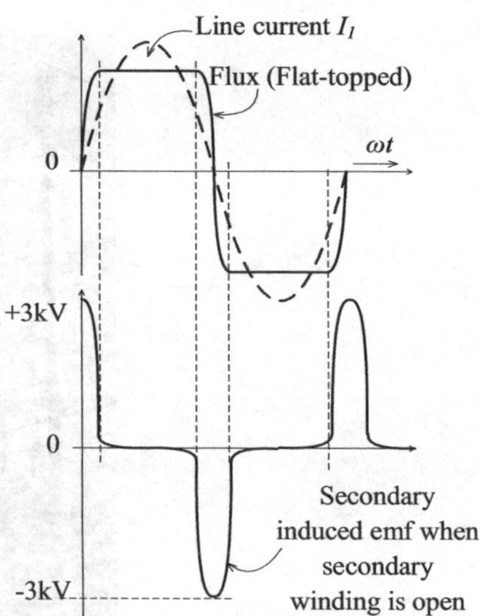

Fig. 5.10: Typical primary current, flux and secondary voltage when CT is open-circuited

5.2.3 Potential Transformer (PT)

Potential transformer is also called as *Voltage transformer*. They are the high precision transformers in which the ratio of primary voltage to secondary voltage is a known constant that is mentioned by the manufacturer on the nameplate. It changes very little with the burden. Also, the secondary voltage is in phase with the primary voltage. The construction of a PT is similar to that of a conventional transformer. The number of turns in the primary are much more than in the secondary winding. The primary winding is made to have higher impedance hence, it carries less current. The insulation between the primary and secondary windings should be sufficient to withstand the high line-to-neutral voltage on HV side.

PTs installed in high-voltage (HV) lines always measure line-to-neutral voltage. One of the primary winding terminal is connected to the ground while the other is connected to the HV line. Thus there is only one porcelain bushing and only one terminal coming out of it. Like CT, one terminal of the secondary winding of a PT is always connected to ground to avoid the danger of fatal shock when touching one of the secondary terminals. In absence of such grounding, the (invisible) distributed capacitance between the primary and secondary windings may produce a very high voltage between the secondary and the ground. Fig. 5.11 shows a typical PTs installed in substation and the symbol of PT.

When the primary of a PT is in energised state, precaution should be taken to *avoid short-circuit of its secondary winding*. This is because, in normal operating condition, the secondary winding of a PT is either connected to a voltmeter or to the actuating coil of a relay, both having high impedance Z_L and therefore, the secondary as well as primary currents are low. But accidently, if the secondary winding gets short-circuited then, Z_L will become zero. Since the primary side is already connected across high line-to-neutral voltage of HV line, it will circulate large short-circuit currents in both primary and secondary windings. This may damage the equipment connected on secondary side or even endanger personal safety.

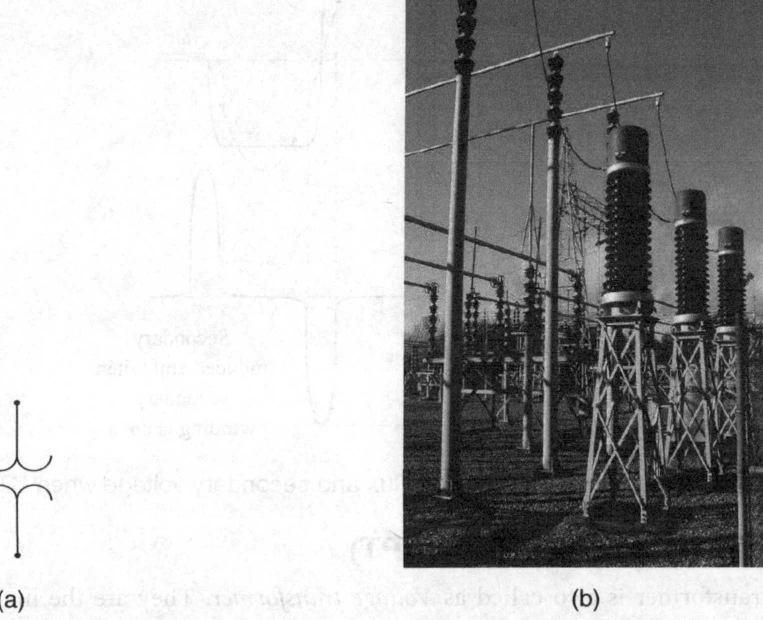

Fig. 5.11: Potential Transformer (a) Symbol, (b) PTs in substation

5.3 PULSE TRANSFORMER

So far, the transformers discussed in this book are designed to operate with continuous and sinusoidally alternating waveforms of voltages and currents. Pulse transformers are the transformers that handle voltages and currents in the form of pulses which are non-continuous signals. The application of pulse transformer is quite common in electronic communication networks, radar systems, televisions, digital computers, electrical control circuits etc. The main functions that a pulse transformer can perform are:

- To change the amplitude of voltage pulse
- To invert the polarity of voltage pulse
- To provide *dc* isolation between the source and load that are connected on the primary and secondary sides of pulse transformer
- To couple different stages of pulse amplifiers and
- To function as an isolation transformer

The pulse transformers used in electronic circuits and electrical control circuits are quite small in size. Both primary and secondary windings have comparatively few turns so that the leakage inductances are minimum. They are designed to operate at high frequencies of the order of 20 kHz – 100 kHz. The pulse transformers employed in electrical control circuits are used to isolate the low-power, low-voltage gate control circuit from the high-power, high-voltage power electronic circuits. Those used in radar systems are relatively bigger in size and voltage rating.

The construction and operation of a pulse transformer is identical to that of conventional transformers. However to allow high frequency, the magnetic core is made from highly permeable material. Ferrite core or the nickel-iron alloy called as 'Permalloy' are commonly used. Fig. 5.12 shows the kind of voltage pulses that are normally given as input to the pulse transformer. The pulse width varies from a fraction of microseconds to about 25 microseconds.

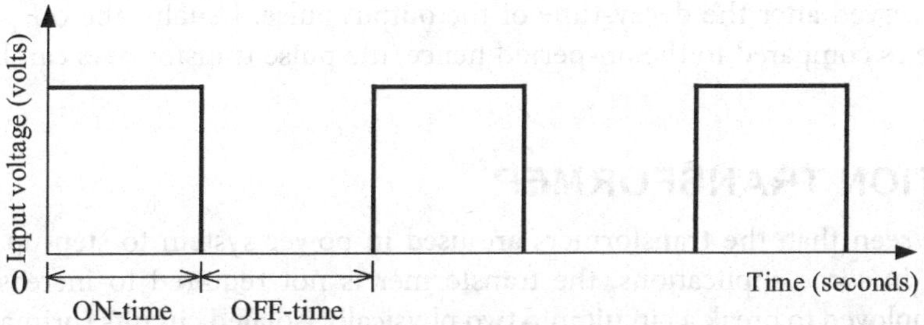

Fig. 5.12: Train of input voltage pulses

It is expected that that a pulse transformer should reproduce the input pulse as accurately as possible on its output side. A typical output pulse produced by it on its secondary side is shown in Fig. 5.13. The distortions in this output pulse can be determined by transient analysis.

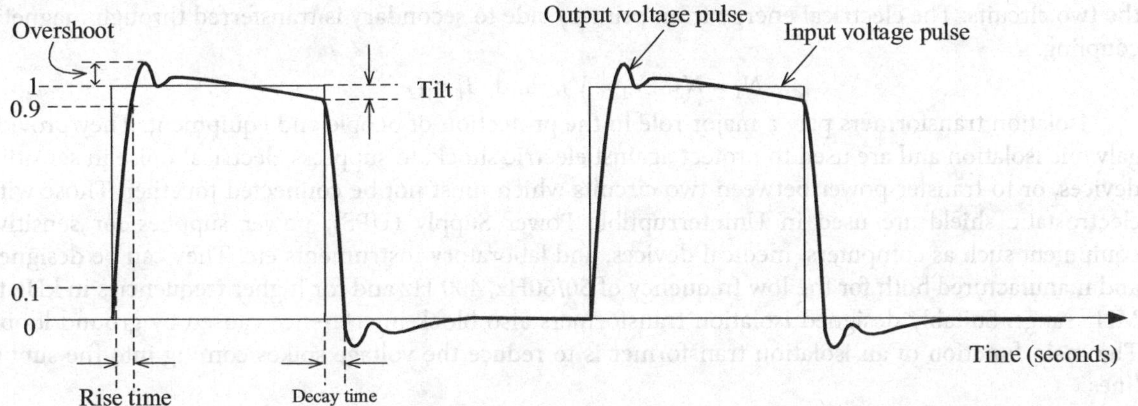

Fig. 5.13: Output voltage pulses of a pulse transformer

The performance analysis of pulse transformer can be carried out by dividing the output response into three parts as:

- The front edge of output voltage pulse
- The middle flat-top region of the pulse and
- The trailing edge of the pulse after its termination.

The Rise-time (t_r) represents the time taken by the output voltage pulse to rise from 10% to 90% of its final magnitude. In order to keep the rise time within limits, the leakage inductances of the transformer should be kept to a minimum. In the middle region, the output voltage pulse is seen to have a downward tilt. This is because, the output voltage cannot remain flat (identical to the input pulse) else, it will mean that the pulse transformer is allowing *dc* to pass through it which is not possible in any transformer. However, this tilt (drop-off) of the output pulse can be kept as small as possible by having high magnetizing inductance of the transformer. In other words, the core should be made from highly permeable material.

From Fig. 5.12 and 5.13, it is observed that when the input pulse is zero, the output pulse does not reduce to zero instantaneously. This happens due to the magnetic energy stored in the transformer inductance when input voltage pulse is not zero. There is a backswing of the output voltage and because of the transformer inductance and stray capacitance, damped oscillations and long-duration negative overshoots are observed after the decay-time of the output pulse. Usually, the off-period of the pulse train is quite large as compared to the on-period hence, the pulse transformers can handle high pulse-power levels also.

5.4 ISOLATION TRANSFORMER

Till now we have seen that, the transformers are used in power system to step-up or step-down the voltage. However in some applications, the transformer is not required to increase or decrease the voltage but it is employed to break a circuit into two physically isolated circuits (primary and secondary) so that the direct current noise signals or the voltage surges can't travel from primary to secondary. Such transformers are called as isolation transformer.

Therefore, in an isolation transformer, the number of turns in primary and secondary windings are equal and transformation ratio K = 1. Hence, it is also called a 1:1 ratio transformer in terms of voltage, current, and turns ratio. An isolation transformer provides isolation physically and electrically between the two circuits. The electrical energy from primary side to secondary is transferred through magnetic coupling.

$$N_1 = N_2, \quad V_1 = V_2, \quad \text{and} \quad I_1 = I_2$$

Isolation transformers play a major role in the protection of people and equipment. They provide galvanic isolation and are used to protect against electric shock, to suppress electrical noise in sensitive devices, or to transfer power between two circuits which must not be connected together. Those with electrostatic shield are used in Uninterruptible Power Supply (UPS), power supplies for sensitive equipment such as computers, medical devices, and laboratory instruments etc. They can be designed and manufactured both for the low frequency of 50/60Hz, 400 Hz and for higher frequencies in kHz to MHz range. Suitably designed isolation transformers also block interference caused by ground loops. The main function of an isolation transformer is to reduce the voltage spikes coming into the supply lines.

Due to lightning, static electricity, or due to sudden change in load, a sudden rise in voltage level takes place that lasts for a very short duration of time (in nanoseconds or microseconds). This transient change in voltage is called as *voltage spike*. The voltage spikes may have high voltages ranging upto several thousand volts. Such high voltage spikes when reaches the load can cause the interruption of services or can damage the electronic equipment. Also, voltage spike causes electromagnetic interference (EMI). It results in malfunctioning of equipment causing loss and damage. If we connect an isolation transformer between the power supply lines then, due to the inductive nature of transformer windings, the voltage spikes get minimized on the secondary side before reaching the load. Fig. 5.14 shows the schematic view of an isolation transformer.

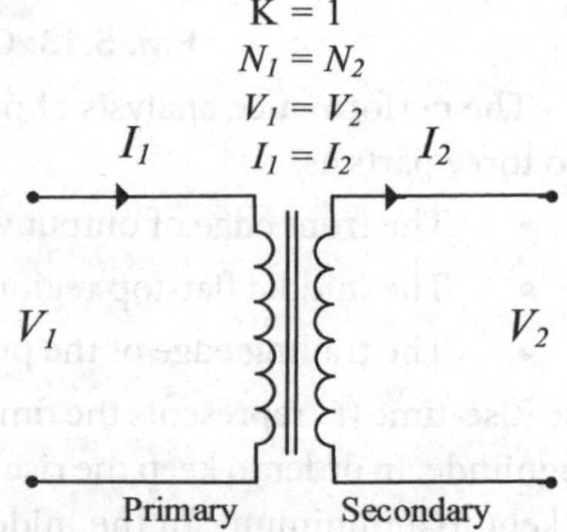

Fig. 5.14: Isolation transformer

5.5 WELDING TRANSFORMER

Arc welding is a process of joining metal to metal using electricity. It involves the creation of enough heat to melt the metals and the melted metals when cooled, results into their binding. The most important device in the arc welding process is the *Welding Transformer*. Welding transformer is basically a step-down transformer with high leakage reactance both in primary and secondary windings. An arc is initiated between the two metal electrodes that are connected to the two terminals of secondary winding. This arc produces sufficient heat to melt the metals to be joined.

Welding transformer differs from a conventional two-winding transformer in construction as well as in performance. In Unit-3 it was informed that, in a conventional transformer, the flux produced by the primary and secondary windings should be able to link with each other's winding to a larger extent so that the leakage flux is minimum and the resulting leakage reactances X_1 and X_2 are also minimum. The effect of this consideration in a conventional transformer is that, with rise in secondary current I_2 (or the rise in load), the change in secondary terminal voltage V_2 is less and hence, the voltage regulation is good. On the contrary, a welding transformer is designed such that, the leakage flux and leakage reactances are purposefully made high in magnitude so that, with rise in secondary current I_2, there is a large drop in secondary terminal voltage V_2. For this to happen, the primary and secondary windings of a welding transformer are either wound on separate limbs of the magnetic core or on the same limb but spaced distance apart. Provision of saw-cut in the magnetic core also helps in increasing the leakage reactance. Schematic diagram of welding transformer and its *V-I* characteristics are shown in Fig. 5.15 (a) and (b) respectively. For initiating the arc, the welding electrodes of a welding transformer are initially short-circuited and then they are separated by a small gap for the conduction of electric current I_2 in the form of arc.

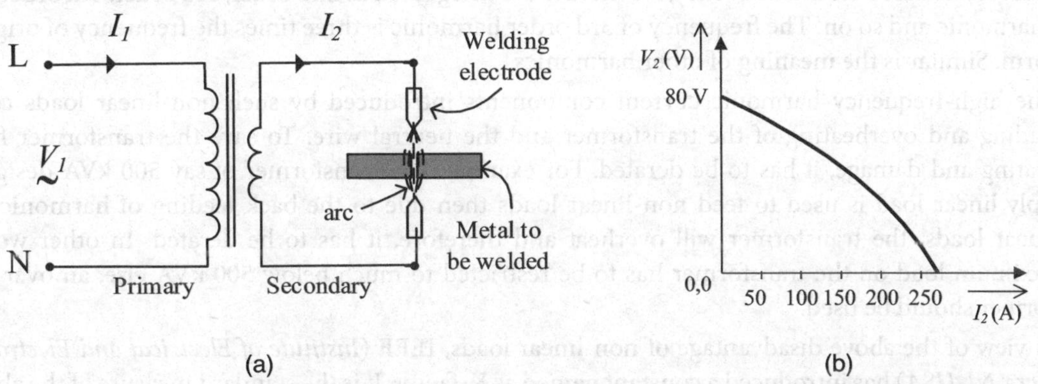

Fig. 5.15: (a) Schematic diagram and (b) V-I characteristic of welding transformer

The magnitude of secondary current I_2 through the arc depends upon the magnitude of voltage in the secondary winding. Once the arc is initiated then, to limit the current magnitude, it is necessary that the voltage should decrease. If the leakage reactances are large then any increase in winding currents will increase the voltage-drop across these reactances and consequently, the output voltage will decrease which is essential to limit the welding current as the weld is practically a short-circuit.

If we use a conventional transformer (having low leakage reactances) in place of a welding transformer for the purpose of arc welding then, due to short-circuit on the secondary side, the short-circuit currents will be significantly high in both windings. They will produce high copper loss resulting into excessive heat in the windings and damage the transformer. In the absence of welding transformer, a conventional two-winding transformer with series inductive reactance capable of carrying large current is also an alternative.

5.6 'K' FACTOR OF A TRANSFORMER

Electrical loads are either linear or non-linear. Linear loads consume electric power from the supply without disturbing the sinusoidal wave-shape of the input current and input voltage. Examples of linear load are the directly line-connected *ac* motors and incandescent lighting. Non-linear loads consume power in a discontinuous manner at designated intervals and hence disrupt the sinusoidal shape of input current. For example, if a diode is connected between an *ac* source and the load, it will allow only positive half-cycle of current waveform. Examples of non-linear loads are electronic lighting ballast, computers, fax and photocopy machines, uninterruptible power supply (UPS), programmable logic controllers (PLC), telecommunication equipment, and induction heating equipment etc.

In the modern era, most of the electrical equipment are controlled by electronic circuits so as to make their operation energy-efficient. Electronically controlled, variable-speed *ac* motor drive is a good example of non-linear load in which the semiconductor switches are continuously switched ON and OFF. The ON and OFF switching results in non-continuous flow of current from the supply lines to the motor. As a result, the input current tends to deviate from its sinusoidal wave-shape. In other words, *harmonics* are introduced in the input current because any waveform that is not purely sinusoidal in shape is said to have harmonics. By Fourier expansion, harmonics are described as the sinusoidal waveforms of higher frequencies and different amplitudes whose presence makes the original current and voltage waveform non-sinusoidal. Based on the frequency, harmonics are designated as 2nd order, 3rd order, 4th order, 5th order harmonic and so on. The frequency of 3rd order harmonic is three times the frequency of original waveform. Similar is the meaning of other harmonics.

The high-frequency harmonic current components introduced by such non-linear loads cause overloading and overheating of the transformer and the neutral wire. To save the transformer from overheating and damage, it has to be derated. For example, if a transformer of say 500 kVA designed to supply linear load is used to feed non-linear loads then due to the back feeding of harmonics by non-linear loads, the transformer will overheat and therefore, it has to be derated. In other words, the maximum load on the transformer has to be restricted to much below 500 kVA else, an over-size transformer should be used.

In view of the above disadvantage of non-linear loads, IEEE (*Institute of Electrical and Electronics Engineers, NJ, USA*) has introduced a constant named as K-factor. It is the standard measure of the ability of a transformer to withstand non-linear loads which produces harmonics. A K-factor rated transformer means a transformer that is designed and manufactured to handle harmonics and operate upto its full-capacity without any damage. The range of constant 'K' varies from 1 to 50. The higher is the K-factor rating of a transformer, the more harmonic distortion it is able to sustain. K-factor rated transformers withstand the heating effects of harmonic currents due to their larger winding coil diameters. They are designed for lower flux density by increasing the size of their magnetic core and thus minimizing the iron loss. K-factor is defined as,

$$\text{K-factor} = K - factor = \sum_{(h=1)}^{\infty} I_h(pu)^2 h^2 \tag{5.13}$$

Where,

$I_h\,(pu)$ = rated current at *hth* order harmonic (in per unit of rated rms load current)

h = the harmonic order

As a rule of thumb, if the load is purely electrical (inductive and resistive) and there are no electronic components in it then, K-1 rated transformer can be used to supply power to such loads. With relative increase in the electronic components compared to electrical components in the load, the K rating of the transformer is increased. For example, for UPS with filter, induction heating equipment and PLC, K-4 rated transformer can be used. For telecommunication equipment and UPS without filter, K-13 rated transformers are used. For mainframe computer loads and variable speed motor drives, the K-20 rated transformers are preferred. For other loads identified as producing very high amounts of harmonics of higher order, K-40 is used.

For systems where the third order and other triplen harmonic components are significant, using a delta-star transformer is also a solution. This is because, triplen harmonics often cause problems for the neutral wire, but when a delta winding is present on the primary side of a two-winding transformer, then triplen harmonics can circulate inside the delta loop instead of the neutral wire.

UNIT SUMMARY

- The transfer of power from primary to secondary winding in a two-winding transformers takes place by mutual induction whereas in an autotransformer, it takes place by mutual induction as well as by direct conduction.
- Autotransformer has higher efficiency and less cost than a two-winding transformer of the same kVA rating.
- Autotransformer does not provide electrical isolation between the primary and secondary sides because it is a single-winding transformer.
- Current transformer is used with low-range ammeter to measure high ac currents whereas, potential transformers are used with low-range voltmeters to measure high ac voltages.
- The secondary winding of a CT should never be open-circuited while its primary is in energised state to avoid the danger of fatal shock to the operator due high induced voltage.
- The secondary winding of a PT should never be short-circuited while its primary is in energised state to avoid the danger of high winding current, excessive heat and damage.
- Pulse transformers are used to transfer the high frequency voltage pulses from primary to secondary with identical reproduction and are mostly employed in Power Electronics circuits and in electronic communication networks.
- Isolation transformers are employed to block the transfer of dc transients (noise signals) from primary to secondary and thus isolate the two circuits from each other physically.
- Welding transformers are purposefully designed to have high leakage flux. This increases the leakage reactance due to which when secondary of welding transformer is initially short-circuited for the initiation of arc, current is limited due to voltage drop across primary leakage reactance.
- A transformer designed to supply linear loads should be derated when employed to supply non-linear loads.
- Transformers with K factor rating are specially designed and fabricated to supply non-linear loads.

EXERCISES

Multiple Choice Questions

1. A dimmerstat can provide output voltage__________ input voltage.
   (a) less than the (b) more than the
   (c) less than and slightly more than (d) more than and slightly less than
2. Copper saved in an autotransformer is ______times the copper required in 2-winding transformer.
   (a) K (b) (1 – K)
   (c) (K – 1) (d) 1/K
3. Autotransformers should not be connected across a dc voltage source because,
   (a) the source may get short-circuited (b) the source may get open-circuited
   (c) the output voltage will be zero (d) Output voltage will be equal to input voltage
4. The primary winding of a CT is always connected in ______ with the supply line and the primary winding of a PT is always connected in ______ with the supply lines.
   (a) series, parallel (b) parallel, series
   (c) series, series (d) parallel
5. A torroidal core CT with only secondary winding is to be designed with a CT ratio of 100/5A. The number of turns of secondary winding should be,
   (a) 100 (b) 20
   (b) 5 (d) 500
6. The two terminals of primary winding of a PT employed in substation are connected to,
   (a) two different HV lines (b) neutral and ground
   (c) HV line and ground (d) HV line and neutral
7. Pulse transformers operate at,
   (a) power frequency of 50 Hz (b) high frequencies between 20 kHz – 100 kHz
   (c) frequencies between 50 Hz – 100 kHz (d) any frequency
8. In isolation transformer, the transformation ratio is,
   (a) K = 1 (b) K < 1
   (c) K > 1
9. The main objective of isolation transformer is to,
   (a) step-down the voltage (b) step-up the voltage
   (c) minimize the voltage spikes (d) minimize the losses
10. Welding transformer is used for,
    (a) arc welding (b) gas welding
    (c) resistance welding (d) laser beam welding

Short Answer Questions

1. Compare an autotransformer with two-winding transformer.
2. Explain how an autotransformer transfers power by mutual induction.

3. Explain: 'Leakage reactance in an autotransformer is less'.
4. Write the applications of single-phase and three-phase autotransformers.
5. Why the primary winding of a CT carries high current whereas the primary winding of a PT carries low current?
6. Why are porcelain bushings required in CT and PT?
7. What is burden?
8. Why are the secondaries of a CT and PT used in substations usually grounded?
9. Why is the core of a pulse transformer made from highly permeable material?
10. What is the meaning of galvanic isolation between primary and secondary winding in an isolation transformer?
11. Explain the significance of K factor.
12. What is IEEE? What are its functions? (Explore from internet)
13. What are harmonics?

Long Answer Questions

1. Derive expressions for the power transferred by autotransformer by mutual induction and conduction.
2. Derive expression for copper saved in an autotransformer compared to a two-winding transformer of same rating.
3. Analyse why a shunt resistor and a series resistor are not used with ammeter and voltmeter to measure high ac currents and high ac voltages respectively.
4. Discuss in detail why the secondary winding of a current transformer should not be open-circuited when its primary is carrying the line current.
5. Discuss in detail why the secondary winding of a voltage transformer should not be short- -circuited when its primary is connected across the HV lines.
6. Explain any one application of pulse transformer.
7. What are the design considerations of a pulse transformer to make the output pulse as accurately as possible identical to the input pulse?
8. Discuss how additional heating is produced in a transformer supplying non-linear loads.
9. Explain how you will identify a suitable transformer on the basis of type of load to which it will be connected.
10. Distinguish between the linear and non-linear loads. Give examples.
11. Discuss welding transformer. Why is it necessary to have high leakage reactance in a welding transformer? How is it ensured?

ANSWERS															
1.	(c)	2.	(a)	3.	(a)	4.	(a)	5.	(b)	6.	(c)	7.	(b)	8.	(a)
9.	(c)	10.	(a)												

PRACTICALS

Experiment No: 1

Aim: To connect the autotransformer in step-up and step-down modes noting the input and output readings

Equipment and Apparatus:

Device	Make	Range	Quantity
1-phase autotransformer (variac)		Input: 0-270 V, 5A	02
Voltmeter		0 - 150V	01
Voltmeter		0 – 300 V	01
Rheostat or lamp load (optional)		750Ω, 5A	01
ICDP Switch			01

Circuit connections:

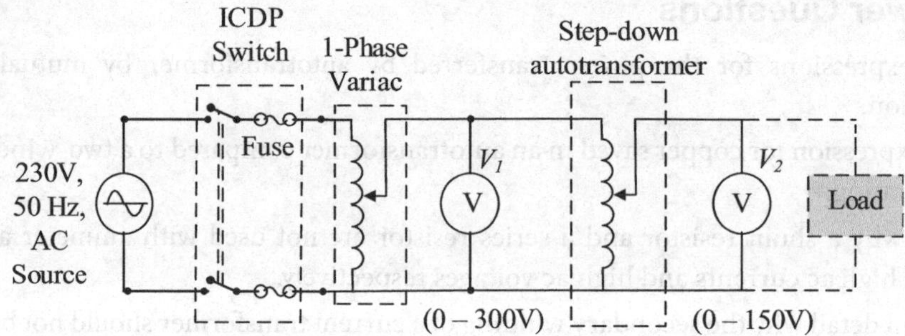

(a) For autotransformer in step-down mode

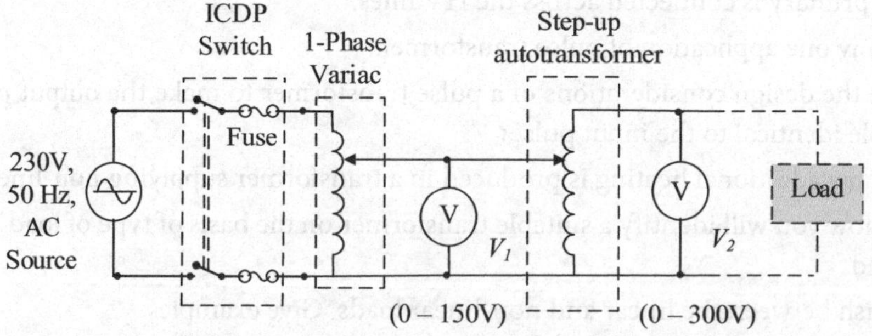

(b) For autotransformer in step-up mode

Fig. 5.16: Connection diagrams for autotransformer in step-up and step-down modes

Theory:

In Section 5.1.1, the single-phase autotransformers are discussed in detail. We know that, autotransformers are the single-winding transformers that could work in step-up as well as in step-down modes as there is no electrical isolation between the primary and secondary sides. When the tap on the winding is made movable, we can get variable output voltage. Such autotransformers are commonly used in institutional laboratories, research centres as well as industries and are commonly called as *dimmerstat* or *variac.*

In industries, it is usually required to use a variac for the testing of electrical equipment whose input voltage rating is less than the voltage made available by power utilities. In view of such application, this experiment gives confidence to students to use autotransformers. Here, we shall connect two variacs. One will be used to vary the input voltage for safe conduction of experiment. The other one will be used as a device on which the experiment is to be performed. Hence to avoid confusion, they will be referred as '*variac*' and '*step-up or step-down autotransformer*' respectively. In above connection diagrams, a load is shown connected across secondary side. However, it is optional and one may keep the secondary side open.

Precautions:

1. Make tight connections as shown and switch on the supply after they are verified by the instructor.
2. Ensure that you are standing on a rubber mat while performing the experiment.
3. Stand at a safe distance from the experimental setup when it is running.
4. Do not bend to take readings. Stand straight and avoid parallax error while taking the readings.
5. Ensure that the variac output voltage setting is at zero setting before switching ON the supply.
6. If you are using a load, make sure that it is turned-OFF at the beginning of the experiment.

Procedure:

(A) For step-down mode of autotransformer:

1. Make the connections as shown above.
2. Ensure that the knob of variacs as well as the knob of step-down autotransformer, both are at zero position before switching the supply ON.
3. Switch ON the input supply.
4. Increase the input voltage using the variacs so that the voltmeter *V1* reads some input voltage say 100V.
5. Turn the knob of step-down autotransformer in clock-wise direction to increase the output voltage *V2*. Note the readings. Observe that, $V2 < V1$.
6. If you have connected a load, turn it ON and ensure that, the load current is not more than the rated current value of the step-down autotransformer.
7. Take more sets of readings by increasing the input voltage *V1* to say 150V, then 200V and repeat steps 4 -7.
8. Switch OFF the load, bring back the knob of step-down autotransformer then, variacs to zero position and switch OFF the supply.
9. Open switch in Fig 5.16 (a), record output voltage. Is autotransformer working?

(B) For step-up mode of autotransformer:

1. Make the connections as shown above.
2. Ensure that the knob of variacs is at zero position and the knob of step-up autotransformer at maximum position before switching the supply ON.
3. Switch ON the input supply.

4. Increase the input voltage using the variacs so that the voltmeter *V1* reads some input voltage say 20V.
5. Turn the knob of step-up autotransformer in counter clock-wise direction to increase the output voltage *V2*. Note the readings. Observe that, *V2 > V1*.
6. If you have connected a load, turn it ON and ensure that, the load current is not more than the rated current value of the step-down autotransformer.
7. Take more sets of readings by increasing the input voltage *V1* to say 25V, then 30V and repeat steps 4 -7.
8. Switch OFF the load, bring back the knob of step-down autotransformer then, variacs to zero position and switch OFF the supply.

Observations:

(A) For step-down mode of autotransformer:

Sr. No.	Input voltage (V_1)	Output voltage (V_2)	Transformation ratio ($K = V_2/V_1$)
1.			
2.			
3.			
4.			

(A) For step-up mode of autotransformer:

Sr. No.	Input voltage (V_1)	Output voltage (V_2)	Transformation ratio ($K = V_2/V_1$)
1.			
2.			
3.			
4.			

Result: Write average value of transformation ratio in both modes

Conclusion:

Discussion:

1. *Distinguish between autotransformer and two-winding transformer.*

   Autotransformer is a single-winding transformer in which the same winding is common for both primary and secondary sides. Thus there is a saving in copper and hence the cost of autotransformer is less than that of a two-winding transformer of same rating. However, the electrical isolation between the two transformers gets lots which is necessarily required in some applications. Due to the presence of direct conducting path between primary and secondary winding, the power transfer in an autotransformer takes place by both methods; electrical conduction and mutual induction. This allows relatively higher power capacity in autotransformers.

As the name imply, a two-winding transformer has two individual and isolated windings for the primary and secondary sides. The transfer of power takes place by mutual induction only. Due to the presence of isolation between primary and secondary sides, the *dc* transients in the primary winding (if any) cannot enter into the secondary circuit. This is the biggest advantage of two-winding transformer.

2. *Write applications of autotransformer.*
   - Testing of electrical appliances/device at reduced voltages
   - Reduced voltage starting of induction motor
   - Speed control of induction motor by varying the input voltage
   - Adjustment of temperature produced by a resistive heater by varying the *ac* input voltage
   - Adjustment of focus light intensity by varying the ac input voltage
   - Step-up and step-down applications
3. *Why is autotransformer not used as distribution transformer?*

   Autotransformers are normally preferred when the transformation ratio (K = V2 / V1) is close to 1. They are used for stepping down the voltage from such as 220kV to 132kV. In distribution network, as it is usually required to step-down the voltage from 11kV to 433V, the required transformation ratio is too low therefore autotransformer is not used on distribution side. There are chances of full high voltage appearing on LV side, which is dangerous and fatal.
4. *Can we operate an autotransformer on dc supply in place of ac?*

   No. Autotransformer is a device to be operated on *ac* supply only. If a *dc* source is connected on its input side then due to zero frequency, the leakage reactance becomes zero and the impedance will be due to the winding resistance only which is normally small. The autotransformer winding will act like a short-circuit across the *dc* source and draw large current leading to damages.
5. What will happen if the magnetic core of a transformer is replaced by non-magnetic material?

   The B-H curve of a non-magnetic material has very less slope. In other words, the permeability of non-magnetic material is very poor. This will require a large mmf and a large magnetizing current for producing the required magnitude of flux in the core. Due to large magnetizing current, the power-factor will be extremely poor. Also, it will result into high copper loss and poor efficiency. The winding will become hot.

Experiment No: 2

Aim: To check the functioning of current transformer (CT)

Equipment and Apparatus:

Device	Make	Range	Quantity
Current transformer		CT ratio: 20/5A	01
1-Phase Variac		Input: 0-270 V, 5A	01
Ammeter		0 – 20A	01
Ammeter		0 – 5A	01
Rheostatic variable load		100Ω, 5A (or as per the CT rating)	01
ICDP Switch			01

Circuit connections:

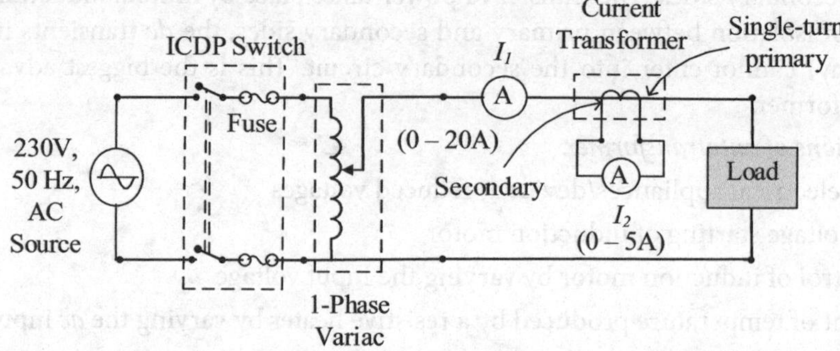

Fig. 5.17: Connection diagram to the functioning of current transformer (CT)

Theory:

In Section 5.2.1 and 5.2.2, the current transformer is discussed in detail. It is extensively used not only in generating stations and electrical substations but also in industrial and commercial electrical power distribution networks. In terms of voltage, it is a step-up transformer where $I_2 < I_1$ and $V_2 > V_1$. Small size, torroidal type CT with single-turn primary is very common in the 11 kV substations as well as in industrial automation system wherever high currents are to be monitored.

Precautions:

1. Make tight connections as shown and switch on the supply after they are verified by the instructor.
2. Ensure that you are standing on a rubber mat while performing the experiment.
3. Stand at a safe distance from the experimental setup when it is running.
4. Do not bend to take readings. Stand straight and avoid parallax error while taking the readings.
5. Ensure that the variac output voltage setting is at zero before switching ON the supply.
6. Ensure that the ammeter is already connected across the secondary winding and that it is not open-circuited.

Procedure:

1. Make the connections as shown above.
2. Ensure that the knob of variac at zero position before switching the supply ON.
3. Ensure that the load is switched OFF.
4. Switch ON the input supply.
5. Increase the input voltage using the variac so as to apply some voltage to the load.
6. Note the primary and secondary currents I_1 and I_2 of CT. Observe that, $I_2 < I_1$.
7. Increase the load and repeat step 6.
8. Switch OFF the load, bring back the knob of variac to zero position and switch OFF the supply.

Observations:

Sr. No.	CT Input current (I_1)	CT Output current (I_2)	CT ratio (I_1/I_2)

Result: Write average value of CT ratio and compare it with that mentioned on the name plate.

Conclusion:

Discussion:

1. *Distinguish between current transformer and power transformer.*

   Current transformer is similar in construction to a power transformer. Both have a magnetic core and two windings. They work on the same principle of electromagnetic induction but there are a few differences. A power transformer can be of single-phase type as well as of three-phase type whereas, the current transformer is in single–phase only.

   The primary winding of a power transformer is normally connected across a substantially constant voltage source while, *the secondary winding is connected across the load* which varies in impedance over a wide range. As a result, the *current flowing in the primary winding of a power transformer is determined by the secondary current* which in turn depends on the magnitude and power-factor of load. Whereas, in a current transformer, *the primary winding is connected in series with the load* whose current is to be measured. A measuring instrument or the actuating coil of a protective relay is connected across the secondary of a CT (also called as burden of CT) and their impedance remains almost constant. Since, the load varies over a wide range a day, the primary current of a CT also varies accordingly and therefore, the *current in the secondary winding of a CT is determined by its primary current* (i.e. load current).

   In a power transformer, due to the fixed input voltage the magnetic flux in the core substantially remains constant at all loads. Whereas in a CT, the magnetic flux in the core varies with primary current (i.e. the load current).

2. *What is burden of a CT?*

   Burden of a CT is the value of load connected across its secondary winding. This load can be in the form of a voltmeter or the actuating coil of a protective relay. It is expressed as an output in voltamperes (VA). The rated burden value is mentioned on the name plate of a CT.

3. *Write applications of current transformer.*

   - For measurement of high *ac* current which cannot be measured directly by the available ammeters of low range.
   - For protection of electrical circuits against high current.
   - For metering and analysing the current

4. *What precaution is necessary while using a CT?*

   Precaution should be taken that the secondary winding is never open-circuited when it is in energized state. Either connect an ammeter, or a relay coil or close the secondary winding circuit.

5. *What are the different types of CT?*

   - *Indoor type CT:* They are used on low voltage circuits.
   - *Outdoor type CT:* They are used in high voltage circuits like substations and switchyards.
   - *Portable type CT:* These CTs are also called as Clamp-on CT, and Split-core CT. The core is split into two parts and can be easily made to encircle any line whose current is to be measured.
   - *Hall Effect Current Sensor:* As shown in Fig. 5.18(a), the Hall effect current sensor has a circular magnetically permeable core, Hall effect device and a signal conditioning circuit. The conductor whose current is to be measured is allowed to pass through the core. A Hall effect device is placed in the gap of the core and at right angles to the flux in the core. For a given magnitude of current (*I*) being measured, a proportional magnetic field is produced around the current-carrying conductor which then concentrates in the core. The magnetic field generated by the input current (*I*) is detected and converted into a voltage signal by the Hall effect device. Since the output of the Hall device is small, it is amplified to a voltage of several volts by an amplifier.

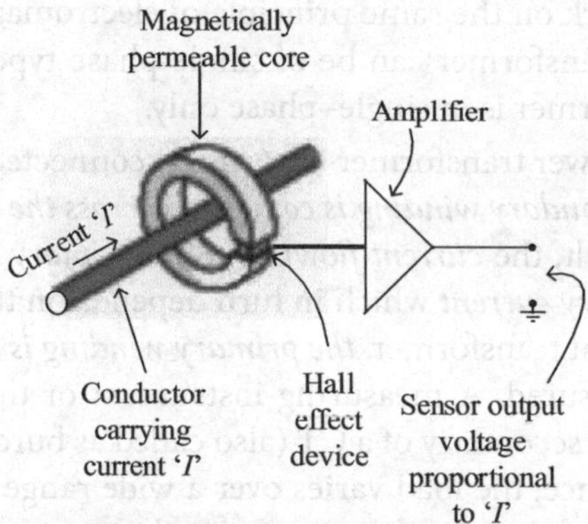

Fig. 5.18: (a) Hall effect current sensor

The Hall effect current sensor is best suited for the measurement of *dc* currents. For *ac* current measurement, inductive sensors are preferred.

Experiment No: 3

Aim: To check the functioning of potential transformer (PT)

Equipment and Apparatus:

Device	Make	Range	Quantity
Potential transformer		1100V/110V	01
1-Phase Variac		Input: 0-270 V, 5A	01
Voltmeter		0 – 300V	01
Voltmeter		0 – 150V	01
ICDP Switch			01

Circuit connections:

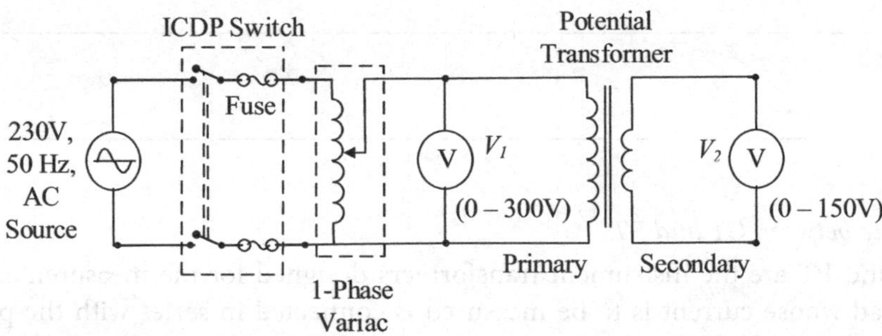

Fig. 5.18: (b) Connection diagram to Check the functioning of potential transformer (PT)

Theory:

In Section 5.2.3, the potential transformer is discussed in detail. It is used for stepping down the system voltage to a convenient value which could be applied across low range voltmeters and relays. This is because, the commercially available relays are designed to operate at low voltages. Obviously, in a potential transformer, $V_2 < V_1$ and $I_2 > I_1$.

Precautions:

1. Make tight connections as shown and switch on the supply after they are verified by the instructor.
2. Ensure that you are standing on a rubber mat while performing the experiment.
3. Stand at a safe distance from the experimental setup when it is running.
4. Do not bend to take readings. Stand straight and avoid parallax error while taking the readings.
5. Ensure that the variac output voltage setting is at zero before switching ON the supply.
6. Ensure that the secondary side of the PT is not short-circuited.

Procedure:

1. Make the connections as shown above.
2. Ensure that the knob of variac at zero position before switching the supply ON.
3. Switch ON the input supply.
4. Increase the input voltage using the variac so as to apply some voltage to the primary of the PT.
5. Note the primary and secondary voltages V_1 and V_2 of PT. Observe that, $V_2 < V_1$.
6. Repeat steps 4 and 5 for different values of input voltage V_1.
7. Bring back the knob of variac to zero position and switch OFF the supply.

Observations:

Sr. No.	PT Input voltage (V_1)	PT Output voltage (V_2)	PT ratio V_1/V_2)
1.			
2.			
3.			
4.			

Result: Write average value of PT ratio and compare it with that mentioned on the name plate.

____

Conclusion:

____

Discussion:

1. *Differentiate between CT and PT.*

   Both CT and PT are the instrument transformers designed for the measurement purpose. In a CT, the load whose current is to be measured is connected in series with the primary winding whereas in a PT, the load whose voltage is to be measured is connected in parallel with the primary winding. In other words, a CT is connected in series with the load and a PT is connected across the load.

   Precaution should be taken that, in energized state, the secondary of a CT is never open-circuited and the secondary of a PT is never short-circuited.

2. *Write applications of potential transformer.*
   - They are used with ammeter and wattmeter to measure the high-voltage transmission line voltage and power.
   - To signal protective relays when a fault occurs on a high-voltage line.

Experiment No: 4

Aim: To check the functioning of isolation transformer

Equipment and Apparatus:

Device	Make	Range	Quantity
1-Phase Isolation transformer		1kVA, 230V	01
1-Phase Variac		Input: 0-270 V, 5A	01
Voltmeter		0 – 300V	02
ICDP Switch			01

Circuit connections:

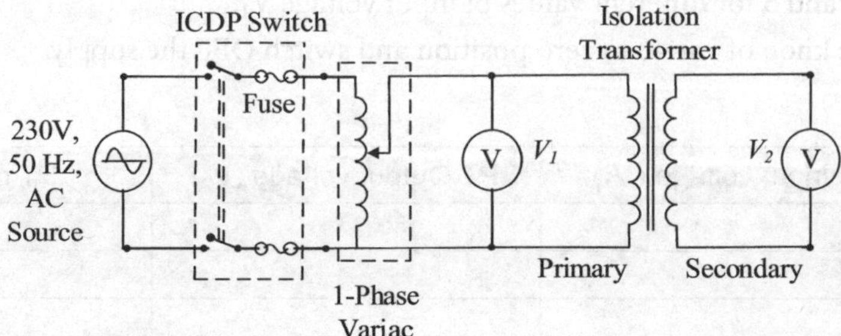

Fig. 5.19: Connection diagram to Check the functioning of isolation transformer

Theory:

In Section 5.4, the isolation transformer is discussed in detail. It is used in industries to transfer the electrical power from an *ac* source to some load (say, equipment or device) or to transfer gating signals from control circuit to power semiconductor circuit, usually for equipment safety reasons. Obviously, in a potential transformer, $V_2 = V_1$ and $I_2 = I_1$.

Precautions:

1. Make tight connections as shown and switch on the supply after they are verified by the instructor.
2. Ensure that you are standing on a rubber mat while performing the experiment.
3. Stand at a safe distance from the experimental setup when it is running.
4. Do not bend to take readings. Stand straight and avoid parallax error while taking the readings.
5. Ensure that the variac output voltage setting is at zero before switching ON the supply.

Procedure:

1. Make the connections as shown above.
2. Ensure that the knob of variac at zero position before switching the supply ON.
3. Switch ON the input supply.
4. Increase the input voltage using the variac so as to apply some voltage to the primary of isolation transformer.
5. Note the primary and secondary voltages V_1 and V_2 of isolation transformer. Observe that, $V_2 = V_1$.
6. Repeat steps 4 and 5 for different values of input voltage V_1.
7. Bring back the knob of variac to zero position and switch OFF the supply.

Observations:

Sr. No.	Input voltage of isolation transformer (V_1)	Output voltage of isolation transformer (V_2)	Transformation ratio ($K = V_2/V_1$)
1.			
2.			
3.			
4.			

Result: Write average value of transformation ratio

____

Conclusion:

____

Discussion:

1. *Compare isolation transformer and autotransformer.*

- Isolation transformer is a two-winding transformer whereas autotransformer is a single-winding transformer.
- The primary and secondary sides of an isolation transformer are electrically isolated from each other. An autotransformer has a primary and a secondary winding that are partly shared.

- The primary and secondary terminal voltages in an isolation transformer are equal whereas an autotransformer is normally used to step-up, to step-down or to provide a variable output voltage.
- The transfer of power from primary to secondary in an isolation transformer is by mutual induction only whereas, in an autotransformer, it takes place by mutual induction as well as direct conduction.

2. *Write applications of isolation transformer.*

- To create isolation between a power source and the powered circuit or powered equipment.
- To change the electrical power flowing between two circuit lines that are not electrically connected.

Experiment No: 5

Aim: To test the operation of pulse transformer

Equipment and Apparatus:

Device	Make	Specification	Quantity
Pulse transformer of suitable make and rating			01
Signal generator or function generator			01
Cathode Ray Oscilloscope (CRO) or Digital Storage Oscilloscope (DSO)			01

Circuit connections:

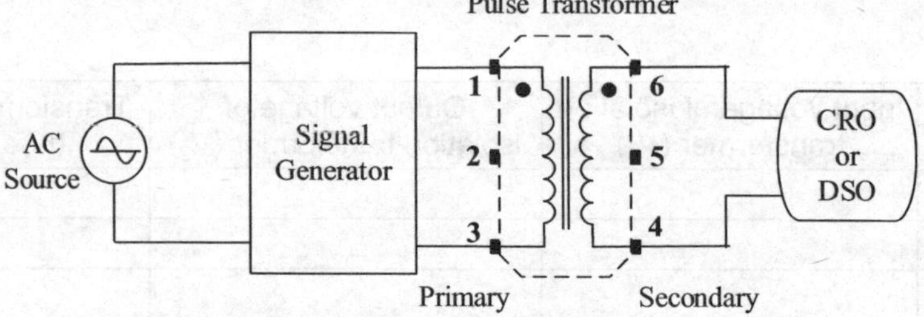

Fig. 5.20: Connection diagram To test the operation of pulse transformer

Theory:

In Section 5.3, the pulse transformer is discussed in detail. Pulse transformers are of two types; (i) power type and (ii) signal type. The high power pulse transformers are mainly used in pulse power technology where the energy from low power source is slowly stored and sharply released to the load.

The signal type pulse transformer delivers a rectangular pulse-like signal or a series of pulses. The number of turns in both primary and secondary are very less to minimize the leakage flux and leakage reactance. They are often used for the transmission of the digital data and in the gate drive circuitry of semiconductor switches. They are also used in digital logic and telecommunication circuits. Pulse transformers normally operate at high frequencies. They are relatively smaller in size and can transfer more power than the same size conventional two-winding transformer. Like isolation transformers, the pulse transformer provides galvanic isolation between the two sides and allows fast control signals to be transmitted without much distortion in the signal wave-shape. Due to high operating frequency (in kHz), the magnetic core of pulse transformer is made from highly permeable material like ferrite, Permalloy and silicon steel laminations. Pulse transformers are also be made of air-core type. Other applications of signal type pulse transformer are line coupling, small isolated power supplies and also as common mode chokes in filtering applications.

Commercially, signal type pulse transformers are also available in encapsulated form with industry-standard-pinout. Example: muRata 78601/1C general purpose pulse transformer (See its data sheet from internet).

Precautions:

1. Read the CRO/DSO manual and understand its control features carefully before operating.
2. Read the data-sheet of pulse transformer being used for its pinout and voltage details.

Procedure:

1. Make the connections as shown above.
2. In this experiment, muRata 78601 IC is considered. Readers can also use some other pulse transformer. The general procedure will remain same.
3. Energize primary pins 1 and 3 at relevant voltage and frequency say 100mV and 1 kHz.
4. Observe the voltage waveforms at input pins (1, 3) and output pins (4, 6) on CRO or DSO.
5. Measure the voltage magnitudes.

Observations:

Sr. No.	Input voltage	Output voltage	Nature of input waveform	Nature of output waveform	Remarks

Result:

____

Conclusion:

____

Discussion:

1. *Why a pulse transformer has less number of turns and low inter-winding capacitance?*
   Pulse transformers have less number of winding turns to minimize the flux leakage and low inter-winding capacitance to ensure that the profile of the signal is maintained on the secondary as accurately as possible.
2. *Write applications of pulse transformer.*

- Pulse transformer is designed for the transmission of voltage pulses between its windings and into the load. They can be used for signal transmission, low-power control circuits, and in high-power switch mode power supplies (SMPS).

- The low-power pulse transformers are used for interfacing the gate control circuit with the power semiconductor switches like power transistors, SCT, traic etc.

KNOW MORE

Dissolved Gas Analysis (DGA) of Transformer Oil

Like a medical doctor, use a syringe and take the oil sample of a transformer to diagnose its internal health and to assess its remnant life.

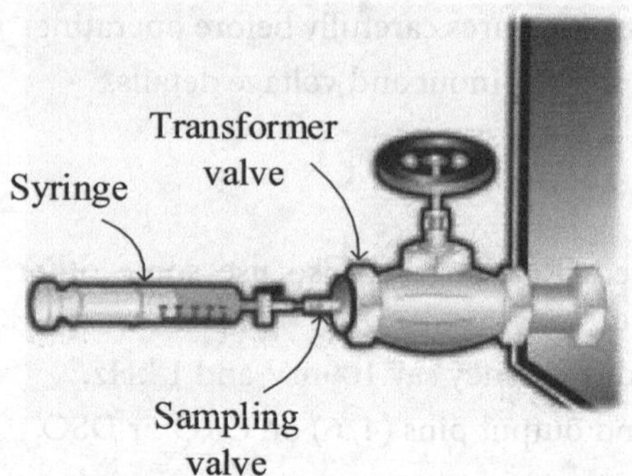

Transformer which has no substitute, plays a major role in power transmission and distribution. Its correct and continuous operation improves the overall efficiency and reliability of the system. Majority of the transformers used in power system are of oil-immersed type. In other words, the complete transformer comprising of magnetic core and windings is immersed in a metal tank that is filled with oil. This oil acts as a dielectric as well as a coolant medium.

When put into service, the iron loss and copper loss (stray loss also) of transformer produces heat and the temperature of all components of the transformer assembly including the transformer oil increases.

This leads to the deterioration of oil. It also leads to the evolution of various hydrocarbon gases in the oil. For example; hydrogen, methane, ethane, ethylene, acetylene and others. Initially, these gases get dissolved in the oil but after oil saturation, they evolve above the oil surface. During abnormal conditions like overload, external short-circuit or internal fault, the temperature increases significantly. The oil may almost boil and liberate excessive gases.

In the 'Know More' section of Unit-4, we have seen that, a protective device called as 'Buchholz relay is mounted on the top of transformer tank and is connected between the transformer tank and the conservator tank. After oil saturation, when high volume of gases evolved in the oil gets liberated from the oil surface, they travel upwards and get trapped in the Buchholz relay. The protection circuit of Buchholz relay gets closed and it disconnects the transformer from the grid thus safeguarding the transformer from explosion.

The Buchholz relay comes into action after the problem has grown up sufficiently and the transformer is almost on the verge of collapse. It responds only to the severe fault conditions requiring immediate removal of transformer from service and resulting in power outages which causes dissatisfaction to the customers. Since the generated gases first dissolve in the oil long before they enter into the Buchholz relay, an analysis of these dissolved gases will help for a preventive action at an initial stage of the fault.

What is DGA?

Dissolved Gas Analysis (DGA) is a diagnostic tool which can tell us about the internal health of the transformer, the rate of growth of the problem and hence estimates the transformer age if the problem keeps growing at the same pace. It is a quantitative and qualitative analysis of oil samples taken from different parts of the transformer for the purpose of taking preventive action much before the fault reaches its critical stage. This is similar to a medical doctor who makes his diagnosis on the basis of pathological test reports and suggests appropriate medicine and care. He doesn't wait for the patient to enter into critical stage and directly admit him in ICU on ventilator.

Due to the presence of any abnormal thermal or electrical stresses on the oil and paper insulation etc., degradation products are formed in the oil which leads to the evolution of dissolved gases. The internal incipient faults like arcing, corona discharge, low energy sparking, overheating of cellulose insulation, overheating due to the concentration of flux, eddy currents in copper, poor contacts etc. are some of the causes that start the degradation process.

How DGA works?

The basic diagnosis is based upon the types and relative quantities of dissolved gases. The gases most commonly analysed are methane (CH_4), ethane (C_2H_6), ethylene (C_2H_4) acetylene (C_2H_2) and hydrogen (H_2). It is found that different types of internal faults result into different gases or combination of gases. After determining the magnitudes of gases in ppm (parts per million), conclusions are drawn on the internal condition of transformer using various methods and standards. The Indian Standard IS-10593 is one of them.

Procedure to conduct DGA

Following are the main steps of DGA.

- Oil sampling
- Extraction of dissolved gases and analysis
- Interpretation of analysis results

Oil samples are normally taken from both the top of the transformer main tank and from the bottom of the main tank. Oil syringes are used to obtain an oil sample from the transformer. This syringe is similar to the medical syringe used by doctors for collecting blood-sample or for injecting medicine into the human body. Alternatively, a closed glass or metal container or tube is also used in accordance with the IEC standard.

Once the oil sample is collected, the dissolved gases are extracted from the oil in a vacuum chamber by stirring it in an air-tight flask. The gases so extracted are a mixture of many gases. Therefore, the individual gases are separated from the mixture by using a device called as '*Gas Chromatograph*'. It is based on a technique called as Chromatography which deals with the separation, identification and quantification of different components in a given mixture. To measure the volume of each gas in ppm, standardisation and calibration of the chromatograph is essential.

Diagnosis using DGA

After the magnitudes of individual gases are known, the next step is to interprete the internal condition of the transformer from these gases. The identified gases are compared with their standard permissible limits considering the transformer rating, period of service, oil-filtration, oil-reclamation and oil-replacement history etc. For this purpose, the various available methods are:

- Key-Gas method
- Total combustible gas content method
- Duval's triangle

- Brown Boveri practice
- Hitachi method
- Roger's ratio method
- Dornenberg ratio method

Every method has its own limitation. Also, it is possible that, the interpretation of the same case by different methods may be different. Hence it is normally said that interpretation about the transformer condition by DGA is not a rule but an Art which depends upon the experience and skill of the person doing it. This problem points towards the need of an artificial intelligence based software. As per Key-Gas method, the following gases when they exceed their permissible limits indicate the type of fault as mentioned against them in Table 5.1.

Table 5.1: Key-Gas method

Sr. No.	Key Gas	Associated Fault
	C_2H_2	Arcing
	H_2	Corona
	C_2H_4	Overheating of oil
	CO	Overheated cellulose insulation

REFERENCES AND SUGGESTED READINGS

1. T. Wildi, Electrical Machines, Drives, and Power Systems, Pearson Education Asia, 2001
2. P. K. Mukherjee, S. Chakravorti, Electrical Machines, Dhanpat Rai Publications, 2010.
3. P. S. Bhimbra, Electrical Machinery, Khanna Publishers, 2009.
4. R. K. Srivastava, Electrical Machines 2e, Cengage Learning, 2013
5. D. P. Kothari, I. J. Nagrath, Electrical Machines, 4/e., Tata McGraw Hill Edn. Pvt. Ltd., 2010
6. J. M. Frank, "Origin, Development, and Design of K-factor Transformers," *IEEE Ind. App. Magazine* September/October 1997.
7. IEEE Std. C57.110 2008, "IEEE Recommended Practice for Establishing Liquid-Filled and Dry-Type Power and Distribution Transformer Capability When Supplying Non-sinusoidal Load Currents," *IEEE Power Engineering Society,* August, 2008.
8. Z. Zhang, X. Tan, "Review of High Power Pulse Transformer Design," *SciVerse ScienceDirect, Physics Procedia,* 32 (2012), pp. 566-574.
9. N. K. Dhote, S. B. Bodkhe, "Proposal for an Improvement in Transformer Diagnosis by DGA using Expert System," *Australasian Universities Power Engineering Conference*, 26-29 September 2004, Brisbane, Australia.

QR Code for Further Reading

CO AND PO ATTAINMENT TABLE

Assessment of attainment of course outcomes (COs) can be done after the completion of the course and the assessment of student performance. They can be mapped with programme outcomes (POs) and for the attainment of POs to analyse the gap. After proper analysis of the gap in the attainment of POs, necessary measures can be taken to overcome the gaps.

Table for CO and PO attainment

Course Outcomes	Expected Mapping with Programme Outcomes (1- Weak Correlation; 2- Medium correlation; 3- Strong Correlation)						
	PO-1	PO-2	PO-3	PO-4	PO-5	PO-6	PO-7
CO-1							
CO-2							
CO-3							
CO-4							
CO-5							

The data filled in the above table can be used for gap analysis.

INDEX

R

S

T

U

V

W

Y

Z

❑❑